AF358798

Theory and Applications of MIMO Radar

Theory and Applications of MIMO Radar

Editors

Guolong Cui
Bin Liao
Yong Yang
Xianxiang Yu

Basel • Beijing • Wuhan • Barcelona • Belgrade • Novi Sad • Cluj • Manchester

Editors

Guolong Cui
University of Electronic
Science and Technology of
China
Chengdu
China

Bin Liao
Shenzhen University
Shenzhen
China

Yong Yang
National University of
Defense Technology
Changsha
China

Xianxiang Yu
University of Electronic
Science and Technology of
China
Chengdu
China

Editorial Office
MDPI
St. Alban-Anlage 66
4052 Basel, Switzerland

This is a reprint of articles from the Special Issue published online in the open access journal *Remote Sensing* (ISSN 2072-4292) (available at: https://www.mdpi.com/journal/remotesensing/special_issues/MIMO_radar).

For citation purposes, cite each article independently as indicated on the article page online and as indicated below:

Lastname, A.A.; Lastname, B.B. Article Title. *Journal Name* **Year**, *Volume Number*, Page Range.

ISBN 978-3-7258-0235-7 (Hbk)
ISBN 978-3-7258-0236-4 (PDF)
doi.org/10.3390/books978-3-7258-0236-4

Contents

About the Editors

Guolong Cui

Guolong Cui (Senior Member, IEEE) received a B.S. in electronic information engineering and an M.S. and Ph.D. in signal and information processing from the University of Electronic Science and Technology of China (UESTC) in Chengdu, China in 2005, 2008, and 2012, respectively. From January 2011 to April 2011, he was a visiting researcher with the University of Naples Federico II in Naples, Italy. From June 2012 to August 2013, he was a post-doctoral researcher with the Department of Electrical and Computer Engineering at the Stevens Institute of Technology in Hoboken, New Jersey, USA. From September 2013 to July 2018, he was an associate professor with UESTC, where he has been a professor since August 2018. His research interests include cognitive radar, array signal processing, multiple-input and multiple-output (MIMO) radar.

Bin Liao

Bin Liao received the B.S. degree and the M.S. degree in Information and Information Processing from Xidian University in 2006 and 2009, respectively, and the Ph.D. degree from the Department of Electrical and Electronic Engineering of the University of Hong Kong in 2013. He is mainly engaged in the research of array signal processing, radar signal processing, wireless communication, and radar and communication sensory integration. He is currently a professor in Shenzhen University and is also a deputy director of Guangdong Key Laboratory of Intelligent Information Processing.

Yong Yang

Yong Yang received the B.S., M.Sc., and Ph.D. degrees in information and communication engineering from the National University of Defense Technology (NUDT), Changsha, China, in 2007, 2009, and 2014, respectively. He is currently a professor with NUDT. His fields of interests include radar target detection and statistical signal processing.

Xianxiang Yu

Xianxiang Yu received the B.S. and Ph.D. degrees from the University of Electronic Science and Technology of China (UESTC), Chengdu, China, in 2014 and 2020, respectively. From 2018 to 2019, he was a Visiting Researcher with Pennsylvania State University, State College, PA, USA. From 2020 to 2022, he was a post-doctoral researcher with UESTC. He is currently an associate professor with UESTC. His research interests include multiple-input multiple-output (MIMO) radar, array signal processing, and waveform optimization.

Preface

The radar detection environment is notably complex due to the multitude of civil and military electronic devices within it, as well as the intricate and non-uniformly time-varying geographic clutter. This complexity significantly undermines the detection performance of traditional radar systems. In contrast, multiple-input multiple-output (MIMO) radars emerge as solutions which possess the capacity to transmit multiple signals through their array of antennas. They adeptly allocate energy tailored to diverse detection scenarios and mission mandates, granting flexibility in designing transmit beampatterns. These potential characteristics lead to pronounced enhancements in target detection precision and superior suppression capabilities against clutter and jamming. As a result, MIMO radars have captured the attention of both researchers and practitioners alike.

This Special Issue assembles the latest advancements in the domain of MIMO radars, with the primary goal of stimulating innovative ideas and fostering enhanced avenues for communication and collaboration among students and researchers engaged in the area of MIMO radars.

Comprising eight papers, this Special Issue places particular emphasis on MIMO radar waveform design, utilizing the potential of waveform diversity to elevate radar performance. Additionally, three papers leverage this diversity to improve target detection and parameter estimation. The remaining papers tackle other important MIMO radar issues, including but not limited to clutter suppression, imaging, and antenna distribution.

We extend our gratitude to the authors who have contributed their insightful content to this Special Issue, acknowledging their exceptional work. Furthermore, we express our appreciation to the individuals and organizations involved in the editorial process of this Special Issue.

Guolong Cui , Bin Liao , Yong Yang, and Xianxiang Yu
Editors

Article

MIMO Radar Transmit Waveform Design for Beampattern Matching via Complex Circle Optimization

Weijie Xiong [1,2]**, Jinfeng Hu** [1,2,*]**, Kai Zhong** [1,2]**, Yibao Sun** [3]**, Xiangqing Xiao** [1,2] **and Gangyong Zhu** [1,2]

[1] Yangtze Delta Region Institute, University of Electronic Science and Technology of China, Quzhou 324003, China

[2] School of Information and Communication Engineering, University of Electronic Science and Technology of China, Chengdu 611731, China

[3] School of Electronic Engineering and Computer Science, Queen Mary University of London, Mile End Road, London E1 4NS, UK

* Correspondence: hujf@uestc.edu.cn

Abstract: In this paper, we study the multiple-input multiple-output (MIMO) radar transmit waveform design method for beampattern matching. The purpose is to design the beampattern to approximate the actual one and minimize the cross-correlation sidelobes under the constant modulus constraint (CMC). Due to the CMC, the problem is non-convex, and the existing methods solve it with relaxation, resulting in performance degradation. Different from these methods, we notice that the CMC is the product of complex circles (CC). Based on this physical characteristic, the direct beampattern matching without relaxation (DBMWR) method is proposed. To be specific, we first formulate the original problem as an unconstrained quartic problem over the CC and then solve it by the proposed method without relaxation. Simulation results show that the proposed method can achieve a balance in terms of accuracy and computation complexity compared with other methods.

Keywords: waveform design; beampattern matching; MIMO radar; constant modulus constraint; complex circle; DBMWR method

Citation: Xiong, W.; Hu, J.; Zhong, K.; Sun, Y.; Xiao, X.; Zhu, G. MIMO Radar Transmit Waveform Design for Beampattern Matching via Complex Circle Optimization. *Remote Sens.* **2023**, *15*, 633. https://doi.org/10.3390/rs15030633

Academic Editor: Danilo Orlando

Received: 2 December 2022
Revised: 8 January 2023
Accepted: 19 January 2023
Published: 20 January 2023

1. Introduction

The multiple-input multiple-output (MIMO) radar waveform design for beampattern matching with the constant modulus constraint (CMC) is a key technology [1–3]. With proper design, the designed beampattern can approximate the desired one, which is essential to enhance the spatial resolution and the target detection capability [4,5]. Besides, the CMC is important to consider since nonlinear amplifiers should operate in saturation mode [6–8]. Therefore, the beampattern matching through waveform design has received wide attention.

The direct solution is difficult to obtain in the problem since both the quartic objective term and the CMC are non-convex. The existing methods mainly solved it with relaxation, which can be divided into two categories: the indirect beampattern matching with relaxation methods and the direct beampattern matching with relaxation methods.

The indirect beampattern matching with relaxation methods first construct the waveform covariance matrix (WCM), then decompose the WCM with relaxation to obtain the waveform. Typically, the problem was relaxed into the semidefinite quadratic programming (SQP) by relaxing the CMC in [9], and then the interior point method (IPM) was utilized to solve the SQP. However, such methods degrade the accuracy of the design due to the CMC relaxation. To improve the accuracy of the design, the cyclic algorithm (CA) was developed by utilizing the matrix singular value decomposition (SVD) in [10]. Nevertheless, the mainlobe fluctuates greatly since only the main eigenvalues are considered and other feature information is ignored. To overcome this shortcoming, the discrete

Fourier transform (DFT)-based method was proposed in [11], which utilizes DFT coefficients and Toeplitz matrices. However, it is not suitable under the condition that the number of antennas is small.

The direct beampattern matching with relaxation methods directly design the waveform with relaxation. Typically, the majorization-minimization (MM)-based method was proposed through generating a tractable surrogate function in [12]. Nevertheless, the performance degrades since it is hard to construct the tractable function, which follows the shape of the original one. To improve the performance, the alternating direction method of multipliers (ADMM)-based method was proposed in [13] by converting the original problem into a bi-convex one with slightly relaxing the CMC. However, only the approximate constant modulus solution is obtained due to the relaxation. To overcome the drawbacks, the phase optimization method based on residual neural network (RNN) method was proposed in [14] by training lots of weight coefficients, which needs huge computational cost.

Due to the relaxation of the above methods, their performance will be degraded. In this paper, we focus on developing a method without relaxation to design the MIMO radar transmit waveform for beampattern matching. To this focus, by noting that the CMC is the product of complex circles (CC), the direct beampattern matching without relaxation (DBMWR) method is proposed. Specifically, we first transform the problem into an unconstrained quartic function over the CC. After that, the DBMWR is used to solve the problem by deriving the actual gradient, the gradient over the CC, the direction of descent, and the step size while assuring non-increase in the objective function. The main contributions are:

- Different from the existing methods, which address the challenge by relaxation, we propose the DBMWR method without relaxation to solve it.
- To reduce the power consumption and improve the accuracy, we further propose the sparse beampattern matching design method based on the DBMWR method.
- Compared with the methods [10,12–14], the proposed method achieves a balance both in terms of computational complexity and accuracy.

Related Works

Several works related to beampattern matching are worth mentioning. Beampattern matching for the MIMO radar system was considered in [9] under an ideal hardware setting; i.e., without CMC. To achieve the CMC, several related works [10,11] considered constructing the WCM to achieve the CMC. However, there is a huge computation complexity to decompose the WCM with relaxation to obtain the waveform. In this situation, works in [12–14] considered optimizing the waveform directly. Nevertheless, these works solve the problem with relaxation which result in performance degradation. For example, in [12,15], the authors proposed the majorization-minimization (MM) method and the block successive upper bound minimization (BSUM) method, respectively, to solve the problem by relaxing the objective function. To avoid the relaxation, the work in [16] also noticed the characteristic of the CC. Nevertheless, it focuses on wideband signals, whereas our work focuses on narrowband signals. Besides, its update strategy is different, especially in deriving the descent direction and the step size.

We arrange the remainder paper as follows. The technical background is introduced in Section 2. The MIMO radar system model and the problem description are established in Section 3. The beampattern matching based on the DBMWR method is described in Section 4. The sparse beampattern matching design is considered in Section 5. Some numerical results shown by figures and tables are obtained in Section 6 The summary of this paper is in Section 7.

In this paper, we use bold-face upper-case and lower-case letters to represent matrices and column vectors, respectively. $\mathbb{C}^{M \times M}$ is the $M \times M$ complex numbers domains. $\otimes$ is the Kronecker product, and $\odot$ is the Hadamard product. $\mathrm{Re}\{\cdot\}$ and $\mathrm{Im}\{\cdot\}$ are the element-wise real and imaginary part of the complex argument, respectively. $(\cdot)^T$, $(\cdot)^H$, and $(\cdot)^*$ are the transpose, the complex conjugate transpose, and the complex conjugate operators. $|\cdot|$

and $\|\cdot\|_2$ are the absolute value and the 2-norm of the complex argument, respectively. $\nabla_{\mathbf{s}}(f(\mathbf{s}))$ is the gradient. $\mathbf{1}_M$ is the all-one vector of length M and $\mathbb{B}_M^N$ is the binary vectors set with N non-zero elements and total M elements.

2. Technical Background

This section briefly introduces MIMO radar systems, MIMO radar waveform design, and CMC.

2.1. MIMO Radar Systems

MIMO radar systems have revolutionized many research areas with their unique abilities. In this subsection, we briefly review MIMO radar systems and their advantages.

MIMO is originally a technology widely used in communication systems. It is a wireless transmission technology that can transmit more data simultaneously by using multiple antennas at the transmitter side and receiver side [17]. MIMO technology has made obvious achievements in communication systems, which has greatly aroused the research interest of experts and scholars in other fields around the world. Considering the similarities between the radar systems and the communication system, MIMO is naturally introduced into the radar systems.

Different from the traditional phased array radar systems that transmit a single waveform, the MIMO radar systems transmit independent waveforms simultaneously through multiple transmitting antennas and centrally process the echo signals through multiple antennas at the receiving end. This waveform diversity enables MIMO radar systems to achieve superior functionality compared to standard phased array radars in several basic aspects, including: (1) greatly enhancing the flexibility of the transmit beampattern design [18], (2) being directly applicable to parameter estimation and target detection [19], and (3) significantly improving parameter identifiability [20,21].

According to the location distribution of the antennas (i.e., the distance between the antennas), the MIMO radar systems are divided into two categories:

- **The colocated MIMO radar systems.** In the colocated MIMO radar systems, the distance between the different transmit and receive antennas is very small, so there are approximately equal observation angles with respect to the target. Besides, the colocated MIMO radar systems are easier to implement because their antenna structure is consistent with the traditional phased array radar. In the paper, we focus on the colocated MIMO radar systems.
- **The distributed MIMO radar systems.** In the distributed MIMO radar systems, the transmitting and receiving antennas are distributed independently over long distances. Since each transmitting and receiving path is approximately independent of the target, it is difficult to realize in practice when performing space and time synchronization.

2.2. MIMO Radar Waveform Design

Waveform design is a unique capability for MIMO radar systems. With proper design, the target detection capability and parameter estimation accuracy can be improved [22]. In this subsection, we will briefly introduce the general classification of the MIMO radar waveform design.

The methods for the MIMO radar waveform design are mainly concluded into three categories. The first category focuses on designing the waveform of the MIMO radar systems by maximizing the mutual information (MI) between the target response and the received echo [23,24].

The second category addresses the design task under signal-dependent clutter. They focus on maximizing the output signal-to-interference-plus-noise ratio (SINR) by optimizing the receive filter and the transmitting waveform jointly [3,25,26].

The third category is the design task we focus on, which solves the MIMO radar beampattern matching problem. The core issue in the problem is to control the transmit power distribution in the spacial and design the beampattern to approximate the actual one and minimize the cross-correlation sidelobes.

2.3. Constant Modulus Constraint

CMC is an essential constraint in MIMO radar systems. In practice, the non-linear power amplifiers in the systems are supposed to operate in saturation mode to maximize their efficiency [27,28]. In this case, the generated waveforms in the MIMO radar transmitters must have a constant modulus. Suppose that x_i is one of the generated waveforms by the MIMO radar systems, then under CMC, it can be expressed as $|x_i| = \xi$, where $\xi > 0$ is the modulus of the waveform x_i.

3. System Model and Problem Description

Consider a colocated MIMO radar system, with M transmit antennas in a uniform linear array (ULA). Let $\mathbf{s}_l = [s_1(l), s_2(l), \ldots, s_M(l)]^T \in \mathbb{C}^{M \times 1}, l = 1, \ldots, L$ denote the transmit waveform at the l-th snapshot with total L snapshots. Then, the transmit waveform matrix is denoted as [29]:

$$\mathbf{S} = \begin{bmatrix} s_1(1) & \cdots & s_1(l) & \cdots & s_1(L) \\ s_2(1) & \cdots & s_2(l) & \cdots & s_2(L) \\ \vdots & \vdots & \vdots & \vdots & \vdots \\ s_M(1) & \cdots & s_M(l) & \cdots & s_M(L) \end{bmatrix} \tag{1}$$

Assuming narrowband signals and non-dispersive propagation, the L snapshots synthesis signal at the direction of θ is denoted as:

$$\mathbf{y} = [\mathbf{I}_L \otimes \mathbf{a}^T(\theta)]\mathbf{s} \tag{2}$$

where

- $\mathbf{s} = \text{vec}(\mathbf{S}) \in \mathbb{C}^{ML \times 1}$ is the column vector of matrix $\mathbf{S}$ in (1).
- $\mathbf{a}(\theta) = [1, e^{-j\pi sin\theta}, \ldots, e^{-j\pi(M-1)sin\theta}]^T \in \mathbb{C}^{M \times 1}$ denotes the steering vector.
- $\mathbf{I}_L$ denotes the $N \times N$ dimensional identity matrix.

In most cases, a set of angles is used to cover the entire spatial region, and the power received at direction θ_r is denoted as:

$$p(\theta_r, \mathbf{s}) = \mathbf{y}^H \mathbf{y} = |[\mathbf{I}_L \otimes \mathbf{a}^T(\theta_r)]\mathbf{s}|^2 = \mathbf{s}^H \mathbf{\Pi}_r \mathbf{s} \tag{3}$$

where

- $\mathbf{y}$ is the synthesis signal in (2)
- $r = 1, \ldots, K$, K is the total number of points to cover the spatial region.
- $\mathbf{\Pi}_r = \mathbf{I}_L \otimes [\mathbf{a}^*(\theta_r)\mathbf{a}^T(\theta_r)]$.

Usually, we define (1) as the transmit beampattern. In addition to considering the beampattern, suppose that there are $\tilde{K}$ ($\tilde{K} \ll K$) targets of interests within the whole space, then the cross-correlation sidelobes term is denoted as [13]:

$$p_c(\tilde{\theta}_a, \tilde{\theta}_b, \mathbf{s}) = ([\mathbf{I}_L \otimes \mathbf{a}^T(\tilde{\theta}_a)]\mathbf{s})^H([\mathbf{I}_L \otimes \mathbf{a}^T(\tilde{\theta}_b)]\mathbf{s}) = \mathbf{s}^H \tilde{\mathbf{\Pi}}_{a,b} \mathbf{s} \tag{4}$$

where

- $a = 1, \ldots, \tilde{K}, b = 1, \ldots, \tilde{K}, a \neq b, \tilde{\theta}_a, \tilde{\theta}_b$ is one of the targets.
- $\tilde{\mathbf{\Pi}} = \mathbf{I}_L \otimes [\mathbf{a}^*(\tilde{\theta}_a)\mathbf{a}^T(\tilde{\theta}_b)]$.

The target of the transmit waveform design for beampattern matching is to approximate the desired beampattern through the designed beampattern and minimize the cross-correlation sidelobe term at multiple targets. Considering the joint optimization, the objective function model is given as [13]:

$$J(\mathbf{s},\delta) = \frac{1}{K}\sum_{r=1}^{K}\omega_r|\delta d_r - p(\theta_r,\mathbf{s})|^2 + \frac{2\omega_c}{\tilde{K}(\tilde{K}-1)}\sum_{a=1}^{\tilde{K}-1}\sum_{b=a+1}^{\tilde{K}}|p_c(\tilde{\theta}_a,\tilde{\theta}_b,\mathbf{s})|^2 \tag{5}$$

where

- $p(\theta_r,\mathbf{s})$ is the power received at direction θ_r in (5) and $p_c(\tilde{\theta}_a,\tilde{\theta}_b,\mathbf{s})$ is the cross-correlation sidelobe terms in (6).
- $\mathbf{d} = [d_1,...,d_r]^T$ denotes the desired beampattern.
- $\omega_r \geq 0(r = 1,...,K)$ and $\omega_c \geq 0$ denote the weighted coefficients.
- δ is a joint optimization parameter to constrain the beampattern to the desired one.

In practice, the CMC maximize the transmitter efficiency by keeping the power amplifier operating in saturation mode. Hence, by considering the CMC, the optimization problems are denoted as:

$$\min_{\mathbf{s},\delta}\quad J(\mathbf{s},\delta)$$
$$s.t.\quad |s(n)| = 1\quad n = 1,...,ML \tag{6}$$

Compactly speaking, the problem in (6) is a bivariate function, which can be alternately optimized by a cyclic optimization algorithm denoted as:

$$\delta^{(l+1)} = \arg\ \min_{\delta} J(\mathbf{s}^{(l)},\delta) \tag{7a}$$

$$\mathbf{s}^{(l+1)} = \arg\ \min_{\mathbf{s}} J(\mathbf{s},\delta^{(l+1)}) \tag{7b}$$

$$|s(n)| = 1\quad n = 1,...,ML$$

where l is the l-th cyclic optimization.

When $\mathbf{s}^{(l)}$ is fixed, the closed form solution for problem (7a) is directly denoted as:

$$\delta^{(l+1)} = \Delta^l \mathbf{d}^T / \|\mathbf{d}\|_2^2 \tag{8}$$

where $\Delta^l = [\mathbf{s}^{(l)H}\Pi_1\mathbf{s}^{(l)},...,\mathbf{s}^{(l)H}\Pi_l\mathbf{s}^{(l)}]^T$.

When $\delta^{(l+1)}$ is fixed, the objective function in (7b) is denoted as:

$$J(\mathbf{s}) = \frac{1}{K}\sum_{r=1}^{K}\omega_r\{|\mathbf{s}^H\Pi_r\mathbf{s}|^2 - 2\delta^{(l+1)}d_r(\mathbf{s}^H\Pi_r\mathbf{s})\} + \frac{2\omega_c}{\tilde{K}(\tilde{K}-1)}\sum_{a=1}^{\tilde{K}-1}\sum_{b=a+1}^{\tilde{K}}|\mathbf{s}^H\tilde{\Pi}_{a,b}\mathbf{s}|^2 \tag{9}$$

After the cyclic optimization, the problem in (9) becomes a single variable function. However, the constraint in the problem remains unchanged (i.e., the CMC), and the problem is finally formulated as:

$$\min_{\mathbf{s}}\quad J(\mathbf{s})$$
$$s.t.\quad |s(n)| = 1\quad n = 1,...,ML \tag{10}$$

Generally speaking, the problem in (10) is solved by most methods with relaxation, leading to performance degradation. In the following, the DBMWR method is developed without relaxation to solve the objective function.

4. The Beampattern Matching Based on the DBMWR Method

In this section, we propose the DBMWR method to optimize the problem (10) without relaxation. To be specific, we first convert the problem into an unconstrained function over the CC, and then we obtain its solution by the proposed DBMWR method.

Generally speaking, for each optimization variable $s(n)$, it lives in a continuous constrained search area given by the CC expressed as:

$$\Xi = \left\{ s \in \mathbb{C} \,\middle|\, \mathrm{Re}\{s\}^2 + \mathrm{Im}\{s\}^2 = 1 \right\} \tag{11}$$

The feasible set of the CMC is the ML times of CC in (11), denoted as:

$$\underbrace{\Xi \times \Xi \cdots \times \Xi}_{ML \ \text{times}} \tag{12}$$

Finally, this product of the CC in (12) is formally defined as:

$$\mathcal{M} = \Xi^{ML} = \{ \mathbf{s} \in \mathbb{C}^{ML} \,||s(n)| = 1 \quad n = 1, \ldots, ML\} \tag{13}$$

Hence, the problem in (10) is reformulated as an unconstrained problem over $\mathcal{M}$ in (13), denoted as:

$$\min_{\mathbf{s} \in \mathcal{M}} \quad J(\mathbf{s}) = \frac{1}{K} \sum_{r=1}^{K} \omega_r \{ |\mathbf{s}^H \mathbf{\Pi}_r \mathbf{s}|^2 - 2\delta^{(l+1)} d_r(\mathbf{s}^H \mathbf{\Pi}_r \mathbf{s}) \} + \frac{2\omega_c}{\tilde{K}(\tilde{K}-1)} \sum_{a=1}^{\tilde{K}-1} \sum_{b=a+1}^{\tilde{K}} |\mathbf{s}^H \tilde{\mathbf{\Pi}}_{a,b} \mathbf{s}|^2 \tag{14}$$

The problem in (14) is unconstrained and can be conveniently minimized by the DBMWR method. In the method, the Euclidean gradient is obtained first, then the gradient over CC is obtained, and, after that, the direction of descent and the step size is calculated, and, finally, the solution is updated by the retract operation. Repeat the above steps until convergence, and the final result is obtained. At i-th inner iteration for updating $\mathbf{s}^{(l+1)}$, the detailed derivation is given as follows:

4.1. Generation of the Euclidean Gradient

The objective functions in (14) can be reformulated as follows:

$$J(\mathbf{s}_i) = \frac{1}{K} \sum_{r=1}^{K} \omega_r \{ f_r(\mathbf{s}_i) f_r^*(\mathbf{s}_i) - g(\mathbf{s}_i) \} + \frac{2\omega_c}{\tilde{K}(\tilde{K}-1)} \sum_{a=1}^{\tilde{K}-1} \sum_{b=a+1}^{\tilde{K}} \{ f_c(\mathbf{s}_i) f_c^*(\mathbf{s}_i) \} \tag{15}$$

where

$$f_r(\mathbf{s}_i) = \mathbf{s}_i^{T} \mathbf{\Pi}_r^{T} \mathbf{s}_i^{*} = \mathbf{s}_i^{H} \mathbf{\Pi}_r \mathbf{s}_i \tag{16a}$$

$$f_r^*(\mathbf{s}_i) = \mathbf{s}_i^{T} \mathbf{\Pi}_r^{*} \mathbf{s}_i^{*} = \mathbf{s}_i^{H} \mathbf{\Pi}_r^{H} \mathbf{s}_i \tag{16b}$$

$$g(\mathbf{s}_i) = 2\delta^{(l+1)} d_r(\mathbf{s}_i^{H} \mathbf{\Pi}_r \mathbf{s}_i) \tag{16c}$$

$$f_c(\mathbf{s}_i) = \mathbf{s}_i^{H} \tilde{\mathbf{\Pi}}_{a,b} \mathbf{s}_i = \mathbf{s}_i^{T} \tilde{\mathbf{\Pi}}_{a,b}^{T} \mathbf{s}_i^{*} \tag{16d}$$

$$f_c^*(\mathbf{s}_i) = \mathbf{s}_i^{H} \tilde{\mathbf{\Pi}}_{a,b}^{H} \mathbf{s}_i = \mathbf{s}_i^{T} \tilde{\mathbf{\Pi}}_{a,b}^{*} \mathbf{s}_i^{*} \tag{16e}$$

To derive the Euclidean gradient, the following definition could be utilized [30]:

$$\nabla(J(\mathbf{s}_i)) = 2\nabla_{\mathbf{s}_i^*} J(\mathbf{s}_i) \tag{17}$$

where $\nabla_{\mathbf{s}_i^*} J(\mathbf{s}_i)$ is the gradient of $J(\mathbf{s}_i)$ is terms of $\mathbf{s}_i^*$.

Based on the definition in (17), the gradient of $f_r(\mathbf{s}_i) f_r^*(\mathbf{s}_i)$ and $f_c(\mathbf{s}_i) f_c^*(\mathbf{s}_i)$ in objective function (15) can be calculated through the following product rules:

$$\begin{aligned} \nabla(f(\mathbf{s}_i) f^*(\mathbf{s}_i)) &= 2\nabla_{\mathbf{s}_i^*}(f(\mathbf{s}_i) f^*(\mathbf{s}_i)) \\ &= 2(\nabla_{\mathbf{s}_i^*} f(\mathbf{s}_i)) f^*(\mathbf{s}_i) + 2(\nabla_{\mathbf{s}_i^*} f^*(\mathbf{s}_i)) f(\mathbf{s}_i) \end{aligned} \tag{18}$$

The gradients in (18) are denoted as:

$$\nabla_{\mathbf{s}_i^*} f_r(\mathbf{s}_i) = \mathbf{s}_i^T \mathbf{\Pi}_r^T = \mathbf{\Pi}_r \mathbf{s}_i \tag{19a}$$

$$\nabla_{\mathbf{s}_i^*} f_r^*(\mathbf{s}_i) = \mathbf{s}_i^T \mathbf{\Pi}_r^* = \mathbf{\Pi}_r^H \mathbf{s}_i \tag{19b}$$

$$\nabla_{\mathbf{s}_i^*} f_c(\mathbf{s}_i) = \mathbf{s}_i^T \tilde{\mathbf{\Pi}}_{a,b}^T = \tilde{\mathbf{\Pi}}_{a,b} \mathbf{s}_i \tag{19c}$$

$$\nabla_{\mathbf{s}_i^*} f_c^*(\mathbf{s}_i) = \mathbf{s}_i^T \tilde{\mathbf{\Pi}}_{a,b}^* = \tilde{\mathbf{\Pi}}_{a,b}^H \mathbf{s}_i \tag{19d}$$

Replacing the identities (19) in (18), the overall Euclidean gradient is denoted as:

$$\nabla(J(\mathbf{s}_i)) = \frac{1}{K} \sum_{r=1}^{K} \omega_r \{\Psi_r - 4\delta^{(l+1)} d_r \mathbf{\Pi}_r \mathbf{s}_i\} + \frac{2\omega_c}{\tilde{K}(\tilde{K}-1)} \sum_{a=1}^{\tilde{K}-1} \sum_{b=a+1}^{\tilde{K}} \Psi_c \tag{20}$$

where

$$\Psi_r = 2\mathbf{\Pi}_r^H \mathbf{s}_i \mathbf{s}_i^H \mathbf{\Pi}_r \mathbf{s}_i + 2\mathbf{\Pi}_r \mathbf{s}_i \mathbf{s}_i^H \mathbf{\Pi}_r^H \mathbf{s}_i \tag{21a}$$

$$\Psi_c = 2\tilde{\mathbf{\Pi}}_{a,b}^H \mathbf{s}_i \mathbf{s}_i^H \tilde{\mathbf{\Pi}}_{a,b} \mathbf{s}_i + 2\tilde{\mathbf{\Pi}}_{a,b} \mathbf{s}_i \mathbf{s}_i^H \tilde{\mathbf{\Pi}}_{a,b}^H \mathbf{s}_i \tag{21b}$$

4.2. Generation of the Gradient Over CC

Since $\mathcal{M}$ is embedded in a Euclidean space, the gradient over CC, namely, $\nabla_{\mathcal{M}}(J(\mathbf{s}))$, is the orthogonal projection from the Euclidean gradient to the tangent space $\mathcal{T}_s\mathcal{M}$, denoted as:

$$\begin{aligned}
\nabla_{\mathcal{M}}(J(\mathbf{s}_i)) &= \mathrm{Proj}_{\mathcal{T}_{\mathbf{s}_i}\mathcal{M}}(\nabla J(\mathbf{s}_i)) \\
&= \nabla J(\mathbf{s}_i) - \mathrm{Re}\{\nabla J(\mathbf{s}_i)^* \odot \mathbf{s}_i\} \odot \mathbf{s}_i
\end{aligned} \tag{22}$$

where

- $\mathcal{T}_{\mathbf{s}_i}\mathcal{M}$ denotes the tangent space composed of all tangent vectors at point $\mathbf{s}_i$ of $\mathcal{M}$, which is formulated as:

$$\mathcal{T}_{\mathbf{s}_i}\mathcal{M} = \{\bar{\mathbf{s}} \in \mathbb{C}^{ML} | \mathrm{Re}\{\bar{\mathbf{s}} \odot \mathbf{s}_i^*\} = 0\} \tag{23}$$

- $\mathrm{Proj}_{\mathcal{T}_{\mathbf{s}_i}\mathcal{M}}(\nabla J(\mathbf{s}_i))$ denotes the orthogonal projection from $\nabla J(\mathbf{s}_i)$ to $\mathcal{T}_{\mathbf{s}_i}\mathcal{M}$ in (23), which is denoted as:

$$\mathrm{Proj}_{\mathcal{T}_{\mathbf{s}_i}\mathcal{M}}(\nabla J(\mathbf{s}_i)) = \nabla J(\mathbf{s}_i) - \mathrm{Re}(\nabla J(\mathbf{s}_i) \odot \mathbf{s}_i^*) \odot \mathbf{s}_i^*. \tag{24}$$

- $\nabla J(\mathbf{s}_i)$ is the Euclidean gradient in (20).

4.3. Derivation of the Descent Direction

In most cases, the Euclidean gradient of $J(\mathbf{s})$ is used as the descent direction, but the speed of this approach is slow in practice [31]. Therefore, we introduce the Polak-Ribiére conjugate gradient algorithm in [32], which contains the second order information and has a faster convergence speed. The descent direction is given by:

$$\eta_i = -\nabla_{\mathcal{M}}(J(\mathbf{s}_i)) + \beta_i^{PR} \mathcal{T}_{\mathbf{s}_{i-1} \to \mathbf{s}_i} \mathcal{M}(\eta_{i-1}) \tag{25}$$

where

- β_i^{PR} is the conjugate parameter in Polak-Ribiére algorithm. For the CC, it is denoted as:

$$\beta_i^{PR} = [\nabla_{\mathcal{M}}(J(\mathbf{s}_i))]^H \frac{\nabla_{\mathcal{M}}(J(\mathbf{s}_i)) - \mathcal{T}_{\mathbf{s}_{i-1} \to \mathbf{s}_i}\mathcal{M}(\nabla_{\mathcal{M}}(J(\mathbf{s}_i)))}{\|\nabla_{\mathcal{M}}(J(\mathbf{s}_{i-1}))\|^2} \tag{26}$$

- $\mathcal{T}_{\mathbf{s}_{i-1}\to\mathbf{s}_i}\mathcal{M}(\cdot)$ in (25) and (26) is a vector transport operation, denoted as:

$$\mathcal{T}_{\mathbf{s}_{i-1}\to\mathbf{s}_i}\mathcal{M}(\boldsymbol{\eta}_{i-1}) = \boldsymbol{\eta}_{i-1} - \mathrm{Re}\{\boldsymbol{\eta}_{i-1}\odot\mathbf{s}_i^*\}\odot\mathbf{s}_i \tag{27}$$

- $\nabla_{\mathcal{M}}(J(\mathbf{s}_i))$ is the gradient over CC in (22)

4.4. Derivation of the Step Size

Here, we introduce the Armijo line search in [33,34] to obtain the step size and ensure non-increase in each iteration of the objective function conveniently. In the algorithm, the smallest integer $m \geq 0$ is found, such that:

$$J(\mathbf{s}_i) - J(\mathbf{s}_i + \tau_{i-1}\beta^m\boldsymbol{\eta}_i) \geq \sigma\tau_{i-1}\beta^m\|\nabla_{\mathcal{M}}J(\mathbf{s}_i)\|^2 \tag{28}$$

where $\tau_{i-1} > 0, \sigma, \beta \in (0,1)$ and $\boldsymbol{\eta}_i$ is the descent direction in (25).

τ_i is the i-th iteration step, denoted as:

$$\tau_i = \tau_{i-1}\beta^m \tag{29}$$

4.5. Updating the Solution

The operation of updating $\mathbf{s}_i$ on tangent space $\mathcal{T}_s\mathcal{M}$ is denoted as:

$$\widetilde{\mathbf{s}}_i = \mathbf{s}_i + \tau_i\boldsymbol{\eta}_i \tag{30}$$

However, the solution can not be simply updated via (30), that is, because the solution lies on the tangent space $\mathcal{T}_s\mathcal{M}$ instead of the CC. Therefore, a mapping operation is needed to retract $\widetilde{\mathbf{s}}_i$ in Equation (30) from the tangent space to the CC:

$$\mathbf{s}_{i+1} = \widetilde{\mathbf{s}}_i \odot \frac{1}{|\widetilde{\mathbf{s}}_i|} \tag{31}$$

According to the above discussion, the DBMWR method for solving (6) is concluded in Algorithm 1.

Algorithm 1 : The DBMWR method for solving (6)

Input: $\mathbf{s}^0 \in \mathcal{M}, \delta^0, \mathbf{d}, \omega_r, \omega_c, \Pi_r$.

1: **Reapeat:**

- **Updating δ^{l+1} with $\mathbf{s}^l$**

2:     Calculate δ^{l+1} according to (8).

- **Updating $\mathbf{s}^{l+1}$ with δ^{l+1}**

3:     $i = 0, \mathbf{s}_i^{l+1} = \mathbf{s}^l$.

4:     **Reapeat:**

5:     Calculate Euclidean gradient by (20).

6:     Calculate gradient over CC by (22).

7:     Calculate descent direction by (25).

8:     Calculate step size by (28).

9:     Obtain $\mathbf{s}_{i+1}^{l+1}$ by (31).

10:     $i = i + 1$.

11:     **Until converge**

12:     $\mathbf{s}^{l+1} = \mathbf{s}_i^{l+1}$.

13:     $l = l + 1$.

14: **Until converge**

Output: $\mathbf{s}^* = \mathbf{s}^l, \delta^* = \delta^l$

4.6. Analysis of the Computation Complexity

The computation complexity is mainly built upon the update of s^{l+1}. For i-th iteration, the complexity for Algorithm 1 is listed as follows:

- Euclidean gradient and gradient over CC: $\mathcal{O}(5K(M^2L^2 + ML) + 2ML)$.
- Descent direction: $\mathcal{O}(4ML)$.
- Step size: $\mathcal{O}((m+1)(2K(M^2L^2 + ML)))$.
- Update the solution: $\mathcal{O}(2ML)$.

Hence, for each iteration, the computational complexity is denoted as: $\mathcal{O}((2m+7)K(M^2L^2 + ML) + 8ML)$.

4.7. Analysis of the Convergence

Generally speaking, finding a proper condition for the step size to ensure the monotonic decrease is difficult for the quartic objective function in (14). In this paper, we introduce the Armijo line search method with variable step size to guarantee a monotonic decrease on the tangent space of the CC at s_i. In [35], Proposition 1.2.1 states that for the gradient based methods, every limit generated point is a stationary point when the step size is chosen by the Armijo line search method. In our setup, we can use the proposition to ensure the reduction of the tangent space and generate a solution $\tilde{s}_i$ on the tangent space of the CC with a reduction on the value of the objective function (i.e., $J(s_i) \geq J(\tilde{s}_i)$). Besides, since the mapping operation to retract $\tilde{s}_i$ from the tangent space to obtain s_{i+1} on the CC is a linear projection, the convergence property on the CC is approximately same as that on the tangent space. Then, the objective function $J(s_i)$ is monotonic decrease and converges to a finite value.

5. The Sparse Beampattern Matching Design

The selection of the sparse antenna positions is a key technology widely used in MIMO radar systems [36,37]. Since the available antennas are placed in a wider transmit field, an additional DOF is introduced to the system. With the increased DOF, the MIMO radar beampattern can achieve better accuracy by using the same number of antennas. In this case, the power consumption is reduced, since fewer antennas can achieve a similar beampattern. Because of the above merits, to further obtain a better performance, we propose a method to jointly optimize the beampattern and the sparse antenna positions.

More specifically, the problem is first formulated as a multi-variables function and then solved by a cyclic optimization through optimizing the beampattern and the sparse antenna positions separately. In each iteration, the beampattern is optimized by the DBMWR method, and the sparse antenna positions are selected by a greedy search method.

5.1. The Problem Formation

Let $\mathbf{p} \in \mathbb{B}_M^N$ denote the sparse antenna positions. Suppose we need to select M antennas in total N grid points. The corresponding signal at the direction of θ is now denoted as:

$$\mathbf{y}_p = [\mathbf{I}_L \otimes (\mathbf{p} \odot \mathbf{a}(\theta))^T]\mathbf{s} \tag{32}$$

Hence, by considering $\mathbf{p}$, the objective function is now denoted as:

$$J_p(\mathbf{p}, \mathbf{s}, \delta) = \frac{1}{K}\sum_{r=1}^{K} \omega_r |\delta d_r - p_p(\theta_r, \mathbf{p}, \mathbf{s})|^2 + \frac{2\omega_c}{\tilde{K}(\tilde{K}-1)} \sum_{a=1}^{\tilde{K}-1}\sum_{b=a+1}^{\tilde{K}} |p_{pc}(\tilde{\theta}_a, \tilde{\theta}_b, \mathbf{p}, \mathbf{s})|^2 \tag{33}$$

where

$$p_p(\theta_r, \mathbf{p}, \mathbf{s}) = |[\mathbf{I}_L \otimes (\mathbf{p} \odot \mathbf{a}(\theta))^T]\mathbf{s}|^2 \tag{34}$$

and

$$p_{pc}(\tilde{\theta}_a, \tilde{\theta}_b, \mathbf{p}, \mathbf{s}) = ([\mathbf{I}_L \otimes (\mathbf{p} \odot \mathbf{a}(\tilde{\theta}_a))^T]\mathbf{s})^H ([\mathbf{I}_L \otimes (\mathbf{p} \odot \mathbf{a}(\tilde{\theta}_b))^T]\mathbf{s}) \tag{35}$$

To minimize the objective function in (33), the cyclic optimization is generated to optimize $(\mathbf{s}, \delta)$ and $\mathbf{p}$ separately. For the fixed $\mathbf{p}$, the problem becomes the same in (10), which can be optimized by the DBMWR method. For a fixed $(\mathbf{s}, \delta)$, the optimization problem is denoted as:

$$
\begin{aligned}
\min_{\mathbf{p}} \quad & J_p(\mathbf{p}) \\
s.t. \quad & \mathbf{p} \in \mathbb{B}_M^N
\end{aligned}
\tag{36}
$$

5.2. The Proposed Greedy Search Framework

A greedy search method is developed in this subsection to optimize the problem in (36) efficiently. In the method, we first generate the feasible solutions set, and then we select the fittest solution. Repeat the above steps until $\mathbf{p}$ obtains stopping criteria. At k-th iteration for updating, the steps are listed as follows:

5.2.1. Generation of Possible Solutions Set

Based on the solution at $(k-1)$-th iteration, the set of possible solutions $\mathbf{p}_S^k$ is generated as follows:

$$
\mathbf{p}_S^k = \left\{ \mathbf{p} \mid H(\mathbf{p}, \mathbf{p}^{k-1}) = 1, \|\mathbf{p}\|_1 < \|\mathbf{p}^{k-1}\|_1 \right\}
\tag{37}
$$

where $H(\mathbf{a}, \mathbf{b})$ is the Hamming distance between $\mathbf{a}$ and $\mathbf{b}$.

More specifically, the generated set $\mathbf{p}_S^k$ composed of the solution that only differs from $\mathbf{p}^{k-1}$ in one bit, which is one less nonzero element.

5.2.2. Updating the Solution

With the current set of vectors $\mathbf{p}_S^k$, we select the fittest solution $\mathbf{p}^k$ based on:

$$
\mathbf{p}^k = \arg \min_{\mathbf{p} \in \mathbf{p}_S^k} J(\mathbf{p})
\tag{38}
$$

According to the above discussion, the developed method for solving (36) is concluded in Algorithm 2.

Algorithm 2 : The greedy search method for solving (36)

Input: $\mathbf{p}^0 = \mathbf{1}_N$, N, M.
  1: **Reapeat:**
  2:     Generate the possible solutions set by (37).
  3:     Updating the solution by (38).
  4: **Until** $\left\| \mathbf{p}^k \right\|_1 = M$.
Output: $\mathbf{p}^* = \mathbf{p}^k$

Finally, the proposed sparse beampattern matching design method to jointly optimize the beampattern and the sparse antenna positions is summarized in Algorithm 3.

Algorithm 3 : The proposed sparse beampattern matching design method for solving (33)

Input: $\mathbf{s}^0 \in \mathcal{M}$, δ^0, $\mathbf{d}$, ω_r, ω_c, $\mathbf{\Pi}_r$, $\mathbf{p}^0 = 1_M$, N, M.

  1: **Reapeat:**

      • **Updating** $(\mathbf{s}^{t+1}, \delta^{t+1})$ **with** $\mathbf{p}^t$

  2:     Solve problem (36) and obtain $(\mathbf{s}^{t+1}, \delta^{t+1})$ based on **Algorithm 1**.

      • **Updating** $\mathbf{p}^{t+1}$ **with** $(\mathbf{s}^{t+1}, \delta^{t+1})$

  3:     Solve problem (6) and obtain $\mathbf{p}^{t+1}$ based on **Algorithm 2**.

  4: **Until converge.**

Output: $\mathbf{s}^* = \mathbf{s}^{t+1}$, $\delta^* = \delta^{t+1}$, $\mathbf{p}^* = \mathbf{p}^{t+1}$

---

6. Numerical Results

In this section, simulation results are shown in the following aspects: (1) analysis of convergence; (2) the comparison of beampattern matching performance between the proposed method and other methods; (3) the capability of the cross-correlation sidelobes controlling; and (4) the performance of the sparse beampattern matching design. The simulation results are run under MATLAB R2019a, CPU Core i7, RAM of 16GB, and GPU GTX2060.

We consider the simulation parameters the same as [13,14] with $M = 10$ transmit antennas, half-wavelength inter-element interval, and $L = 32$ snapshots. The angle is set in the range of $(90°, 90°)$ with $1°$ space. The weighted coefficients are $\omega_r = 1$ for $r = 1, \ldots, K$, and $\omega_c = 0$.

We further consider two cases of desired beampatterns, Case 1 and Case 2, which are the same as [13,14]:

- Case 1 considers the desired beampattern with one mainlobe at the direction of $\theta = 0°$, and the width of it is $\Delta = 60°$, denoted as:

$$\mathbf{d}_1(\theta) = \begin{cases} 1, & \theta \in (-30, 30), m = 1, 2, 3 \\ 0, & \text{otherwise} \end{cases} \tag{39}$$

- Case 2 considers the desired beampattern with three mainlobes at the direction of $\theta_1 = -40°$, $\theta_2 = 0°$, and $\theta_3 = 40°$, and each of them has a width of $\Delta = 20°$, denoted as:

$$\mathbf{d}_2(\theta) = \begin{cases} 1, & \theta \in (\theta_m - \frac{\Delta}{2}, \theta_m + \frac{\Delta}{2}), m = 1, 2, 3 \\ 0, & \text{otherwise} \end{cases} \tag{40}$$

6.1. Analysis of Convergence

Figure 1 shows normalized objective function versus the iteration numbers with different desired beampatterns. As shown in the figure, when we consider Case 1 as the desired beampattern, the objective function decreases sharply in the first 40 iterations and starts convergence after 150 iterations. When we consider Case 2 as the desired beampattern, the objective function decreases sharply in the first 40 iterations and starts convergence after 110 iterations. Besides, the total computation time when considering Case 1 is 30.2 s, and the total computation time when considering Case 2 is 38.0 s.

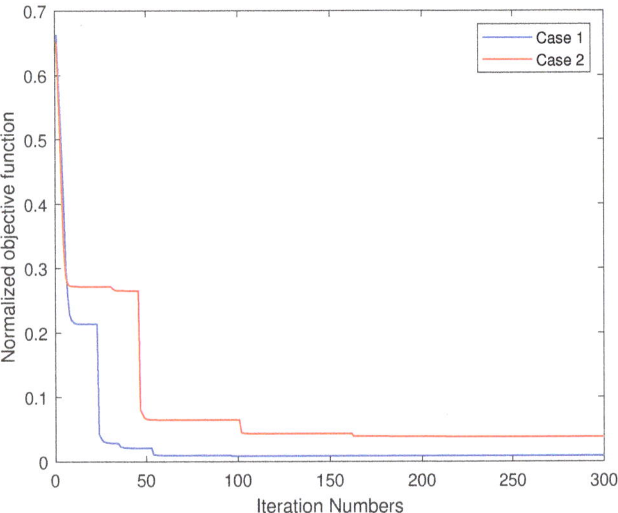

Figure 1. The normalized objective function versus iteration numbers with different desired beampatterns Case 1 and Case 2.

6.2. The Comparison of Beampattern Matching Performance between the Proposed Method and Other Methods

In this subsection, we consider the MM-based method in [12], the CA method in [10], the ADMM-based method in [13], and the RNN method in [14] for comparison. We further consider Case 1 as the desired beampattern with one mainlobe in Figure 2 and Case 2 as the desired beampattern with three mainlobes in Figure 3, respectively.

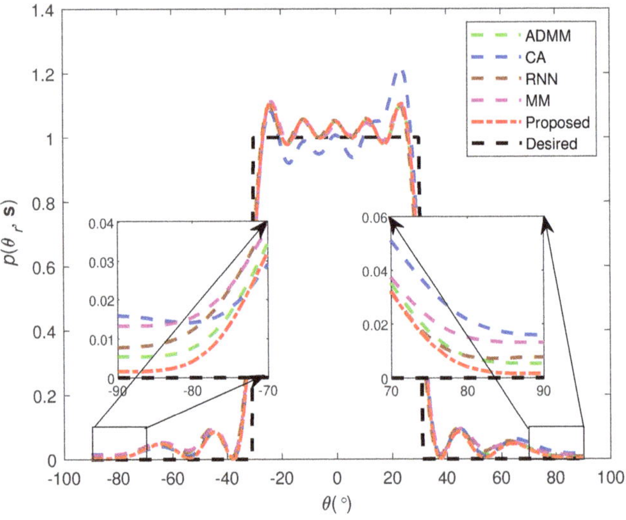

Figure 2. The beampattern matching comparison between different methods when considering Case 1 as the desired beampattern.

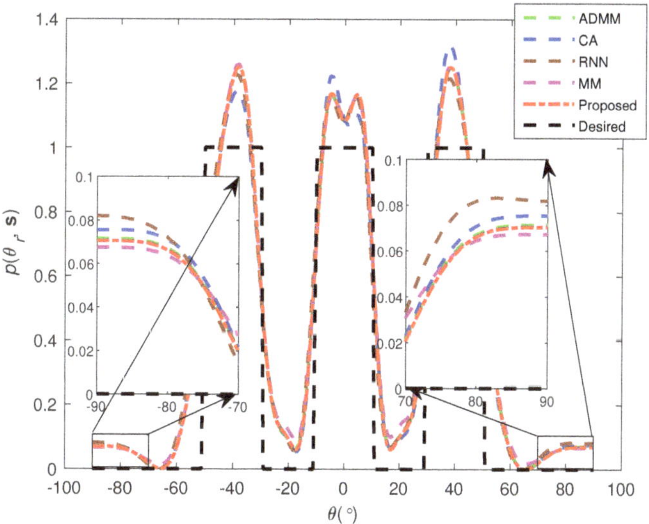

Figure 3. The beampattern matching comparison between different methods when considering Case 2 as the desired beampattern.

As can be seen in Figures 2 and 3, the beampattern generated by the proposed method has the lowest sidelobe value in whole angle space, meaning that the proposed method has the best sidelobe suppression ability and the highest beampattern matching accuracy.

We further introduce the mean square error (MSE) in (41) to show the beampattern matching performance between different methods more clearly. The MSE is the square error between the designed beampattern and the desired one. Hence the lower MSE value means better matching performance. Tables 1 and 2 show the MSE, σ, computation complexity, and convergence time for different methods when considering Case 1 and Case 2 as the desired beampattern, respectively. As can be seen from the tables, the proposed method has the lowest MSE values and reasonable computation time. Although the MSE values are similar between the ADMM-based method and the proposed method, the ADMM-based method is not strictly CMC. Hence, with relaxation, the method may have a better performance where we introduce σ in (42) to demonstrate it. As can be seen from the tables, the σ in the ADMM-based method has a small value (i.e., not strictly CMC), while the value of it is zero (i.e., strictly CMC) in other methods.

Table 1. The MSE, σ, computation complexity, and convergence time (seconds) for different methods when considering Case 1 as the desired beampattern.

Methods	MSE	σ	Computation Complexity	Convergence Time (s)
RNN in [14]	0.0098	0	-	430
CA in [10]	0.0102	0	$\mathcal{O}(M^{3.5} + M^3)$	182
MM in [12]	0.0095	0	$\mathcal{O}(M^2 L)$	6.25
ADMM in [13]	0.0089	0.0018	$\mathcal{O}(M^3 L^3)$	444
Proposed method	**0.0086**	**0**	$\mathcal{O}(M^2 L^2)$	30.2

Table 2. The MSE, σ, computation complexity, and convergence time (seconds) for different methods when considering Case 2 as the desired beampattern.

Methods	MSE	σ	Computation Complexity	Convergence Time (s)
RNN in [14]	0.0390	0	-	450
CA in [10]	0.0391	0	$\mathcal{O}(M^{3.5} + M^3)$	176
MM in [12]	0.0442	0	$\mathcal{O}(M^2 L)$	6.18
ADMM in [13]	0.0382	0.0015	$\mathcal{O}(M^3 L^3)$	440
Proposed method	**0.0381**	0	$\mathcal{O}(M^2 L^2)$	38.0

The mean square error (MSE) is defined based on [38]:

$$\mathbf{MSE} = \frac{1}{K} \sum_{r=1}^{K} \left(d_r - p(\theta_r, \mathbf{s})/\delta^2 \right)^2 \tag{41}$$

where d_r is the desired waveform and $p(\theta_r, \mathbf{s})/\delta^2$ is the normalized waveform.

The σ is defined based on:

$$\sigma = \max(\mathbf{s}) - \min(\mathbf{s}) \tag{42}$$

where $\min(\mathbf{s})$ is the minimum amplitude in the designed waveform, and $\max(\mathbf{s})$ is the maximum amplitude in the designed waveform.

6.3. The Capability of the Cross-Correlation Sidelobes Controlling

We consider that ω_c is not zero and Case 2 as the desired beampattern in the following simulations while other simulation parameters keep the same. Table 3 shows the results of the cross-correlation sidelobes term (i.e., $\frac{2\omega_c}{\tilde{K}(\tilde{K}-1)} \sum_{a=1}^{\tilde{K}-1} \sum_{b=a+1}^{\tilde{K}} |p_c(\tilde{\theta}_a, \tilde{\theta}_b, and\mathbf{s})|^2$ in (5)) under different values of ω_c). As shown in the table, the cross-correlation sidelobes term is minimized when ω_c is not zero. Besides, the optimal performance is obtained when $\omega_c = 1$.

Table 3. The cross-correlation sidelobes term under different values of ω_c by the proposed method.

	$\omega_c = 0$	$\omega_c = 0.001$	$\omega_c = 0.01$	$\omega_c = 0.1$	$\omega_c = 1$	$\omega_c = 10$
The values of the cross-correlation sidelobes term	3.012×10^4	538.0	23.03	0.2541	2.577×10^{-4}	1.313×10^{-3}

Next, the beampattern matching performance comparison with $\omega_c = 1$ and $\omega_c = 0$ is illustrated in Figure 4. The designed beampattern generated considering the cross-correlation sidelobes term (i.e., MSE = 0.0380) is similar to the beampattern obtained without considering the cross-correlation sidelobes term (i.e., MSE = 0.0381), which is because the cross-correlation sidelobes term is quite small compared with the beampattern matching term. Nevertheless, the cross-correlation behavior is much better when using $\omega_c = 1$ than that of using $\omega_c = 0$, which is because the generated waveforms under the case of $\omega_c = 1$ are almost uncorrelated with each other.

As a result, the proposed method has great cross-correlation sidelobes control capability and can generate uncorrelated waveforms.

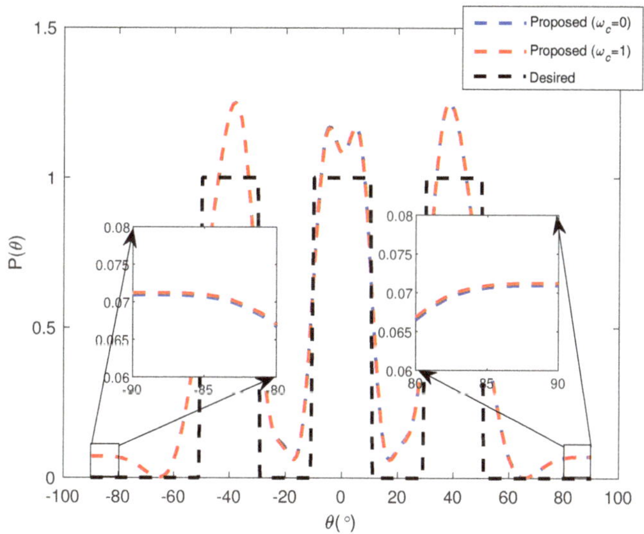

Figure 4. The beampattern matching performance comparison with $\omega_c = 0$ and $\omega_c = 1$.

6.4. The Performance of the Sparse Beampattern Matching Design

In this subsection, we test the performance of the proposed sparse beampattern matching design method. We consider Case 2 as the desired beampattern with $M = 10$ antennas and $N = 10, 13, 15$ and 20 grid points in the following simulations while other simulation parameters keep the same.

Figure 5 shows beampattern matching performance under different numbers of grid points. It can be seen that the sidelobes of the beampattern with sparse antenna positions selection are significantly reduced compared to the beampattern without sparse antenna positions selection (i.e., $N = 10$). Besides, as the number of grid points N increases, the sidelobe values of the designed beampatterns decrease. Table 4 shows the corresponding MSE values under different numbers of grid points. As can be seen from the table, the MSE values are lower after sparse antenna positions selection and decrease with the increase in the grid points N, which is consistent with the results shown in Figure 5. The results shown in Figure 5 and Table 4 demonstrate that the proposed sparse beampattern matching design method can greatly enhance the beampattern matching performance by selecting proper sparse antenna positions. Besides, with the increase in the grid points N, more DOF is provided to the system for beampattern designing, which is helpful to increase the beampattern matching performance. Finally, Figure 6 shows the corresponding sparse antenna positions.

Table 4. The corresponding MSE values with $M = 10$ antennas and a different number of grid points.

Grid Points	$N = 10$	$N = 13$	$N = 15$	$N = 20$
MSE	0.0381	0.0240	0.0221	0.0191

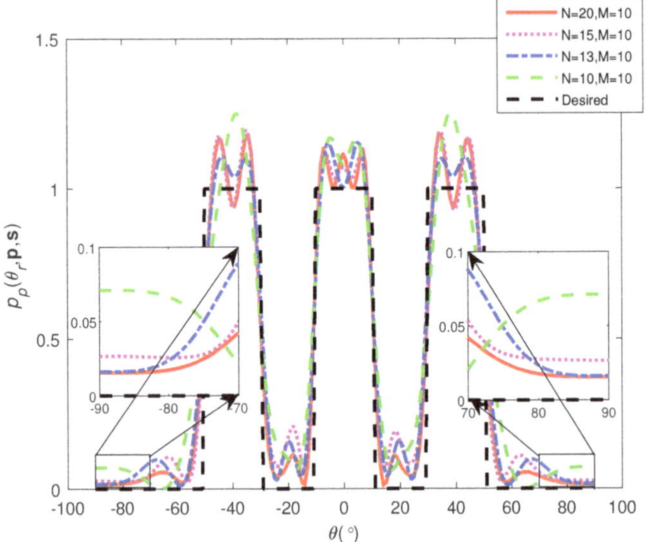

Figure 5. The beampattern matching performance comparison with $M = 10$ antennas and a different number of grid points.

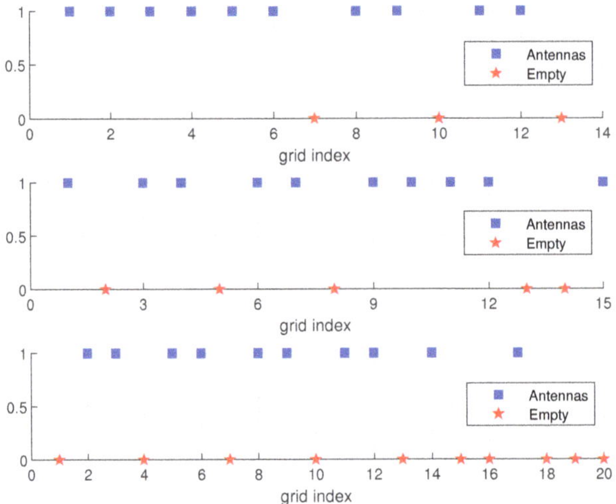

Figure 6. The corresponding sparse antenna positions with $M = 10$ antennas and a different number of grid points.

We further compare the proposed method with the method in [39], which jointly optimize the WCM and the sparse antenna positions. However, the method in [39] is an indirect optimization method that cannot obtain the waveform directly. To obtain the waveform from the WCM, the CA method in [10] is used. The result of the beampattern matching comparison between the proposed method and the method in [39] after the sparse antenna position selection is illustrated in Figure 7. As shown in the figure, the proposed method has lower sidelobes compared with the method in [39]. Similar to the above, the comparison in terms of MSE between the two methods is illustrated in Table 5. As can be seen from the table, the proposed method has lower MSE values compared with the method in [39]. Therefore, the proposed method has better accuracy in terms of the sparse beampattern matching design than the method in [39].

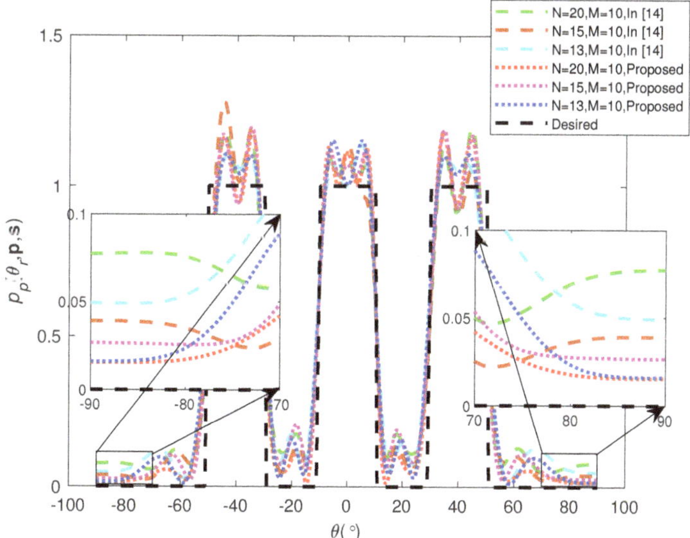

Figure 7. The beampattern matching performance comparison between the method in [39] and the proposed method with $M = 10$ antennas and a different number of grid points.

Table 5. The corresponding MSE values between the proposed method and the method in [39] with $M = 10$ antennas and a different number of grid points.

Grid Points/MSE	$N = 13$	$N = 15$	$N = 20$
The method in [39]	0.0266	0.0235	0.0249
The proposed method	**0.0240**	**0.0221**	**0.0191**

7. Conclusions

In this paper, we consider the MIMO radar waveform design for beampattern matching problem. Unlike most existing methods, which approach this challenge by relaxation, the novel optimization method DBMWR is developed without relaxation by noticing the new intrinsic of CMC. More precisely, the problem is first transformed into an unconstraint quartic polynomial over the CC, and, after that, it is solved through the proposed method. Results show that the DBMWR method obtains a balance in terms of computation complexity and accuracy compared with the existing methods.

Author Contributions: Conceptualization, W.X. and J.H.; methodology, W.X., Y.S., and K.Z.; software, W.X. and G.Z.; validation, W.X., J.H. and K.Z.; formal analysis, W.X. and G.Z.; investigation, W.X. and X.X.; resources, W.X. and X.X.; data curation, W.X.; writing—original draft preparation, W.X.; writing—review and editing, W.X., J.H., and K.Z.; visualization, K.Z.; supervision, J.H.; project administration, W.X.; funding acquisition, J.H. All authors have read and agreed to the published version of the manuscript.

Funding: This research was funded by the Science and Technology Project of Sichuan Province under Grant 2023YFG0176, the National Natural Science Foundation of China under Grant 62231006 and the Municipal Government of Quzhou under Grant 2022D009.

Data Availability Statement: Not applicable.

Acknowledgments: Thanks to the editor and all reviewers for their valuable comments.

Conflicts of Interest: The authors declare no conflict of interest.

Abbreviations

The following abbreviations are used in this manuscript:

MIMO	Multiple-Input Multiple-Output
CMC	Constant Modulus Constraint
DBMWR	Direct Beampattern Matching Without Relaxation
CC	Complex Circle
DOF	Degrees Of Freedom
WCM	Waveform Covariance Matrix
IPM	Interior Point Method
SQP	Semidefinite Quadratic Programming
CA	Cyclic Algorithm
SVD	Singular Value Decomposition
DFT	Discrete-Fourier-Transform
MM	Majorization-Minimization
ADMM	Alternating Direction Method Of Multipliers
RNN	Residual Neural Network
ULA	Uniform Linear Array

References

1. Liao, B. Fast Angle Estimation for MIMO Radar With Nonorthogonal Waveforms. *IEEE Trans. Aerosp. Electron. Syst.* **2018**, *54*, 2091–2096. [CrossRef]
2. Gemechu, A.Y.; Cui, G.; Yu, X. Spectral-Compatible Transmit Beampattern Design With Minimum Peak Sidelobe for Narrowband MIMO Radar. *IEEE Trans. Veh. Technol.* **2022**, *71*, 11900–11910. [CrossRef]
3. Wang, L.; Wang, W.Q.; So, H.C. Covariance Matrix Estimation for FDA-MIMO Adaptive Transmit Power Allocation. *IEEE Trans. Signal Process.* **2022**, *70*, 3386–3399. [CrossRef]
4. Arora, A.; Tsinos, C.G.; Shankar, M.R.B.; Chatzinotas, S.; Ottersten, B. Efficient Algorithms for Constant-Modulus Analog Beamforming. *IEEE Trans. Signal Process.* **2022**, *70*, 756–771. [CrossRef]
5. Cui, G.; Yu, X.; Foglia, G.; Huang, Y.; Li, J. Quadratic Optimization With Similarity Constraint for Unimodular Sequence Synthesis. *IEEE Trans. Signal Process.* **2017**, *65*, 4756–4769. [CrossRef]
6. Yu, X.; Alhujaili, K.; Cui, G.; Monga, V. MIMO Radar Waveform Design in the Presence of Multiple Targets and Practical Constraints. *IEEE Trans. Signal Process.* **2020**, *68*, 1974–1989. [CrossRef]
7. Aubry, A.; De Maio, A.; Govoni, M.A.; Martino, L. On the Design of Multi-Spectrally Constrained Constant Modulus Radar Signals. *IEEE Trans. Signal Process.* **2020**, *68*, 2231–2243. [CrossRef]
8. Domouchtsidis, S.; Tsinos, C.G.; Chatzinotas, S.; Ottersten, B. Constant Envelope MIMO-OFDM Precoding for Low Complexity Large-Scale Antenna Array Systems. *IEEE Trans. Wirel. Commun.* **2020**, *19*, 7973–7985. [CrossRef]
9. Stoica, P.; Li, J.; Xie, Y. On Probing Signal Design For MIMO Radar. *IEEE Trans. Signal Process.* **2007**, *55*, 4151–4161. [CrossRef]
10. Stoica, P.; Li, J.; Zhu, X. Waveform Synthesis for Diversity-Based Transmit Beampattern Design. *IEEE Trans. Signal Process.* **2008**, *56*, 2593–2598. [CrossRef]
11. Lipor, J.; Ahmed, S.; Alouini, M.S. Fourier-Based Transmit Beampattern Design Using MIMO Radar. *IEEE Trans. Signal Process.* **2014**, *62*, 2226–2235. [CrossRef]
12. Shi, S.; Wang, Z.; He, Z.; Cheng, Z. Constrained waveform design for dual-functional MIMO radar-Communication system. *Signal Process.* **2020**, *171*, 107530. .: 10.1016/j.sigpro.2020.107530. [CrossRef]
13. Cheng, Z.; He, Z.; Zhang, S.; Li, J. Constant Modulus Waveform Design for MIMO Radar Transmit Beampattern. *IEEE Trans. Signal Process.* **2017**, *65*, 4912–4923. [CrossRef]
14. Zhang, W.; Hu, J.; Wei, Z.; Ma, H.; Yu, X.; Li, H. Constant modulus waveform design for MIMO radar transmit beampattern with residual network. *Signal Process.* **2020**, *177*, 107735. [CrossRef]
15. Raei, E.; Alaee-Kerahroodi, M.; M. R., B.S. MIMO Radar Transmit Beampattern Matching Based on Block Successive Upper-bound Minimization. In Proceedings of the 2022 30th European Signal Processing Conference (EUSIPCO), Belgrade, Serbia, 29 August–2 September 2022; pp. 1901–1905. [CrossRef]
16. Alhujaili, K.; Monga, V.; Rangaswamy, M. Transmit MIMO Radar Beampattern Design via Optimization on the Complex Circle Manifold. *IEEE Trans. Signal Process.* **2019**, *67*, 3561–3575. [CrossRef]
17. Naeem, M.; De Pietro, G.; Coronato, A. Application of Reinforcement Learning and Deep Learning in Multiple-Input and Multiple-Output (MIMO) Systems. *Sensors* **2022**, *22*, 309. [CrossRef]

18. Wang, W.Q.; So, H.C.; Farina, A. FDA-MIMO Signal Processing for Mainlobe Jammer Suppression. In Proceedings of the 2019 27th European Signal Processing Conference (EUSIPCO), A Coruna, Spain, 2–6 September 2019; pp. 1–5. [CrossRef]
19. Forsythe, K.; Bliss, D. Waveform Correlation and Optimization Issues for MIMO Radar. In Proceedings of the Conference Record of the Thirty-Ninth Asilomar Conference onSignals, Systems and Computers, Pacific Grove, CA, USA, 29 October–1 November 2005; pp. 1306–1310. [CrossRef]
20. Li, J.; Stoica, P. MIMO Radar Diversity Means Superiority. In *MIMO Radar Signal Processing*; John Wiley & Sons: Hoboken, NJ, USA, 2009; pp. 1–64. [CrossRef]
21. Li, J.; Stoica, P.; Xu, L.; Roberts, W. On Parameter Identifiability of MIMO Radar. *IEEE Signal Process. Lett.* **2007**, *14*, 968–971. [CrossRef]
22. Basit, A.; Wang, W.Q.; Wali, S.; Yaw Nusenu, S. Transmit beamspace design for FDA–MIMO radar with alternating direction method of multipliers. *Signal Process.* **2021**, *180*, 107832. [CrossRef]
23. Chen, Y.; Nijsure, Y.; Yuen, C.; Chew, Y.H.; Ding, Z.; Boussakta, S. Adaptive Distributed MIMO Radar Waveform Optimization Based on Mutual Information. *IEEE Trans. Aerosp. Electron. Syst.* **2013**, *49*, 1374–1385. [CrossRef]
24. Yang, Y.; Blum, R.S. Waveform Design for MIMO Radar Based on Mutual Information and Minimum Mean-Square Error Estimation. In Proceedings of the 2006 40th Annual Conference on Information Sciences and Systems, Princeton, NJ, USA, 22–24 March 2006; pp. 111–116. [CrossRef]
25. Cui, G.; Li, H.; Rangaswamy, M. MIMO Radar Waveform Design With Constant Modulus and Similarity Constraints. *IEEE Trans. Signal Process.* **2014**, *62*, 343–353. [CrossRef]
26. Jiu, B.; Liu, H.; Wang, X.; Zhang, L.; Wang, Y.; Chen, B. Knowledge-Based Spatial-Temporal Hierarchical MIMO Radar Waveform Design Method for Target Detection in Heterogeneous Clutter Zone. *IEEE Trans. Signal Process.* **2015**, *63*, 543–554. [CrossRef]
27. Kim, S.; Kim, J.; Chung, C.; Ka, M.H. Derivation and Validation of Three-Dimensional Microwave Imaging Using a W-Band MIMO Radar. *IEEE Trans. Geosci. Remote. Sens.* **2022**, *60*, 1–16. [CrossRef]
28. Yu, X.; Yao, X.; Yang, J.; Zhang, L.; Kong, L.; Cui, G. Integrated Waveform Design for MIMO Radar and Communication via Spatio-Spectral Modulation. *IEEE Trans. Signal Process.* **2022**, *70*, 2293–2305. [CrossRef]
29. Cui, G.; Fu, Y.; Yu, X.; Li, J. Robust Transmitter–Receiver Design in the Presence of Signal-Dependent Clutter. *IEEE Trans. Aerosp. Electron. Syst.* **2018**, *54*, 1871–1882. [CrossRef]
30. Alhujaili, K.; Monga, V.; Rangaswamy. Quartic Gradient Descent for Tractable Radar Slow-Time Ambiguity Function Shaping. *IEEE Trans. Aerosp. Electron. Syst.* **2020**, *56*, 1474–1489. [CrossRef]
31. Xiong, K.; Iu, H.H.C.; Wang, S. Kernel Correntropy Conjugate Gradient Algorithms Based on Half-Quadratic Optimization. *IEEE Trans. Cybern.* **2021**, *51*, 5497–5510. [CrossRef]
32. ElMossallamy, M.A.; Seddik, K.G.; Chen, W.; Wang, L.; Li, G.Y.; Han, Z. RIS Optimization on the Complex Circle Manifold for Interference Mitigation in Interference Channels. *IEEE Trans. Veh. Technol.* **2021**, *70*, 6184–6189. [CrossRef]
33. Zhong, K.; Hu, J.; Pan, C.; Yu, X.; Li, X. MIMO Radar Beampattern Design Based on Manifold Optimization Method. *IEEE Commun. Lett.* **2022**, *26*, 1086–1090. [CrossRef]
34. Zhong, K.; Hu, J.; Cong, Y.; Cui, G.; Hu, H. RMOCG: A Riemannian Manifold Optimization-Based Conjugate Gradient Method for Phase Only Beamforming Synthesis. *IEEE Antennas Wirel. Propag. Lett.* **2022**, *21*, 1625–1629. [CrossRef]
35. Bertsekas, D. *Nonlinear Programming*; Athena Scientific: Nashua, NH, USA, 1999.
36. Ding, S.; Tong, N.; Zhang, Y.; Hu, X.; Zhao, X. Cognitive Antenna Selection in MIMO Imaging Radar. *IEEE Trans. Geosci. Remote. Sens.* **2021**, *59*, 9829–9841. [CrossRef]
37. Zhang, H.; Shi, J.; Zhang, Q.; Zong, B.; Xie, J. Antenna Selection for Target Tracking in Collocated MIMO Radar. *IEEE Trans. Aerosp. Electron. Syst.* **2021**, *57*, 423–436. [CrossRef]
38. Cui, G.; Yu, X.; Carotenuto, V.; Kong, L. Space-Time Transmit Code and Receive Filter Design for Colocated MIMO Radar. *IEEE Trans. Signal Process.* **2017**, *65*, 1116–1129. [CrossRef]
39. Bose, A.; Khobahi, S.; Soltanalian, M. Efficient waveform covariance matrix design and antenna selection for MIMO radar. *Signal Process.* **2021**, *183*, 107985. [CrossRef]

Technical Note

Radar Phase-Coded Waveform Design with Local Low Range Sidelobes Based on Particle Swarm-Assisted Projection Optimization

Xiang Feng *, Zhanfeng Zhao, Fengcong Li, Wenqing Cui and Yinan Zhao

School of Information Science and Engineering, Harbin Institute of Technology, Weihai 264209, China
* Correspondence: fengxiang@hit.edu.cn

Abstract: In modern electronic warfare, cognitive radar with knowledge-aided waveforms would show significant flexibility in anti-interference. In this paper, a novel method, named particle swarm-assisted projection optimization (PSAP), is introduced to design phase-coded waveforms with multi-level low range sidelobes, which mainly considers the stability for randomized initialization under the unimodular constraint. Firstly, the mathematical problem corresponding to avoid the range sidelobe masking from multiple non-cooperative targets or interference is formulated by giving different threat levels. Then, based on the alternating direction decomposition idea, the original problem is divided into triple-variable ones where these non-linear approximations can be solved via alternating projections along with FFT. Furthermore, the PSAP method with swarm intelligence, learning factor, and particle-assisted projection could ensure the optimization convergence in a parallel way, which could relax the non-convex constraint and enhance the global exploiting performance. Finally, simulations for several typical scenarios and numerical results are all provided to assess the waveforms generated by PSAP and other prevalent ones.

Keywords: cognitive radar; waveform design; range sidelobe suppression; particle swarm-assisted projection; FFT

Citation: Feng, X.; Zhao, Z.; Li, F.; Cui, W.; Zhao, Y. Radar Phase-Coded Waveform Design with Local Low Range Sidelobes Based on Particle Swarm-Assisted Projection Optimization. *Remote Sens.* **2022**, *14*, 4186. https://doi.org/10.3390/rs14174186

Academic Editor: Okan Yurduseven

Received: 7 July 2022
Accepted: 23 August 2022
Published: 25 August 2022

Publisher's Note: MDPI stays neutral with regard to jurisdictional claims in published maps and institutional affiliations.

1. Introduction

Waveform diversity aided by high-performance radar hardware has received considerable attention and even made a great step forward to cognitive radar (CR) [1–3]. Most radars transmit a modulated waveform and then use some type of matched filter (MF) to enhance the signal-to-noise ratio (SNR) of the return echo. In mathematical sense, the MF output is usually the convolution between the received signal and the time-reversed replica of the transmitted signal, which is also regarded as the aperiodic auto-correlation [4–6]. Generally, to suppress range sidelobes which might obscure small targets of interest (especially, some dot targets or low RCS target) and further improve the anti-interference ability, the transmitting waveforms with desirable auto-correlation property are imperative by using some prior information [7–10]. Moreover, in engineering, the constant modulus (CM) of waveform (in most cases, i.e., unimodular property) could maximize the transmitter's efficiency, but also makes the mathematical problem of generating waveform be more non-convex [11–13].

In the past 10 years, to achieve waveforms with ideal range sidelobes, minimizing integrated sidelobe level (ISL) and weighted integrated sidelobe level (WISL) have been developed as the classical metrics [14–16]. Therein, typical methods, such as cyclic algorithm new (CAN) [4], iterative spectral approximation algorithm (ISAA) [12], coordinate descent method (CD) [15], majorization minimization (MM) [16], weighted CAN (WeCAN) [17], alternating direction method of multipliers (ADMM) [18,19], etc., have also been presented and have received much attention. Note that, for mathematical problems under convex constraints, CANs could give some asymptotic convergence and make a big difference, while for the non-convex case CANs might stagnate into a suboptimum or local area [20,21]. Authors in [22] discussed the successive application of MM and phase gradient algorithm to synthesize the waveforms with low sidelobes. Additionally, authors adopted the limited memory Broyden–Fletcher–Goldfarb–Shanno algorithm (L-BFGS) to solve a fourth order polynomial formula and design unimodular sequences, but their ideas might incur invalidation for waveforms with large size [23]. Meanwhile, waveform optimization based on simulated annealing and stochastic neighborhood searching mechanism have also been presented, but these heuristic algorithms may be restricted to the modulus intricacy and large size of the waveforms [24,25].

Note that, when discussing the non-convex optimization with random initialization, these algorithms above would lead to a different terminus and only guarantee a local convergence [26]. Especially, to design the random phase-coded waveforms, how to tackle the random initialization for non-convex case has always been the key issue in engineering. Parallel optimization based on swarm intelligence maybe a good choice to improve the robustness. Unlike the optimization methods regarding each phase-coded unit, the waveform sample-based projection optimization using FFT will make a difference. In this paper, we use the idea of swarm particle intelligence [27–29] and combine alternating projection and particle swarm intelligence together to improve the global exploiting. To this end, our particle swarm-assisted projection optimization (PSAP) method is introduced to design the waveforms with desirable range sidelobes. Firstly, the mathematical problem is formulated to tackle the multiple non-cooperative targets or interference. Furthermore, using the alternating direction idea, the original problem is divided into some triple-variable ones considering different constraints. Then, the spectrum approximation in the sense of F-norm can be transformed into multi-variable alternating optimization cases. Finally, with the help of particle swarm intelligence, phase retrieval, learning factor and accelerated projections, PSAP method and its accelerated version have been formulated.

The remainder of paper is organized as follows. In Section 2, the system model is shown, and the formulated optimization problem for minimizing sidelobes is derived. In Section 3, PSAP as a novel alternative optimization mechanism based on swarm intelligence and FFT is presented. In Section 4, the performance of the proposed algorithms is evaluated and a series of numerical examples are also provided. Finally, in Section 5, the concluding remarks are provided.

2. The Signal Model and Problem Description

In this section, we discuss the range sidelobes masking effect from some strong RCS scatters which might obscure small targets of interest (such as dot target or low RCS target), and mainly focus on the waveform design for the static target detection in masking scenario and ignore the Doppler effect. For general description, the pre-modulated transmitting waveform in time domain has the discrete form of base band sequence with N code elements, i.e., $s = [s_1 \quad s_2 \quad \dots \quad s_N]^T$. The received signal is down converted to base band and undergoes the MF at the receiver [30]. The vector format of autocorrelation function which can be regarded as the MF output at the zero Doppler shift, has been listed as follows

$$\alpha(s) = [\alpha_{-N+1}(s) \quad \dots \quad \alpha_{-1}(s) \quad \alpha_0(s) \quad \alpha_1(s) \quad \dots \quad \alpha_{N-1}(s)]^T \tag{1}$$

$$\alpha_n(s) = \sum_{\tilde{n}=1}^{N-n} s_{\tilde{n}} s_{\tilde{n}+n}^* = \alpha_{-n}^*(s) \tag{2}$$

As shown in [12], suppose that a strong point scatterer (or the interference) with echo power $\sigma_t^2(q)$ exists in the q-th range cell, and a weak target of interest with echo power $\sigma_t^2(r)$ exists in the r-th range cell, the noise plus range sidelobe interference for the r-th range cell can be represented as

$$\sigma_I^2(r) = \Omega(N - 1 - |q - r|) |\sigma_t(q)\alpha_{q-r}(s)/N|^2 + \sigma_n^2, \tag{3}$$

where $\Omega(n) = \begin{cases} 1 & n \geq 0 \\ 0 & n < 0 \end{cases}$, σ_n^2 is the power of the thermal noise, and $\alpha_{q-r}(s)$ denotes the $(q\text{-}r)$-th lag of the aperiodic autocorrelation of (1). Then the target Signal power to Interference plus Noise Ratio (i.e., SINR) can be expressed as

$$\text{SINR} = \frac{\sigma_t^2(r)}{\sigma_I^2(r)} == \frac{\sigma_t^2(r)}{\Omega(N-1-|q-r|)\left|\sigma_t(q)\alpha_{q-r}(s)/N\right|^2 + \sigma_n^2} \tag{4}$$

here, of all the variables in (4), $\alpha_{q-r}(s)$ is the only one under the control of radar transmitter.

Despite the phase-coded waveform s with arbitrary amplitude, the CM waveform could maximize the transmitter's efficiency [13,16,31]. Let $x = \begin{bmatrix} x_1 & x_2 & \dots & x_N \end{bmatrix}^T \in \mathbb{C}^{N \times 1}$ denote CM phase-coded one with N discrete phase-coded units, i.e., $x_n = e^{j\psi_n}$ where ψ_n denotes the n-th phase-unit extracted from $[0, 2\pi]$. $(\cdot)^T$, $(\cdot)^*$ means the operation of transpose, complex conjugate, respectively. Similarly, we use $\alpha_n(x) = \sum\limits_{\tilde{n}=1}^{N-n} x_{\tilde{n}} x_{\tilde{n}+n}^* = \alpha_{-n}^*(x)$ to denote the range sidelobes of MF output in lieu of $\alpha_n(s)$ in (1) and (2). In modern electronic countermeasures scenarios, range sidelobes units occupied by some powerful interference or extended-scatters which mask the interesting target need to be suppressed [8,10,17]. Namely, local low range sidelobes is more convenient for weak targets detection and anti-masking effect as discussed in (3) and (4). Suppose that a powerful dot-scatter exists in the q-th range unit and inevitably impacts the target detection of the r-th one. With the help of some prior information R_r, i.e., $R_r = \{\pm|n-q| : n \in \hat{Z}_r\} \backslash \{N-1, -N+1\}$ where $\hat{Z}_r$ denotes the area with a foreseeable target, we could further describe the range interval R_r to be suppressed. We use an indicating vector $z = [z_1, z_2 \dots z_{2N}]^T$ to formulate the area-mapping of these pre-suppressed range sidelobes interval, i.e., $z_n = 1$ when $n \in R_r$, and $z_n = 0$ when $n \notin R_r$. The classical weighted ISL given some prior information has been discussed in [4,7,17], i.e.,

$$\text{WISL}(x) = \min_x 2 \sum_{n=1}^{N-1} w_n |\alpha_n(x)|^2, w_n \geq 0 \tag{5}$$

Borrowed the idea of (5), we denote $\{\omega_n\}_{n=1}^{2N}$ as the weight corresponding to each unit in $z = [z_1, z_2 \dots z_{2N}]^T$, i.e., $\omega_n = \delta_1 \ll 1$ when $z_n = 1$, and $\omega_n = 1$ when $z_n = 0$. We further define $\tilde{x}$ as the desirable waveform with ideal local low sidelobes, i.e., $\lim\limits_{x \to \tilde{x}} \alpha(\tilde{x}) = \omega \odot \alpha(x)$, $\odot$ denotes the Hadamard element-wise product. Let v denote the approximation vector with $v \to \omega \odot \alpha(x)$. To design waveform with desirable property, namely, we should make $\alpha(x)$ and $\alpha(\tilde{x})$ be more approximate. Here we use the norm-metric $\|\alpha(x) - v\|^2$ to denote the approximation level of them. Finally, the objective function can be formulated as

$$\min_x \|\alpha(x) - \alpha(\tilde{x})\|^2 = \min_x \|\alpha(x) - v\|^2 \tag{6}$$

According to the "Parseval-type equality" in [17], i.e., $\|\mathbf{F}\mathbf{C}^T x\|^2 = \|x\|^2$, the objective function of (6) is equivalent to

$$\min_x \|\mathbf{F}\mathbf{C}^T \alpha(x) - \mathbf{F}\mathbf{C}^T v\|^2 \tag{7}$$

$$\mathbf{F}\mathbf{C}^T v = \tilde{f} \odot \tilde{f}^* \in \mathfrak{R}^{2N \times 1}, \quad \mathbf{F}\mathbf{C}^T \alpha(x) = f \odot f^* \in \mathfrak{R}^{2N \times 1} \tag{8}$$

where $\mathbf{C}$ denotes the extend or cutoff matrix with $\mathbf{C} = \begin{bmatrix} \mathbf{I}_{N \times N} & \mathbf{0}_{N \times N} \end{bmatrix}$, $\tilde{f}$ and $f = \mathbf{F}\mathbf{C}^T x$ refers to the frequency spectrum of $\tilde{x}$ and the designing one, respectively. The DFT matrix $\mathbf{F} \in \mathbb{C}^{2N \times 2N}$ is constituted by unity exponential factor, i.e., $\hat{f}_k^H = \begin{bmatrix} e^{-jw_k} & \dots & e^{-j2N \cdot w_k} \end{bmatrix}$ with $w_k = 2\pi k/2N$. Next, the optimization problem in (6) can be transformed into the spectrum approximation as following

$$\min_{x}\left\|(\mathbf{FC}^T x)\odot(\mathbf{FC}^T x)^* - \tilde{f}\odot\tilde{f}^*\right\|^2 \triangleq \min_{x}\sum_{k=1}^{2N}\left|\left(\left|\left(\hat{f}_k^H\mathbf{C}^T x\right)\right|^2 - \left(\sqrt{\left(\hat{f}_k^H\mathbf{C}^T v\right)}\right)^2\right)\right|^2 \tag{9}$$
$$\text{s.t.}\quad |x_n| = 1,\; n = 1, 2, \ldots, N$$

Note that the problem of (9) is a quartic function of $\{x_n\}_{n=1}^{N}$, using the auxiliary phase vector $\boldsymbol{\phi} = [\phi_1 \quad \phi_2 \quad \cdots \quad \phi_{2N}]^T$, (9) can be justified 'almost equivalent' to the following quadratic function of $\{x_n\}_{n=1}^{N}$ [17], i.e.,

$$\min_{x,\phi}\sum_{k=1}^{2N}\left|\sqrt{\left(\hat{f}_k^H\mathbf{C}^T v\right)}\cdot e^{j\phi_k} - \left(\hat{f}_k^H\mathbf{C}^T x\right)\right|^2 \tag{10}$$

and the ideal frequency spectrum vector $\tilde{f}$ can be further expressed as

$$\tilde{f} = \left(\mathbf{FC}^T v\right)^{1/2}\odot\exp(j\boldsymbol{\phi}) \tag{11}$$

As $v \in \mathbb{C}^{2N\times 1} \to \omega\odot\alpha(x)$, (10) implies that once given x, the ideal v satisfies

$$\min_{v}\|v - \omega\odot\alpha(x)\|^2 \tag{12}$$
$$\text{s.t.}\quad |x_n| = 1, n = 1, 2, \ldots, N$$

namely, $v \triangleq \omega\odot\alpha(x)$, so that $\sqrt{\left(\hat{f}_k^H\mathbf{C}^T v\right)}$ will be constant once given x, then (10) in the vector format has

$$\min_{\phi}\left\|\left(\mathbf{FC}^T v\right)^{1/2}\odot\exp(j\boldsymbol{\phi}) - \left(\mathbf{FC}^T x\right)\right\|^2 \tag{13}$$
$$\text{s.t.}\begin{cases} v \triangleq \omega\odot\alpha(x) \\ \phi_n \in [0, 2\pi], n = 1, 2, \ldots, 2N \end{cases}$$

For brevity, defining a novel operator $\text{diag}(\cdot)$ which rearranges the column vector to be a square diagonal matrix, i.e., $\left(\mathbf{FC}^T v\right)^{1/2}\odot\exp(j\boldsymbol{\phi}) = \text{diag}(\exp(j\boldsymbol{\phi}))\left(\mathbf{FC}^T v\right)^{1/2}$, and the objective function (13) has

$$\min_{\phi}\left\|\text{diag}(j\boldsymbol{\phi})\left(\mathbf{FC}^T v\right)^{1/2} - \mathbf{FC}^T x\right\|^2 \tag{14}$$
$$\text{s.t.}\begin{cases} v \triangleq \omega\odot\alpha(x) \\ \phi_n \in [0, \pi], n = 1, 2, \ldots, N \end{cases}$$

Similarly, given $\boldsymbol{\phi}$ and v, the quadratic optimization problem of (10) has

$$\min_{x}\left\|\left(\left(\mathbf{FC}^T v\right)^{1/2}\odot\exp(j\boldsymbol{\phi})\right) - \mathbf{FC}^T x\right\|^2 \tag{15}$$
$$\text{s.t.}\quad |x_n| = 1, n = 1, 2, \ldots, N$$

Let $J = \left(\mathbf{FC}^T v\right)^{1/2} \in \Re^{2N\times 1}$, then $\text{diag}(\exp(j\boldsymbol{\phi}))J = \text{diag}(J)\exp(j\boldsymbol{\phi})$, (14) can be rewritten as

$$\min_{\phi}\left\|\text{diag}(J)\exp(j\boldsymbol{\phi}) - \mathbf{FC}^T x\right\|^2 \tag{16}$$

Meanwhile, $\text{diag}(J)$ is also an invertible matrix, the estimated $\boldsymbol{\phi}$ in (16) could be given by

$$\boldsymbol{\phi} = ang\left(\left((\text{diag}(J))^H\text{diag}(J)\right)^{-1}(\text{diag}(J))^H\mathbf{FC}^T x\right) \tag{17}$$

where $ang(\cdot) = \mathrm{im}(\ln(\mathrm{diag}(\cdot)))$ represents the phase extractor from its vector argument, and $\mathrm{im}(\cdot)$ denotes the imaginary component extraction operator. Similarly, recall "Parseval-type equality" again, the objective function in (15) could be expressed as:

$$
\begin{aligned}
&\min_{x} \left\| \mathbf{CF}^{H}\left(\left(\mathbf{FC}^{T}v\right)^{1/2} \odot \exp(\mathrm{j}\boldsymbol{\phi}) \right) - x \right\|^{2} \\
&= \left\| \mathbf{CF}^{H}\left(\mathrm{diag}(\exp(\mathrm{j}\boldsymbol{\phi})) \left(\mathbf{FC}^{T}v\right)^{1/2} \right) - x \right\|^{2} \\
&\text{s.t.} \quad |x_{n}| = 1, n = 1, 2, \ldots, N
\end{aligned}
\tag{18}
$$

On account of CM property, we only retain the phase section of the estimated x of (18). Then the designed waveform by the phase-retrieval operation of [20] has

$$
x = \exp\left(\mathrm{j} \cdot ang\left(\mathbf{CF}^{H}\left(\mathrm{diag}(\exp(\mathrm{j}\boldsymbol{\phi})) \left(\mathbf{FC}^{T}v\right)^{1/2} \right) \right) \right)
\tag{19}
$$

3. The Proposed Particles Swarm Assisted Projection Framework

In recent years, CAN [4], RISAAP [5], PONLP [12], CD [15], WeCAN [17], and ADMM [18] are all presented to deal with the Non-deterministic Polynomial-time hard (NP-hard) problem. They have the common trait, i.e., iterative optimization mechanism. Therein, each iteration of them requires to solve a non-convex problem under unimodular constraint, no matter by virtue of handling the bisection gradient optimization or FFT-based one. However, the initialized random phases where each phase unit is distributed in $[0, 2\pi]$, would incur a different terminus when using different random initialization. Namely, each Monte-Carlo trial could obtain different solution and remain some local convergence. To tackle these, we borrow the idea of parallel optimization to combine the particles swarm intelligence and projection optimization together, where the novel particles projection mechanism will avoid the local area in the statistical sense. Next, we present the PSAP framework in lieu of the traditional evolution mechanism of PSO or DE [28,29], and thus could enhance the global exploiting for non-convex phase-coded problem. The detailed descriptions of PSAP have been listed as follows

Step 1. Formulate the waveform set rather than one single sequence, i.e., $X = [x_{1} \ldots \ x_{m} \ldots \ x_{M}] \in \mathbb{C}^{M \times N}$ with $x_{m} = [x_{m}(1) \ldots \ x_{m}(n) \ x_{m}(N)]^{T} \in \mathbb{C}^{N \times 1}$. Note that each sequence has $x_{m}(n) = e^{\mathrm{j}\psi_{n}}$, where ψ_{n} denotes the independent phase-coded variables extracted from $[0, 2\pi]$. Additionally, the waveform set could also be initialized by Frank or Barker sequence.

Step 2. Define a novel metric to assess the sidelobe performance of waveform in the range interval R_{l} as

$$
fitness(x_{m}) = \frac{1}{num\_R_{r}} \sum_{k \in R_{r}} \alpha_{k}(x_{m})
\tag{20}
$$

where num_R_{r} denotes the number of pre-suppressed sidelobes units.

Step 3. Using (20) as the fitness function to evaluate each waveform x_{m} of set X, and select the best fitness function and its corresponding waveform p^{t}.

Step 4. For the t-th iteration, select $\hat{M} < M$ waveforms from waveform set X to formulate the novel subset. Here we should update the subset by some rules that if the best fitness function and its corresponding waveform have not been incorporated, then use it to replace the worst one in current subset.

Step 5. For each waveform x_{m} at the t-th iteration, utilize the oversampling FFT to get $x_{m} \to \widehat{f}_{m}$, and then formulate the relaxing factor $\delta(x_{m})$, relational factor d_{m}, and also projection vector v_{m} by using (22)~(24) respectively, as follows

$$
\widehat{f}_{m} = (\mathbf{F}\omega \odot \alpha(x_{m}))^{(1/2)} \odot \left(\exp\left(\mathrm{j} \cdot ang\left(\mathbf{FC}^{T}x_{m} \right) \right) \right)
\tag{21}
$$

$$\delta(x_m) = \sqrt{\frac{\left\|(\mathbf{F}\omega \odot \alpha(x_m))^{(1/2)} \odot \left(\exp\left(\mathbf{j}\cdot ang\left(\mathbf{FC}^T x_m\right)\right)\right) - x_m\right\|}{\left\|\left(\exp\left(\mathbf{j}\cdot ang\left(\mathbf{C}^T\mathbf{F}\widehat{f}_m\right)\right)\right) - x_m\right\|}} \tag{22}$$

$$d_m = \delta(x_m)\cdot\left(\exp\left(\mathbf{j}\cdot ang\left(\mathbf{C}^T\mathbf{F}\widehat{f}_m\right)\right) - x_m\right) \tag{23}$$

$$v_m = w\cdot d_m + c1\cdot\text{rand}\cdot(y_m - x_m) + c2\cdot\text{rand}\cdot(p - x_m) \tag{24}$$

where d_m denotes the relational factor of the m-th waveform at the t-th iteration to the $t+1$-th iteration, rand denotes one random value of $[0,1]$, $w \in [1,1.5]$ represent the inertia factor, $c1 \in [0.5,1]$ is the learning factor related to the best waveform y_m which is selected from initial iteration to the current iteration $\{x_m^1 \quad \dots \quad x_m^t\}$. $c2 \in [0,0.5]$ represents the learning factor related to the best waveform p^t at the current t-th iteration.

Step 6. Use the multi-particles projection of (25)~(26) to achieve unimodular waveforms at the $t+1$-th iteration, as follows

$$\widehat{x}_m = x_m + v_m \tag{25}$$

$$\hat{x}_m = \left(\exp\left(\mathbf{j}\cdot ang\left(\widehat{x}_m\right)\right)\right) \tag{26}$$

where $\hat{x}_m$ with CM property has been obtained by the phase retrieval operation in (26).

Step 7. Consider the remaining subset, update them by

$$\phi_m = ang\left(\text{diag}\left(\left(\left(\mathbf{FC}^T(\omega\odot\alpha(x_m))\right)^{1/2}\right)^{-1}\right)\mathbf{FC}^T x_m\right) \tag{27}$$

$$\hat{x}_m = \exp\left(\mathbf{j}\cdot ang\left(\mathbf{CF}^H\left(\text{diag}(\exp(\mathbf{j}\phi_m))\left(\mathbf{FC}^T(\omega\odot\alpha(x_m))\right)^{1/2}\right)\right)\right) \tag{28}$$

Step 8. Merge the subset and remaining part, use the following selection rules to get y_m, i.e., $y_m = \hat{x}_m$ when $fitness(\hat{x}_m) < fitness(x_m)$, or not $y_m = x_m$, and select the best fitness function and its corresponding waveform p^t.

Step 9. Repeat step 3~8 until $\left|fitness(p^t) - fitness(p^{t-1})\right| < \varepsilon$ or num > iter_num, then output p^t. Otherwise, update $t = t+1$ and continue iterating.

Furthermore, we could incorporate the gradient steepest idea into PSAP. The objective function of (15) can be expressed as:

$$\min_x J(x) = \left\|\mathbf{FC}^T x - \sqrt{\left|\mathbf{FC}^T\text{diag}(\ddagger)\alpha(x)\right|}\odot ang\left(\widetilde{f}\right)\right\|^2 \tag{29}$$

Let $\breve{\psi} = ang\left(\widetilde{f}\right)$, then the two-variable optimization problem has

$$\min_{\psi,x} J\left(x, \breve{\psi}\right) = \left\|\mathbf{FC}^T x - \sqrt{\left|\mathbf{FC}^T\text{diag}(\ddagger)\alpha(x)\right|}\odot\breve{\psi}\right\|^2 \tag{30}$$

Given the latest iterative x, $\breve{\psi}$ can be achieved by the follow gradient optimization, and its gradient matrix has

$$\nabla_{\psi} J(x) =$$
$$\frac{\partial\left\|abs\left(\mathbf{F}\begin{bmatrix} x \\ \mathbf{0}_{N\times 1}\end{bmatrix}\right)\odot\exp\left(\mathbf{j}ang(\mathbf{FC}^T x)\right) - \sqrt{\left|\mathbf{FC}^T\text{diag}(\ddagger)\alpha(x)\right|}\odot\breve{\psi}\right\|^2}{\partial\breve{\psi}} \tag{31}$$
$$= \mathbf{j}\cdot\text{diag}\left(abs\left(\mathbf{FC}^T x\right)\right)\text{diag}\left(\sqrt{\left|\mathbf{FC}^T\text{diag}(\ddagger)\alpha(x)\right|}\odot\breve{\psi}\right)\cdot$$
$$\left(\exp\left(\mathbf{j}\left(ang\left(\mathbf{FC}^T x\right) - \breve{\psi}\right)\right) - \exp\left(\mathbf{j}\left(\breve{\psi} - ang\left(\mathbf{FC}^T x\right)\right)\right)\right)$$

$$\frac{\partial^2 \left\| abs\left(\mathbf{FC}^T x\right) \odot \exp\left(j ang\left(\mathbf{FC}^T x\right)\right) - \sqrt{\left|\mathbf{FC}^T \mathrm{diag}(\ddagger)\alpha(x)\right| \odot \breve{\psi}} \right\|^2}{\partial \breve{\psi} \partial \breve{\psi}^T} \tag{32}$$
$$= \mathrm{diag}\left(aba\left(\mathbf{FC}^T x\right)\right) \mathrm{diag}(v) \cdot \mathrm{diag}\left(\exp\left(j\left(ang\left(\mathbf{FC}^T x\right) - \breve{\psi}\right)\right) + \exp\left(j\left(\breve{\psi} - ang\left(\mathbf{FC}^T x\right)\right)\right)\right)$$

Let (32) equal to 0, then $\breve{\psi}$ has:

$$\breve{\psi} = ang\left(\mathbf{FC}^T x\right) + \pi \cdot \xi \tag{33}$$

where $\xi \in \mathbb{Z}$, as known, only if ξ is the even value, the hessian matrix of (33) can be positive, and (30) will get the minimum. Moreover, given $\breve{\psi}$, the phase vector of x has:

$$ang(x) = ang\left(\mathbf{CF}^H\left(\sqrt{\left|\mathbf{FC}^T \mathrm{diag}(\ddagger)\alpha(x)\right|} \odot \breve{\psi}\right)\right) + \pi \cdot \xi \tag{34}$$

Finally, the detailed descriptions of particle swarm-assisted projection with optimizing mechanism (named as PSAPOM) have been listed as follow:

Step 1~Step 3 are similar to PSAP.

Step 4. Define $\bar{x}^{(1)} = p^t$, then calculate the gradient direction $d_0 = -g_0 = -\nabla_\psi J(\bar{x}^{(1)})$, then search the best length ϑ which satisfies

$$J\left(\mathrm{diag}(\exp(j\vartheta d_k))\bar{x}^{(k)}\right) \geq J\left(\mathrm{diag}(\exp(j\eta d_k))\bar{x}^{(k)}\right), \quad \forall \eta \geq 0 \tag{35}$$

define $\bar{x}^{(k+1)} = \mathrm{diag}(\exp(j\vartheta d_k))\bar{x}^{(k)}$, then

$$g_{k+1} = \nabla_\psi J\left(\bar{x}^{(k+1)}\right) \tag{36}$$

$$d_{k+1} = -\left(g_{k+1} + (g_{k+1} - g_k)^T g_{k+1} d_k / \|g_k\|^2\right) \tag{37}$$

if $\left| fitness(\bar{x}^{(k+1)}) - fitness(\bar{x}^{(k)}) \right| < 10^{-5}$ or $k > 100$, output the initiation $p^t = \bar{x}^{(k+1)}$;

Step 5. For the current iteration, select $\hat{M} < M$ waveforms from set X to formulate the novel subset. Here we should update the subset by some rules that if the best fitness function and its corresponding waveform have not been incorporated, then use it to replace the worst one in current subset.

Step 6. For each waveform x_m of subset, utilize the oversampling FFT to get $x_m \to \widehat{f}_m$, and then formulate the relaxing factor $\delta(x_m)$, relational factor d_m, and also projection vector v_m by using (39)~(41) respectively, as follows

$$\widehat{f}_m = (\mathbf{F}\omega \odot \alpha(x_m))^{(1/2)} \odot \left(\exp\left(j \cdot ang\left(\mathbf{FC}^T x_m\right)\right)\right) \tag{38}$$

$$\delta(x_m) = \sqrt{\frac{\left\| (\mathbf{F}\omega \odot \alpha(x_m))^{(1/2)} \odot \left(\exp\left(j \cdot ang\left(\mathbf{FC}^T x_m\right)\right)\right) - x_m \right\|}{\left\| \left(\exp\left(j \cdot ang\left(\mathbf{CF}^H \widehat{f}_m\right)\right)\right) - x_m \right\|}} \tag{39}$$

$$d_m = \delta(x_m) \cdot \left(\left(\exp\left(j \cdot ang\left(\mathbf{CF}^H \widehat{f}_m\right)\right)\right) - x_m\right) \tag{40}$$

$$v_m = w \cdot d_m + c1 \cdot \mathrm{rand} \cdot (y_m - x_m) + c2 \cdot \mathrm{rand} \cdot (p^t - x_m) \tag{41}$$

where d_m denotes the relational factor of the m-th waveform at the t-th iteration to the $t + 1$-th iteration, rand denotes one random value in $[0, 1]$, $w \in [1, 1.5]$ represent the inertia factor, $c1 \in [0.5, 1]$ is the learning factor related to the best waveform y_m which is chosen from initial iteration to the current iteration $\{x_m^1 \quad \cdots \quad x_m^t\}$. $c2 \in [0, 0.5]$ represents the learning factor related to the best waveform p^t at the current t-th iteration.

Step 7. Use the multi-particles projection to achieve unimodular waveforms at $t + 1$-th iteration,

$$\widehat{x}_m = x_m + v_m \tag{42}$$

$$\hat{x}_m = \left(\exp\left(j \cdot ang\left(\widehat{x}_m \right) \right) \right) \tag{43}$$

where $\hat{x}_m$ with CM property has been obtained by the phase retrieval operation of (43).

Step 8. For the remaining subset, update them by

$$\phi_m = ang\left(\mathrm{diag}\left(\left(\left(\mathbf{FC}^T(\omega \odot \alpha(x_m)) \right)^{1/2} \right)^{-1} \right) \mathbf{FC}^T x_m \right) \tag{44}$$

$$\hat{x}_m - \exp\left(j \cdot ang\left(\mathbf{CF}^H\left(\mathrm{diag}(\exp(j\phi_m))\right)\left(\mathbf{FC}^T(\omega \odot \alpha(x_m)) \right)^{1/2} \right) \right) \tag{45}$$

Step 9. Merge the subset and the remaining part, use the following rules to get y_m, i.e.,

$$\begin{aligned} if \quad & \mathrm{fitness}(\hat{x}_m) < \mathrm{fitness}(x_m) \\ & y_m = \hat{x}_m \\ else \quad & y_m = x_m \end{aligned} \tag{46}$$

then select the best fitness function and its corresponding waveform p^t.

Step 10. Repeat above-mentioned steps 5~9 until num $>$ iter_num or $\left| fitness(p^t) - fitness(p^{t-1}) \right| < \varepsilon$, then output p^t. Otherwise, update $t = t + 1$ and continue iterating.

4. Simulations and Performance Analysis

Note that the non-convex optimization problem under different initialization is usually providing a different terminus, and hard to obtain the global solution within polynomial time [5,11,26]. For the phase-coded CM waveform design, selecting an effective technique has always been the focus [12–14]. In this section, to further assess PSAP's performance, we firstly assume code length of waveform $N = 150$, then the initialized waveform set has

$$set^0 = \begin{bmatrix} x_1^0 & \cdots & x_m^0 & \cdots & x_M^0 \end{bmatrix} \in \mathbb{C}^{N \times M} \tag{47}$$

$$x_m^0 = \begin{bmatrix} x_m^0(1) \cdots & x_m^0(n) & x_m^0(N) \end{bmatrix}^T \in \mathbb{C}^{N \times 1} \tag{48}$$

where $M = 20$, $x_m(n) = e^{j\psi_n}$. In addition, the inertia weight, individual learning factor as well as group learning factor are set as $w = 1.5$, $c_1 = 0.5$, and $c_2 = 0.5$, respectively. The iterations number iter_num is set as 2000, threshold value of $\left| fitness(p^t) - fitness(p^{t-1}) \right|$ is set as $\varepsilon = 10^{-10}$.

Next, we take the scenario of suppressing one single area for comparison, i.e., single interval $R_r = [2 : 30]$. PSAP will be compared with WeCAN, ISAA, RISAAP, and PONLP, by 20 Monte Carlo (MC) trials. Here, we define the averaging autocorrelation sidelobe level (Aver-ACL) and local PSL (LPSL) in suppressed regions as the metric, as follows

$$\text{Aver-ACL} = \frac{1}{num\_R_r} \sum_{k \in R_r} 20 \cdot \log_{10} \frac{|\alpha_k(x)|}{|\alpha_0(x)|} (\text{dB}) \tag{49}$$

$$\text{LPSL} = 20 \cdot \log_{10}\left(\max\left(\frac{|\alpha_k(x)|}{N} \right) \right) (\text{dB}), k \in R_r \tag{50}$$

for the sake of comparison, all methods or algorithms would be initialized by random phase-coded sequence. Performance comparisons have been shown in Table 1 and Figure 1. Simulations are performed on a PC with 3.40 GHz i7 CPU.

Table 1. Performance comparison for suppressing single area sidelobes of different algorithms.

Algorithms	Single Area		
	Aver-ACL (dB)	LPSL (dB)	Time Consumption (s)
RISAAP	-228.4560	-216.5490	0.14
ISAA	-158.0292	-144.8821	0.35
WeCAN	-48.5622	-35.6573	5.11
PONLP	-139.6091	-130.2762	0.48
PSAP	-333.4510	-321.8462	0.76

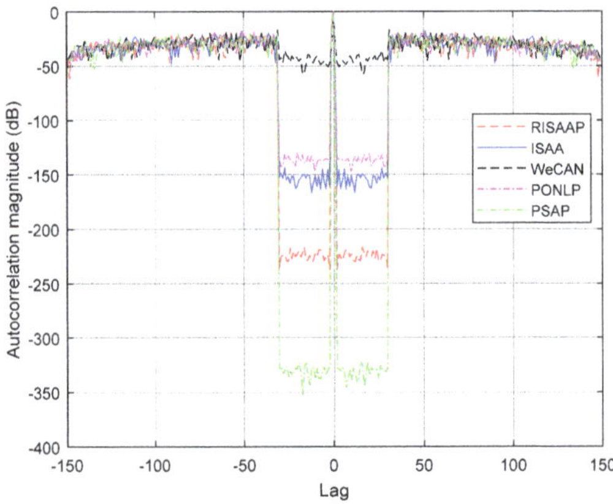

Figure 1. ACL comparison for suppressing sidelobes in a single area.

In Table 1 and Figure 1, our proposed PSAP with parallel optimization mechanism has achieved the best performance, PONLP with stochastic gradient optimization has achieved only -139 dB of Aver-ACL and -130 dB of LPSL, while WeCAN achieves -48 dB of Aver-ACL and -35 dB of LPSL. Moreover, ISAA and RISAAP using the mechanism of alternating projection has obtained -228 dB and -158 dB of Aver-ACL, -216 dB and -144 dB of LPSL, respectively. We need to declare that, with $20 \cdot \log_{10}(\cdot)$ as the mathematical metric referring to [4,5,12,17], these methods or algorithms conduct their iterating or optimizing with the same stop criteria/thresholds (iter_num is set as 2000, threshold value is set as $\varepsilon = 10^{-10}$). Here, the phenomenon with -228 dB or -333 dB may be unnecessary to have so low sidelobe levels for practical engineering, but these low values in mathematical sense would demonstrate some quickly converging or optimizing level of our PSAP framework even for the future electronic countermeasure scenario.

To further discuss the computation complexity of them, here we mainly demonstrate the number of iterations in convergence graph (as shown in Figure 2), and the CPU time consumption in Table 1. In Figure 2, we use $\log\left|fitness(x^t) - fitness(x^{t-1})\right|$ as the convergence metric of the y-coordinate. Figure 2 has shown obvious converging difference. PSAP uses the idea of particles swarm intelligence, i.e., $M = 20$, to establish thus cooperative optimization (in CPU model), also occupies much more time than RISAAP, ISAA and PONLP. WeCAN has consumed the longest time 5.11 s. By 20 MC trails, we can see that PSAP has achieved the best robustness performance, which is own to the parallel cooperative mechanism. In addition, these trails and simulations are all based on CPU; when given the GPU condition, PSAP will occupy much less time than others. As seen in Table 1 and Figure 2, WeCAN has slow convergence and PONLP with the steepest descent gradient might stagnate into the local area. ISAA and RISAAP have oscillations in Figure 2 which might attribute to the alternating projection between multi-local areas for the non-convex case.

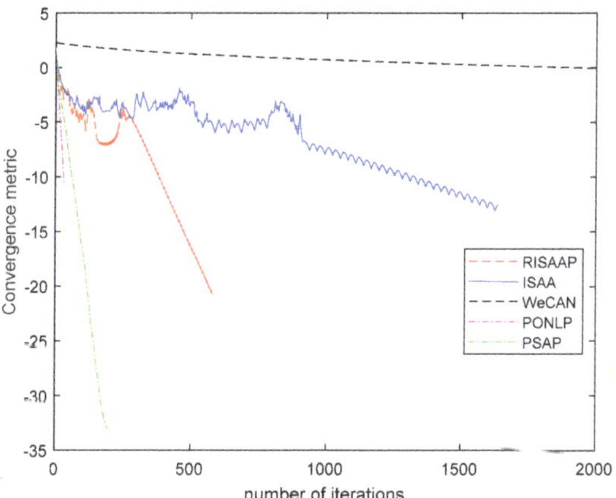

Figure 2. Convergence comparison of different methods.

In addition, we also discuss the Aver-ACL and LPSL comparisons with different code lengths N (i.e., 150, 200, 250, 300) which are shown in Figure 3. No matter $N = 150, 200, 250,$ and 300, our proposed PSAP could obtain the best Aver-ACL and LPSL in different code lengths. Meanwhile, the results of these methods in Figure 3 have shown some similarity that WeCAN might lose the performance for engineering application.

Meanwhile, we also consider the sidelobe suppression in multiple-area case, i.e., $R_l = [2:10] \cup [30:40]$. Here, we assign different suppressing levels with $\delta_1 = 0, \delta_2 = 10^{-4}$ where the former corresponds to the range sidelobe area next to the mainlobe, and the second refers to the farther one. Note that, the interferences near the mainlobe would also produce more effect than the farther one, and affect the detecting performance. Namely, we arrange two suppressing areas with different weights to demonstrate the suppressing levels for different non-cooperative targets. When discussing $\delta_1 = 0, \delta_2 = 10^{-4}$, the former $\delta_1 = 0$ with relative-low weights means more emphasis on the first area. In Table 2 and Figure 4, we could see that the first area has achieved more excellent performance than the second one, which is due to the weights of the different non-cooperative targets.

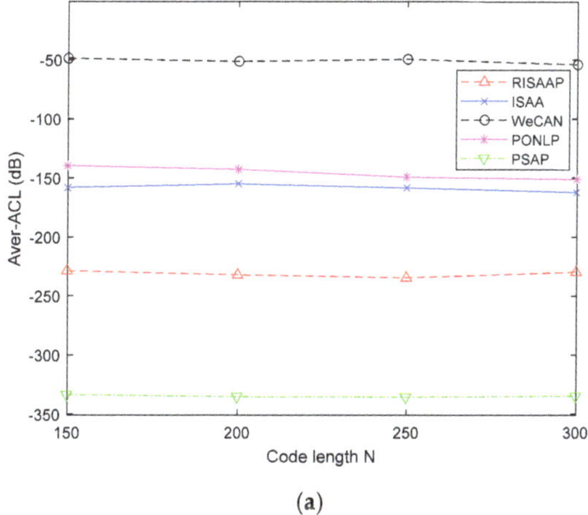

(**a**)

Figure 3. *Cont.*

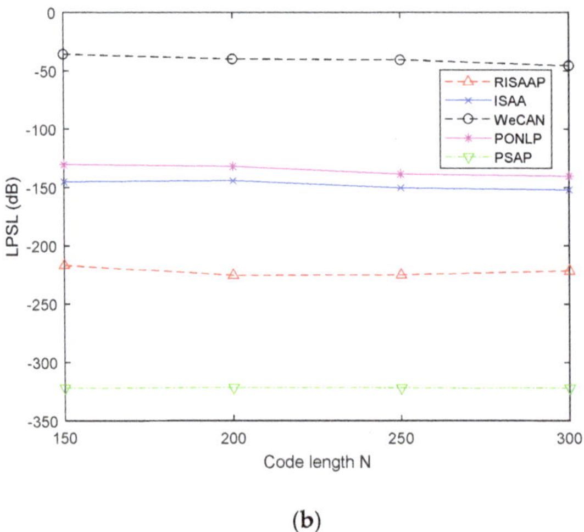

(b)

Figure 3. Aver-ACL and LPSL comparison with different code lengths N. (**a**) Aver-ACL comparison of RISAAP, ISAA, WeCAN, PONLP, and PSAP. (**b**) LPSL comparison of RISAAP, ISAA, WeCAN, PONLP, and PSAP.

Table 2. Performance comparison for suppressing multiple area sidelobes with different weights.

Algorithms	Multiple Area				
	Aver-ACL in 1# Area (dB)	LPSL in 1# Area (dB)	Aver-ACL in 2# Area (dB)	LPSL in 2# Area (dB)	Time Consumption (s)
RISAAP	−233.2183	−220.6892	−163.5196	−163.5110	0.16
ISAA	−119.9062	−117.8060	−100.1328	−93.8627	0.34
WeCAN	−61.2209	−47.9854	−41.1983	−34.9465	5.94
PONLP	−147.6981	−141.6530	−116.1224	−112.5260	0.45
PSAP	−336.3241	−329.4050	−169.5424	−169.5420	0.84

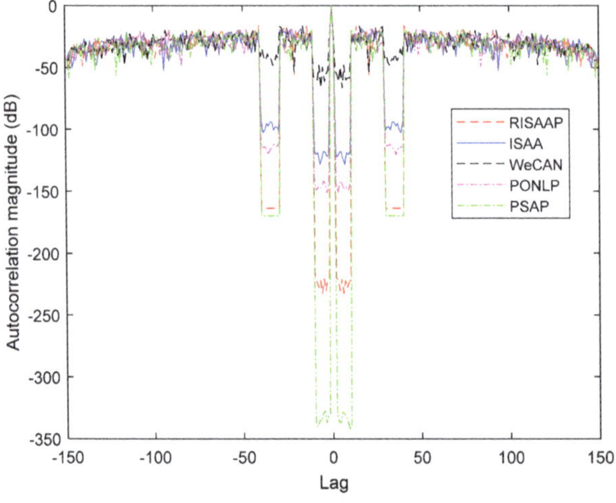

Figure 4. ACL comparison for suppressing sidelobes for multiple areas.

In Table 2 with recorded time-consumption data, obviously, our proposed PSAP with 0.84 s has achieved the best Aver-ACL and LPSL using 20 MC trials, but with more time than RISAAP (with 0.16 s), ISAA (with 0.34 s) and PONLP (with 0.45 s). Moreover, WeCAN (with 5.94 s) has lost the performance and cost more time than others.

Moreover, we also consider the multiple-case, i.e., $R_l = [2:10] \cup [30:40]$ and arrange two suppressing areas with same weights $\delta_1 = 0$, $\delta_2 = 0$. In Table 3 and Figure 5, we could see that these two areas have achieved same excellent performance. Table 3 has shown same characteristics as shown in Tables 1 and 2. Our proposed PSAP has achieved the best result by 20 MC trials. WeCAN has lost its performance.

Table 3. Performance comparison for suppressing multiple area sidelobes with same weights.

Algorithms	Multiple Area			
	Aver-ACL in 1# Area (dB)	LPSL in 1# Area (dB)	Aver-ACL in 2# Area (dB)	LPSL in 2# Area (dB)
RISAAP	−238.1425	−232.5312	−236.9185	−235.2602
ISAA	−137.7802	−134.5042	−134.4622	−131.5260
WeCAN	−59.4995	−50.2199	−53.3124	−48.0641
PONLP	−151.2146	−142.0220	−149.6574	−142.9440
PSAP	−340.8439	−326.2900	−336.8021	−323.6540

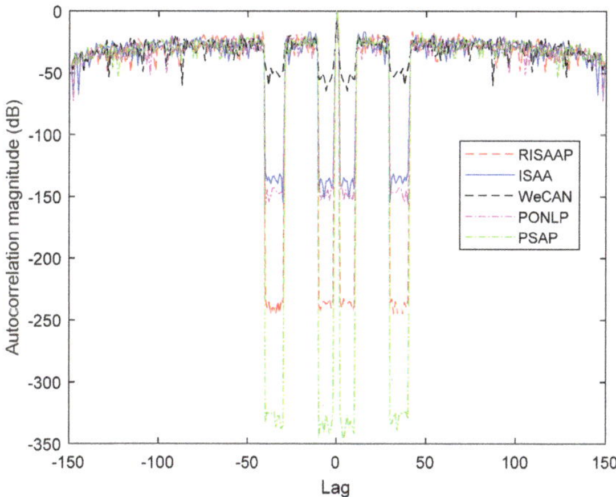

Figure 5. ACL comparison of suppressing sidelobes for multiple areas.

Note that the initialization of algorithms is indeed significant no matter for cyclic algorithms or alternating projection. As the Gradient Descent (GD) mechanism could accelerate the exploiting for local optimization, and we combine the GD and PSAP together to formulate the PSAPOM and enhance the global robustness. To further assess the performance of PSAPOM and PSAP, despite the random phase-coded sequence, here we assume that these algorithms have been initialized by the Frank-coded or Barker-coded sequence, respectively. For the length $N = \tilde{N}^2$ ($N = 196$), the Frank sequence is given by:

$$x(n\tilde{N} + \tilde{n} + 1) = e^{j2\pi n\tilde{n}/\tilde{N}}; n, \tilde{n} = 0, 1, 2 \ldots, \tilde{N} - 1 \tag{51}$$

we also assume another waveform sequence ($N = 169$) formulated by the 13 Barker sequence. As shown in Figure 6 and Table 4, for both the Frank sequence and Barker sequence, PSAPOM also achieved the same performance as PSAP, but its time consumption, 0.4210 s and 0.2502 s, was less than PSAP.

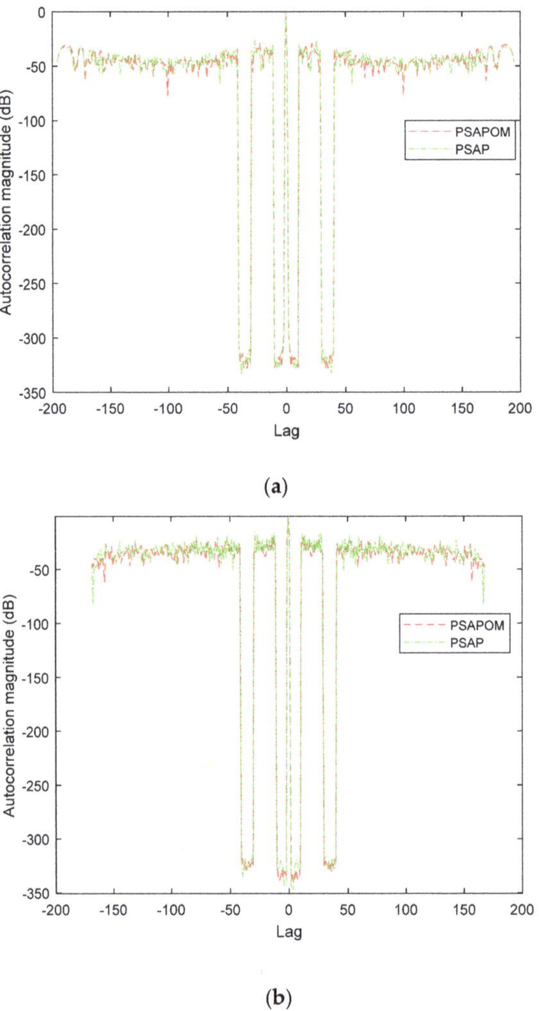

Figure 6. ACL performance comparison for suppressing multiple sidelobe areas. (**a**) Description of PSAPOM and PSAP initialized by Frank-coded sequence; (**b**) Description of PSAPOM and PSAP initialized by the 13 Barker-coded sequence.

Table 4. Algorithm (initialized by the Frank or Barker sequence) comparison for suppressing range sidelobes.

Algorithms	Frank Sequence		Barker Sequence	
	Aver-ACL in 1# Area (dB)	Aver-ACL in 2# Area (dB)	Aver-ACL in 1# Area (dB)	Aver-ACL in 2# Area (dB)
PSAPOM	−321.1162	−320.7214	−326.3520	−323.2342
PSAP	−321.4633	−321.2361	−325.7421	−323.1652

In these comparisons, we could draw a basic conclusion that our improved PSAP algorithms have a remarkable performance compared to WeCAN, PONLP, ISAA, and RISAAP, which will have a large influence on future practical applications. Moreover, given the sophisticated scenarios, i.e., the unimodular constraint and multiple suppressing areas, our PSAP and PSAPOM have shown a more powerful convergence than these methods. Meanwhile, our simulations have demonstrated that PSAP and PSAPOM have excellent robustness and stability. We may attribute these traits to FFT leverage and swarm particle intelligence due to alternating projection mechanisms.

5. Conclusions

In this paper, a PSAP framework (such as PSAP and PSAPOM) is introduced to design a unimodular CM phase-coded waveform with local low range sidelobes. Therein, PSAP with learning factor and particle-assisted projection could improve the convergence in the non-convex case. Numerical trails and simulations have also provided plenty of analysis to assess the waveforms generated by PSAP, WeCAN, ISAA, RISAAP, and PONLP. Regarding statistical performance, PSAP and PSAPOM via swarm intelligence and parallel optimization idea have achieved outstanding results. Additionally, in this paper, as we only discussed the masking scenarios of detecting static targets despite the Doppler effect of moving targets, in our further research, we will continue designing other phase-coded waveforms considering the non-zero Doppler effect. Moreover, we will also use GPU to accelerate the distributed parallel optimization.

Author Contributions: Conceptualization, X.F. and F.L.; Data curation, X.F.; Formal analysis, F.L., W.C. and Z.Z.; Funding acquisition, Z.Z.; Investigation, F.L. and Y.Z.; Methodology, X.F., F.L., W.C., Z.Z. and Y.Z.; Project administration, X.F. and Y.Z.; Resources, Z.Z.; Supervision, Z.Z.; Validation, F.L. and W.C.; Writing—original draft, X.F. and W.C.; Writing—review and editing, X.F. and Y.Z. All authors have read and agreed to the published version of the manuscript.

Funding: This research was funded by the National Natural Science Foundation of China, grant number 42127804.

Data Availability Statement: Not applicable.

Acknowledgments: We also appreciate the anonymous reviewers.

Conflicts of Interest: The authors declare no conflict of interest.

References

1. Blunt, S.D.; Mokole, E.L. Overview of radar waveform diversity. *IEEE Aerosp. Electron. Syst. Mag.* **2016**, *31*, 2–42. [CrossRef]
2. Guolong, C.; Xianxiang, Y.; Jing, Y.; Yue, F.; Lingjiang, K. An Overview of Waveform Optimization Methods for Cognitive Radar. *J. Radars* **2019**, *8*, 537–557.
3. Li, Y.; Vorobyov, S.A. Fast algorithms for designing unimodular waveform(s) with good correlation properties. *IEEE Trans. Signal Process.* **2017**, *66*, 1197–1212. [CrossRef]
4. Stoica, P.; He, H.; Li, J. New algorithms for designing unimodular sequences with good correlation properties. *IEEE Trans. Signal Process.* **2009**, *57*, 1415–1425. [CrossRef]
5. Feng, X.; Zhao, Y.; Zhou, Z.; Zhao, Z.-F. Waveform design with low range sidelobe and high Doppler tolerance for cognitive radar. *Signal Process.* **2017**, *139*, 143–155. [CrossRef]
6. Kajenski, P.J. Design of low-sidelobe phase-coded waveforms. *IEEE Trans. Aerosp. Electron. Syst.* **2019**, *55*, 2891–2898. [CrossRef]
7. Bu, Y.; Yu, X.; Yang, J.; Fan, T.; Cui, G. A new approach for design of constant modulus discrete phase radar waveform with low WISL. *Signal Process.* **2021**, *187*, 108145. [CrossRef]
8. Fan, W.; Liang, J.; Yu, G.; So, H.C.; Lu, G. Minimum local peak sidelobe level waveform design with correlation and/or spectral constraints. *Signal Process.* **2020**, *171*, 107450. [CrossRef]
9. Bolhasani, M.; Mehrshahi, E.; Ghorashi, S.A.; Alijani, M.S. Constant envelope waveform design to increase range resolution and SINR in correlated MIMO radar. *Signal Process.* **2019**, *163*, 59–65. [CrossRef]
10. Thakur, A.; Saini, D.S. MIMO radar sequence design with constant envelope and low correlation side-lobe levels. *AEU-Int. J. Electron. Commun.* **2021**, *136*, 153769. [CrossRef]
11. Song, J.; Babu, P.; Palomar, D.P. Optimization methods for designing sequences with low autocorrelation sidelobes. *IEEE Trans. Signal Process.* **2015**, *63*, 3998–4009. [CrossRef]
12. Li, F.; Zhao, Y.; Qiao, X. A waveform design method for suppressing range sidelobes in desired intervals. *Signal Process.* **2014**, *96*, 203–211. [CrossRef]
13. Ge, P.; Cui, G.; Karbasi, S.M.; Kong, L.; Yang, J. A template fitting approach for cognitive unimodular sequence design. *Signal Process.* **2016**, *128*, 360–368. [CrossRef]

14. Aubry, A.; De Maio, A.; Jiang, B.; Zhang, S. Ambiguity function shaping for cognitive radar via complex quartic optimization. *IEEE Trans. Signal Process.* **2013**, *61*, 5603–5619. [CrossRef]
15. Kerahroodi, M.A.; Aubry, A.; De Maio, A.; Naghsh, M.M.; Modarres-Hashemi, M. A coordinate-descent framework to design low PSL/ISL sequences. *IEEE Trans. Signal Process.* **2017**, *65*, 5942–5956. [CrossRef]
16. Esmaeili-Najafabadi, H.; Leung, H.; Moo, P.W. Unimodular waveform design with desired ambiguity function for cognitive radar. *IEEE Trans. Aerosp. Electron. Syst.* **2019**, *56*, 2489–2496. [CrossRef]
17. He, H.; Li, J.; Stoica, P. *Waveform Design for Active Sensing Systems: A Computational Approach*; Cambridge University Press: Cambridge, UK, 2012.
18. Cheng, Z.; Liao, B.; He, Z.; Li, J.; Han, C. A nonlinear-ADMM method for designing MIMO radar constant modulus waveform with low correlation sidelobes. *Signal Process.* **2019**, *159*, 93–103. [CrossRef]
19. Liu, Y.; Jiu, B.; Liu, H. ADMM-based transmit beampattern synthesis for antenna arrays under a constant modulus constraint. *Signal Process.* **2020**, *171*, 107529. [CrossRef]
20. Patton, L.K.; Rigling, B.D. Phase retrieval for radar waveform optimization. *IEEE Trans. Aerosp. Electron. Syst.* **2012**, *48*, 3287–3302. [CrossRef]
21. Zhang, X.; Wang, X. Waveform design with controllable modulus dynamic range under spectral constraints. *Signal Process.* **2021**, *189*, 108285. [CrossRef]
22. Wu, Z.J.; Zhou, Z.Q.; Wang, C.X.; Li, Y.-C.; Zhao, Z.-F. Doppler resilient complementary waveform design for active sensing. *IEEE Sens. J.* **2020**, *20*, 9963–9976. [CrossRef]
23. Wang, Y.C.; Dong, L.; Xue, X.; Yi, K.-C. On the design of constant modulus sequences with low correlation sidelobes levels. *IEEE Commun. Lett.* **2012**, *16*, 462–465. [CrossRef]
24. Xue-Bin, S.; Zhan-Min, L.; Cheng-Lin, Z.; Zheng, Z. Cognitive UWB Pulse Waveform Design Based on Particle Swarm Optimization. *Adhoc Sens. Wirel. Netw.* **2012**, *16*, 215–228.
25. Wang, S. Efficient heuristic method of search for binary sequences with good aperiodic autocorrelations. *Electron. Lett.* **2008**, *44*, 731–732. [CrossRef]
26. Zhao, D.; Wei, Y.; Liu, Y. Design of unimodular sequence train with low central and recurrent autocorrelations. *IET Radar Sonar Navig.* **2018**, *13*, 45–49. [CrossRef]
27. Tanweer, M.R.; Suresh, S.; Sundararajan, N. Self regulating particle swarm optimization algorithm. *Inf. Sci.* **2015**, *294*, 182–202. [CrossRef]
28. Zhang, H.; Xie, J.; Zong, B. Bi-objective particle swarm optimization algorithm for the search and track tasks in the distributed multiple-input and multiple-output radar. *Appl. Soft Comput.* **2021**, *101*, 107000. [CrossRef]
29. Lin, R.; Soltanalian, M.; Tang, B.; Li, J. Efficient design of binary sequences with low autocorrelation sidelobes. *IEEE Trans. Signal Process.* **2019**, *67*, 6397–6410. [CrossRef]
30. Levanon, N.; Mozeson, E. *Radar Signals*; John Wiley & Sons: Hoboken, NJ, USA, 2004.
31. Klauder, J.R.; Price, A.C.; Darlington, S.; Albersheim, W.J. The theory and design of chirp radars. *Bell Syst. Tech. J.* **1960**, *39*, 745–808. [CrossRef]

 *remote sensing*

Article

MIMO Radar Waveform Design for Multipath Exploitation Using Deep Learning

Zixiang Zheng [1], Yue Zhang [1,*], Xiangyu Peng [1], Hanfeng Xie [1], Jinfan Chen [1], Junxian Mo [1] and Yunfeng Sui [2]

[1] School of Electronic and Communication Engineering, Sun Yat-sen University, Shenzhen 518107, China; zhengzx26@mail2.sysu.edu.cn (Z.Z.)
[2] Research Center, Second Research Institute of CAAC, Chengdu 610041, China
* Correspondence: zhangyue8@mail.sysu.edu.cn

Abstract: This paper investigates the design of waveforms for multiple-input multiple-output (MIMO) radar systems that can exploit multipath returns to enhance target detection performance. By making reasonable use of multipath information in the waveform design, MIMO radar can effectively improve the signal-to-interference and noise ratio (SINR) of the receiver under a constant modulus (CM) constraint. However, optimizing the waveform design under these constraints is a challenging nonlinear and non-convex problem that cannot be easily solved using traditional methods. To overcome this challenge, we proposed a novel waveform design method for MIMO radar in multipath scenarios based on deep learning. Specifically, we leveraged the powerful nonlinear fitting ability of neural networks to solve the non-convex optimization problem. First, we constructed a deep residual network and transform the CM constraint into a phase sequence optimization problem. Next, we used the constructed waveform optimization design problem as the loss function of the network. Finally, we used the adaptive moment estimation (Adam) optimizer to train the network. Simulation results demonstrated that our proposed method outperformed existing methods by achieving better SINR values for the receiver.

Keywords: MIMO radar; multipath exploitation; waveform design; deep learning; SINR

Citation: Zheng, Z.; Zhang, Y.; Peng, X.; Xie, H.; Chen, J.; Mo, J.; Sui, Y. MIMO Radar Waveform Design for Multipath Exploitation Using Deep Learning. *Remote Sens.* **2023**, *15*, 2747. https://doi.org/10.3390/rs15112747

Academic Editor: Stefano Tebaldini

Received: 28 March 2023
Revised: 5 May 2023
Accepted: 22 May 2023
Published: 25 May 2023

1. Introduction

Radar systems often encounter multipath effects when detecting low-altitude targets. In such scenarios, the received returns consist not only of the backscattered line-of-sight (LOS) component, but also of the multipath returns component [1–3]. However, the presence of multipath returns can cause the received echo signal to fluctuate and even cancel, which will reduce the performance of target detection and parameter estimation. The early research has been devoted to suppressing the multipath returns [4–7]. In [4], multipath returns are regarded as clutter and suppressed.

However, the principle of multipath generation suggests that both the direct and multipath returns are coherent and contain target energy [8]. If the energy of the multipath returns can be accurately estimated and accumulated, it can improve the detection and tracking performance of the target [9–11].

Multiple-input multiple-output (MIMO) radar is a new type of radar system in which the transmitting antennas can transmit mutually independent signals [12–16]. By adaptively adjusting the MIMO radar transmission waveform for different tactical needs, the radar detection performance can be significantly improved in complex environments [17,18].

MIMO channel models can be broadly categorized into two types based on the methods used for their establishment. The first type is deterministic channel models [19–21], where precise information about the channel is obtained, and the wireless propagation is deemed as a deterministic process. This enables the determination of the spatio-temporal characteristics of any point in space. One study [19] employed a multilayer artificial neural network method to predict channel characteristics, thereby overcoming the low computational efficacy of traditional deterministic channel models. In contrast, the other type is statistical channel models [22–25], wherein the channel is regarded as a stochastic process, and a probability distribution is used to describe its temporal and spatial changes.

Depending on the specific application requirements of the radar, the waveform optimization design criteria for MIMO radar can be divided into four types. The first type of optimization criterion aims to maximize the signal-to-interference and noise ratio (SINR) [26–29]. Through appropriate waveform design, MIMO radar can maximize the SINR to improve target detection performance. In [29], an optimization design problem for maximizing the SINR based on constant modulus, similarity, and spectrum constraints was presented, and a semi-definite programming method was used to solve the problem. The second type of optimization criterion aims to maximize mutual information [30,31]. By maximizing the amount of mutual information between the corresponding target and the received waveform, the echo can exhibit more target characteristics. The third type of optimization criterion is the pattern matching problem [32,33]. The goal of this type of problem is to concentrate the transmitted beam energy of the MIMO radar in a specified airspace while minimizing the transmitted energy of the side lobes. The last type of optimization criterion is the orthogonal waveform design problem [34–36]. By reducing the correlation between the transmitted waveforms, the performance of matched filtering can be improved.

Combining high degrees of freedom in the MIMO radar waveform design with multipath exploitation [37–39] has shown great potential for further improving target detection performance. Most existing waveform design methods using multipath returns focus on slow time domain weight optimization for fixed waveform. In [37], the authors proposed to improve the detection performance of moving targets in multipath scenarios by optimizing the weights between different coherent processing intervals (CPIs), which they performed by using the orthogonal frequency division multiplexing (OFDM) waveform. On the other hand, in [38], the authors proposed an OFDM MIMO adaptive waveform design algorithm based on the mutual information criteria, thus aiming to select the best OFDM waveform by maximizing the mutual information between the state and measurement vectors. This particular type of slow time domain MIMO radar waveform can present challenges in scenarios that demand low range domain sidelobes, due to its high range domain sidelobes. Additionally, complex signal processing algorithms are required for slow time domain waveforms, which can increase system cost and overall complexity.

In recent years, there has been an increasing interest among radar technicians in exploring fast time domain waveform design methods that make use of multipath information. In [40], the transmit waveform and receive filter of a MIMO radar system were jointly designed to maximize the SINR of the receiver for multipath exploitation. In [41], the robust joint design of MIMO radar transmit waveform and receive filter banks was considered under the uncertainty of multipath returns information. This method addressed the limitation of requiring accurate prior knowledge of the multipath returns information in the method presented in [40]. The optimization problem discussed in the literature is non-convex and, therefore, cannot be solved directly. Existing research primarily relies on algorithms such as semi-definite relaxation algorithms to solve non-convex problems indirectly by relaxing the objective function or the constraint to a more tractable form. Oftentimes, these methods experience degradation in performance due to the relaxation process.

The constant modulus (CM) constraint is a common requirement for MIMO radar waveform design to avoid distortion of the transmit signal in near-saturated operating modes of the high frequency amplifier [42]. The objective of MIMO radar waveform design for multipath exploitation is to maximize the SINR of the receiver while adhering to the CM constraint of the transmit waveform. However, this problem is non-linear and non-convex, thereby making it difficult to solve with traditional optimization algorithms.

Deep learning, as a natural non-linear system [43], can effectively solve such problems. In this paper, we proposed a method to design MIMO radar fast time domain waveforms for multipath exploitation using deep learning. We used a residual neural network to directly solve the non-convex optimization problem, rather than indirectly solving it by relaxing the CM constraint or SINR function. Firstly, we transformed the CM constraint problem into an unconstrained phase optimization problem and took a random phase sequence as the input of the network. Next, the reciprocal of the SINR function of the received signal model in the multipath scenario was used as the loss function of the network. After training the network using an adaptive moment estimation (Adam) optimizer, the output of the network became the phase sequence of the designed waveform.

The simulation results demonstrated that the proposed algorithm made full use of the multipath energy, thus resulting in the waveform with higher SINR performance and detection probability compared to existing methods.

In summary, the main contributions of this article can be summarized as follows:

1. A MIMO radar signal model was constructed for multipath scenarios;
2. The MIMO radar waveform design problem for multipath exploitation was modeled as a maximizing SINR problem with the constant modulus constraint on the transmit waveform;
3. Our proposed MIMO radar waveform design algorithm employed deep learning that utilized the non-linear fitting ability of neural networks to directly solve the non-convex waveform optimization problem.

The remainder of this paper is organized as follows. In Section 2, the multipath signal model for the MIMO radar is established. In Section 3, the MIMO radar waveform optimization design problem is formulated, and the waveform design algorithm based on deep learning is presented. Section 4 provides the simulation results to demonstrate the effectiveness of the proposed algorithm and the superiority of the designed waveform. Finally, Section 5 draws the conclusions.

Notations: In this paper, we use italic letters for scalars, bold italic lowercase letters for vectors, and bold italic uppercase letters for matrices. The superscripts $(\cdot)^T$, $(\cdot)^H$ and $(\cdot)^*$ denote the transpose, conjugate transpose, and conjugate, respectively. $\mathbb{C}^{N \times N}$ denotes the sets of $N \times N$ complex matrices. $\text{vec}(\cdot)$ denotes stacking the matrix by column. The symbol $\otimes$ denotes the Kronecker product. $\text{E}(\cdot)$ represents the calculation expectation, and $\text{tr}(\cdot)$ denotes the trace of a square matrix. $\|\cdot\|$ denotes the Frobeneous norm. $\text{Re}\{a\}$ and $\text{Im}[a]$ denote the real part and imaginary part of the vector a, respectively.

2. Signal Model

A colocated MIMO radar consisting of N_T transmitting antennas and N_R receiving antennas was considered in the multipath scenario. The transmit waveform of the lth snapshot can be expressed as

$$s_l = [s_l(1), s_l(2), \ldots, s_l(N_T)]^T, \tag{1}$$

where $l = 1, 2, \ldots, L$, L denotes the number of samples in the fast time domain. The transmit signal matrix can be represented by

$$S = [s_1, s_2, \ldots, s_L] \in \mathbb{C}^{N_T \times L}. \tag{2}$$

It is assumed that the unit spacing in the transmitting antenna array and the receiving antenna array is set at half wavelength, that is, $d = \lambda/2$. The synthetic signal of the lth snapshot at the azimuth θ can be expressed as

$$y_l = a_t^T(\theta)s_l, l = 1, 2, \ldots, L, \tag{3}$$

where $a_t(\theta) = \left[1, e^{-\frac{j2\pi d \sin\theta}{\lambda}}, \ldots, e^{-\frac{j2\pi(N_T-1)d\sin\theta}{\lambda}}\right]^T$ denotes the transmit steering vector.

The signal received by the MIMO radar receiver of the lth snapshot can be modeled as

$$z_l = \alpha A(\theta)s_l = \alpha a_r(\theta)a_t^T(\theta)s_l, l = 1, 2, \ldots, L, \tag{4}$$

where α is the complex reflection coefficient, and $a_r(\theta) = \left[1, e^{-\frac{j2\pi d\sin\theta}{\lambda}}, \ldots, e^{-\frac{j2\pi(N_R-1)d\sin\theta}{\lambda}}\right]^T$ denotes the receive steering vector $A(\theta) = a_r(\theta)a_t^T(\theta)$.

As shown in Figure 1, it is assumed that there is a reflective surface in the scenario that results in multipath returns. When the radar system transmits signals to detect a target, the returns received by the radar array include not only direct returns, but also multipath returns from the reflector. The transmit–receive path of direct returns is A $\rightarrow$ D $\rightarrow$ A. Multipath returns include the multipath of the transmission process and the multipath of the reception process, namely, A $\rightarrow$ B $\rightarrow$ D $\rightarrow$ A and A $\rightarrow$ D $\rightarrow$ B $\rightarrow$ A. In addition, there are returns from the clutter area, namely A $\rightarrow$ C $\rightarrow$ A. The received signal model is given by

$$y = y_d + y_m + y_c + y_n, \tag{5}$$

where y_d, y_m, y_c, and y_n represent the direct returns, multipath returns, clutter returns, and noise, respectively.

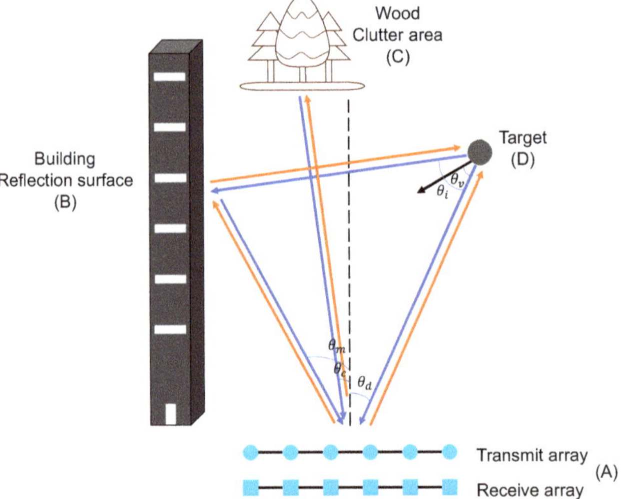

Figure 1. Multipath scenario diagram.

2.1. Direct Returns Model

Assuming that the target velocity and the angle between the target motion direction and LOS are v and θ_v, respectively, the Doppler frequency of the direct returns can be expressed as

$$F_d = \frac{2v\cos\theta_v}{\lambda}, \tag{6}$$

where λ is the wavelength of the transmit waveform. The shift matrix is defined as G_l, which is represented as

$$G_l(m, n) = \begin{cases} 1, & \text{if } m - n + l = 0 \\ 0, & \text{if } m - n + l \neq 0 \end{cases}. \tag{7}$$

The direct returns model of the pth pulse can be expressed as

$$\mathbf{Y}_{d,p} = \alpha \mathbf{a}_r(\theta_d)\mathbf{a}_t^T(\theta_d)\mathbf{S}\mathbf{G}_0 e^{j2\pi F_d(p-1)} \in \mathbb{C}^{N_R \times 2L}, \tag{8}$$

where α is the complex scattering coefficient of the target, and $\mathbf{a}_t(\theta_d)$ and $\mathbf{a}_r(\theta_d)$ denote the transmit steering vector and receive steering vector in the target azimuth θ_d, respectively. Assuming that the radar transmits P pulses in a CPI, we have

$$\mathbf{y}_d = \left[\mathbf{y}_{d,1}^T, \mathbf{y}_{d,2}^T, \dots, \mathbf{y}_{d,P}^T\right]^T, \tag{9}$$

where $\mathbf{y}_{d,p} = \text{vec}(\mathbf{Y}_{d,p})$. The direct return model vector can be expressed as

$$\mathbf{y}_d = \alpha \left(\mathbf{f} \otimes \mathbf{G}_0^T \otimes \left(\mathbf{a}_r(\theta_d)\mathbf{a}_t^T(\theta_d)\right)\right)\mathbf{s}, \tag{10}$$

where $\mathbf{s} = \text{vec}(\mathbf{S})$, $\mathbf{f} = \left[1, e^{j2\pi F_d}, \dots, e^{j2\pi F_d(P-1)}\right]^T$.

2.2. Multipath Returns Model

The Doppler frequency of the multipath returns can be given by

$$F_m = \frac{F_d}{2} + \frac{\cos(\theta_i - \theta_v)F_d}{2\cos\theta_v}, \tag{11}$$

where θ_i is the angle between $\overline{DB}$ and $\overline{DA}$.

For the multipath returns, the transmit steering vector and the receive steering vector will point in different azimuths due to the reflective surface. Taking into account both the receiving and transmitting multipath, the received signal model of the pth pulse can be represented as

$$\mathbf{Y}_{m,p} = \rho\alpha\left(\mathbf{a}_r(\theta_m)\mathbf{a}_t^T(\theta_d) + \mathbf{a}_r(\theta_d)\mathbf{a}_t^T(\theta_m)\right)\mathbf{S}\mathbf{G}_{l_m}e^{j2\pi F_m(p-1)} \in \mathbb{C}^{N_R \times 2L}, \tag{12}$$

where ρ and θ_m are the complex reflection coefficient of the surface and the arrival azimuth of multipath returns, respectively, and l_m is the relative fast time delay of the multipath path with respect to the direct path.

Considering the CPI, the multipath returns model can finally be expressed as

$$\mathbf{y}_m = \rho\alpha\left(\mathbf{f}' \otimes \mathbf{G}_{l_m}^T \otimes \left(\mathbf{a}_r(\theta_d)\mathbf{a}_t^T(\theta_m) + \mathbf{a}_r(\theta_m)\mathbf{a}_t^T(\theta_d)\right)\right)\mathbf{s}, \tag{13}$$

where $\mathbf{f}' = \left[1, e^{j2\pi F_m}, \dots, e^{j2\pi F_m(P-1)}\right]^T$.

2.3. Clutter

In radar operations, there will inevitably be reflections from other objects in the scene that are not desired, which will have a noticeable effect on the received returns. The clutter returns of the pth pulse can be expressed as

$$\mathbf{Y}_{c,p} = \beta \mathbf{a}_r(\theta_c)\mathbf{a}_t^T(\theta_c)\mathbf{S}\mathbf{G}_{l_c} \in \mathbb{C}^{N_R \times 2L}, \tag{14}$$

where β is the complex backscattering coefficient of the clutter area, θ_c is the azimuth of the clutter area, and l_c is the relative fast time delay of the clutter path with respect to the direct path. The clutter signal model can be further expressed as

$$\mathbf{y}_c = \beta\left(\mathbf{1}_P \otimes \mathbf{G}_{l_c}^T \otimes \left(\mathbf{a}_r(\theta_c)\mathbf{a}_t^T(\theta_c)\right)\right)\mathbf{s}, \tag{15}$$

where $\mathbf{1}_P$ denotes a row unit vector of length P.

2.4. Noise

There is also signal-independent interference in the received data, which is caused by system noise. Typically, signal-independent noise is modeled as a Gaussian distribution.

3. Methods

3.1. Problem Formulation

The SINR is a commonly used optimization criterion that is closely related to target detection and parameter estimation performance. In our approach, we maximized the SINR in multipath scenarios to optimize the waveform design. Unlike common methods that treat multipath returns as clutter and suppress them, we leveraged the information from the multipath returns to improve the SINR of the output.

In order to calculate the SINR, the signal power of the received direct returns, multipath returns, clutter, and noise should be calculated separately and expressed as P_d, P_m, P_c, and P_n.

Based on the direct returns model, the signal power of the direct returns can be expressed as

$$
\begin{aligned}
P_d &= \frac{1}{L} \sum_{p=1}^{P} \sum_{l=1}^{L} E\left[\left\| \alpha e^{j2\pi F_d(p-1)} A(\theta_d) s_l \right\|^2 \right] \\
&= \sum_{p=1}^{P} E\left[\left\| e^{j2\pi F_d(p-1)} \right\|^2 \right] E\left[\|\alpha\|^2 \right] \frac{1}{L} \sum_{l=1}^{L} s_l^H a_t(\theta_d) a_r^T(\theta_d) a_r^*(\theta_d) a_t^H(\theta_d) s_l \\
&= \sum_{p=1}^{P} E\left[\left\| e^{j2\pi F_d(p-1)} \right\|^2 \right] E\left[\|\alpha\|^2 \right] \mathrm{tr}\left(a_t^H(\theta_d) \sum_{l=1}^{L} s_l s_l^H a_t(\theta_d) \right) \\
&= \sum_{p=1}^{P} E\left[\left\| e^{j2\pi F_d(p-1)} \right\|^2 \right] \frac{E\left[\|\alpha\|^2 \right]}{L} \left\| a_t^H(\theta_d) S \right\|^2,
\end{aligned}
\tag{16}
$$

where $A(\theta_d) = a_r(\theta_d) a_t^T(\theta_d)$.

The signal power of the multipath returns consists of two parts, namely, the signal power of the transmitting multipath returns and the signal power of the receiving multipath returns, which can be expressed as

$$
\begin{aligned}
P_m &= \frac{1}{L} \sum_{p=1}^{P} \sum_{l=1}^{L} E\left[\left\| \rho\alpha e^{j2\pi F_m(p-1)} B(\theta_{md}) s_l \right\|^2 \right] + \frac{1}{L} \sum_{p=1}^{P} \sum_{l=1}^{L} E\left[\left\| \rho\alpha e^{j2\pi F_m(p-1)} B(\theta_{dm}) s_l \right\|^2 \right] \\
&= \sum_{p=1}^{P} E\left[\|\alpha\|^2 \right] E\left[\|\rho\|^2 \right] E\left[\left\| e^{j2\pi F_m(p-1)} \right\|^2 \right] \frac{1}{L} \sum_{l=1}^{L} s_l^H a_t(\theta_d) a_r^T(\theta_m) a_r^*(\theta_m) a_t^H(\theta_d) s_l \\
&\quad + \sum_{p=1}^{P} E\left[\|\alpha\|^2 \right] E\left[\|\rho\|^2 \right] E\left[\left\| e^{j2\pi F_m(p-1)} \right\|^2 \right] \frac{1}{L} \sum_{l=1}^{L} s_l^H a_t(\theta_m) a_r^T(\theta_d) a_r^*(\theta_d) a_t^H(\theta_m) s_l \\
&= \sum_{p=1}^{P} \eta\, \mathrm{tr}\left(a_t^H(\theta_d) \sum_{l=1}^{L} s_l s_l^H a_t(\theta_d) \right) + \sum_{p=1}^{P} \eta\, \mathrm{tr}\left(a_t^H(\theta_m) \sum_{l=1}^{L} s_l s_l^H a_t(\theta_m) \right) \\
&= \sum_{p=1}^{P} \frac{\eta}{L} \left(\left\| a_t^H(\theta_d) S \right\|^2 + \left\| a_t^H(\theta_m) S \right\|^2 \right),
\end{aligned}
\tag{17}
$$

where $\eta = E\left[\|\alpha\|^2 \right] E\left[\|\rho\|^2 \right] E\left[\left\| e^{j2\pi F_m(p-1)} \right\|^2 \right]$, $B(\theta_{md}) = a_r(\theta_m) a_t^T(\theta_d)$, and $B(\theta_{dm}) = a_r(\theta_d) a_t^T(\theta_m)$.

The signal power of the clutter is given by

$$
\begin{aligned}
P_c &= \frac{1}{L} \sum_{p=1}^{P} \sum_{l=1}^{L} E\left[\|\beta A(\theta_c)s_l\|^2\right] \\
&= \sum_{p=1}^{P} E\left[\|\beta\|^2\right] \frac{1}{L} \sum_{l=1}^{L} s_l^H a_t(\theta_c) a_r^T(\theta_c) a_r^*(\theta_c) a_t^H(\theta_c) s_l \\
&= \sum_{p=1}^{P} E\left[\|\beta\|^2\right] \mathrm{tr}\left(a_t^H(\theta_c) \sum_{l=1}^{L} s_l s_l^H a_t(\theta_c) \right) \\
&= \sum_{p=1}^{P} \frac{E\left[\|\beta\|^2\right]}{L} \left\| a_t^H(\theta_c)S \right\|^2,
\end{aligned}
\tag{18}
$$

where $A(\theta_c) = a_r(\theta_c) a_t^T(\theta_c)$.

The noise power can be expressed as

$$
P_n = PN_T L \sigma_n^2,
\tag{19}
$$

where σ_n^2 represents the variance of Gaussian white noise.

To sum up, the output SINR can be expressed as

$$
\begin{aligned}
\mathrm{SINR} &= \frac{P_d + P_m}{P_c + P_n} \\
&= \frac{\sum_{p=1}^{P} \mu_1 \left\| a_t^H(\theta_d)S \right\|^2 + \sum_{p=1}^{P} \mu_2 \left\| a_t^H(\theta_m)S \right\|^2}{\sum_{p-1}^{P} \mu_3 \left\| a_t^H(\theta_c)S \right\|^2 + PN_T L \sigma_n^2},
\end{aligned}
\tag{20}
$$

where $\mu_1 = \dfrac{E\left[\|\alpha\|^2\right] \left(E\left[\left\|e^{j2\pi F_d(p-1)}\right\|^2\right] + E\left[\left\|e^{j2\pi F_m(p-1)}\right\|^2\right] E\left[\|\rho\|^2\right] \right)}{L}$, $\mu_2 = \dfrac{E\left[\left\|e^{j2\pi F_m(p-1)}\right\|^2\right] E\left[\|\rho\alpha\|^2\right]}{L}$, and $\mu_3 = \dfrac{E\left[\|\beta\|^2\right]}{L}$.

The aim of this paper was to optimize the waveform design by maximizing the output SINR with the constant modulus constraint. Therefore, the waveform optimization problem in this paper can be summarized as

$$
\max_{S} \quad \frac{\sum_{p=1}^{P} \mu_1 \left\| a_t^H(\theta_d)S \right\|^2 + \sum_{p=1}^{P} \mu_2 \left\| a_t^H(\theta_m)S \right\|^2}{\sum_{p=1}^{P} \mu_3 \left\| a_t^H(\theta_c)S \right\|^2 + PN_T L \sigma_n^2}
\tag{21}
$$

$$
\mathrm{s.t.} \quad |s(l)| = 1, l = 1, \ldots, MN_T.
$$

3.2. The Proposed Design Method

The problem of optimizing MIMO radar waveform to maximize SINR with a CM constraint is challenging due to the non-convex nature of both the SINR function and the CM constraint. This is a high-dimensional and non-convex problem that cannot be solved optimally using conventional methods. However, deep learning models are highly suitable for solving this problem, since they are high-dimensional and non-linear systems.

As the residual network can effectively tackle the issues of gradient disappearance and explosion while increasing the number of network layers [44], we constructed a deep residual network to use its non-linear fitting ability to solve the nonconvex problem of maximizing SINR with the CM constraint. The optimized training network designed in this paper is displayed in Figure 2 and comprises five modules: input, forward propagation, output, loss function, and Adam optimizer.

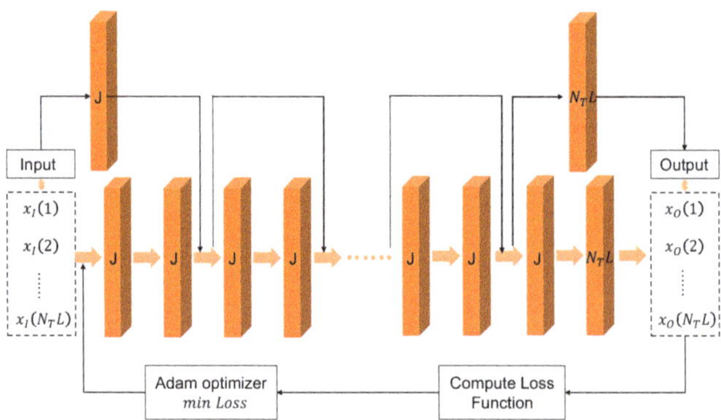

Figure 2. Constructed residual network.

3.2.1. Input and Output

A randomly generated normalized phase sequence of length $N_T L$ was used as the input of the network and can be represented as

$$x_I = [x_I(1), \ldots, x_I(i), \ldots, x_I(N_T L)] \in \mathbb{C}^{N_T L \times 1}, \tag{22}$$

where $x_I(i) \in [0,1], i = 1, 2, \ldots, N_T L$.

The output of the network is a normalized phase sequence and can be represented as

$$x_O = [x_O(1), \ldots, x_O(i), \ldots, x_O(N_T L)] \in \mathbb{C}^{N_T L \times 1}, \tag{23}$$

where $x_O(i) \in [0,1], i = 1, 2, \ldots, N_T L$.

3.2.2. Forward Propagation Module

The constructed residual network consists of 10 residual blocks, each comprising two fully connected neural network layers. The first layer of the first residual block is a neural network with weight matrix dimension of $N_T L \times J$, where the number of neurons is J. The second layer is a neural network with weight matrix dimension of $J \times J$ and J neurons.

The first layer of the last residual block is a neural network with weight matrix dimension of $J \times J$, with J neurons. The second layer is a neural network with weight matrix dimension of $J \times N_T L$, with J neurons.

To achieve dimension matching of the residual network, there is a network with weight matrix dimension of $N_T L \times J$ and J neurons between the input and output of the first residual block, and there is a network with weight matrix dimension of $J \times N_T L$ and J neurons between the input and output of the last residual block.

The other residual blocks in between consist of two layers of neural networks with weight matrix dimension of $J \times J$ and J neurons.

3.2.3. Loss Function

The output of forward propagation is a normalized phase sequence $\hat{x}_O \in \mathbb{C}^{N_T L \times 1}$. Assuming $\hat{\varphi} = 2\pi \times \hat{x}_O$, the original signal from one forward propagation output can be expressed as $\hat{s} = e^{j\hat{\varphi}} \in \mathbb{C}^{N_T L \times 1}$. By the inverse process of vectorizing the waveform columns, an output signal matrix of dimension $N_T \times L$ can be obtained, where the lth column is the signal of the lth snapshot.

As the residual network cannot directly handle complex problems, an algebraic transformation of the original objective function is necessary. Assuming $\hat{s} = \cos\hat{\varphi} + j\sin\hat{\varphi}$ and $a_t^T(\theta) = \text{Re}\{a_t^T(\theta)\} + j * \text{Im}\{a_t^T(\theta)\}$, the synthetic signal of the lth snapshot of MIMO radar in azimuth θ can be expressed as

$$
\begin{aligned}
y_\theta(l) &= \boldsymbol{a}_t^T(\theta)\hat{\boldsymbol{s}}_l \\
&= \left(\mathrm{Re}\{\boldsymbol{a}_t^T(\theta)\} + j * \mathrm{Im}\{\boldsymbol{a}_t^T(\theta)\}\right)(\cos\hat{\varphi}_l + j * \sin\hat{\varphi}_l) \\
&= \mathrm{Re}\{\boldsymbol{a}_t^T(\theta)\}\cos\hat{\varphi}_l - \mathrm{Im}\{\boldsymbol{a}_t^T(\theta)\}\sin\hat{\varphi}_l + j * \left[\mathrm{Re}\{\boldsymbol{a}_t^T(\theta)\}\sin\hat{\varphi}_l + \mathrm{Im}\{\boldsymbol{a}_t^T(\theta)\}\cos\hat{\varphi}_l\right].
\end{aligned}
\tag{24}
$$

According to (24), we have

$$
\mathrm{Re}\{y_\theta(l)\} = \mathrm{Re}\left\{\boldsymbol{a}_t^T(\theta)\right\}\cos\hat{\varphi}_l - \mathrm{Im}\left\{\boldsymbol{a}_t^T(\theta)\right\}\sin\hat{\varphi}_l,
\tag{25}
$$

$$
\mathrm{Im}\{y_\theta(l)\} = \mathrm{Re}\left\{\boldsymbol{a}_t^T(\theta)\right\}\sin\hat{\varphi}_l + \mathrm{Im}\left\{\boldsymbol{a}_t^T(\theta)\right\}\cos\hat{\varphi}_l.
\tag{26}
$$

The direct returns, multipath returns, and clutter of the lth snapshot of the pth pulse can be expressed as

$$
\begin{aligned}
y_{d,p}(l) =\alpha e^{j2\pi F_d(p-1)}\Big(&\mathrm{Re}\{\boldsymbol{a}_t^T(\theta_d)\}\cos\hat{\varphi}_l - \mathrm{Im}\{\boldsymbol{a}_t^T(\theta_d)\}\sin\hat{\varphi}_l \\
&+ j * \left[\mathrm{Re}\{\boldsymbol{a}_t^T(\theta_d)\}\sin\hat{\varphi}_l + \mathrm{Im}\{\boldsymbol{a}_t^T(\theta_d)\}\cos\hat{\varphi}_l\right]\Big),
\end{aligned}
\tag{27}
$$

$$
\begin{aligned}
y_{m,p}(l) = \rho\alpha e^{j2\pi F_m(p-1)}\Big(&\mathrm{Re}\{\boldsymbol{a}_t^T(\theta_d)\}\cos\hat{\varphi}_l - \mathrm{Im}\{\boldsymbol{a}_t^T(\theta_d)\}\sin\hat{\varphi}_l \\
&+ j * \left[\mathrm{Re}\{\boldsymbol{a}_t^T(\theta_d)\}\sin\hat{\varphi}_l + \mathrm{Im}\{\boldsymbol{a}_t^T(\theta_d)\}\cos\hat{\varphi}_l\right] \\
&+ \mathrm{Re}\{\boldsymbol{a}_t^T(\theta_m)\}\cos\hat{\varphi}_l - \mathrm{Im}\{\boldsymbol{a}_t^T(\theta_m)\}\sin\hat{\varphi}_l \\
&+ j * \left[\mathrm{Re}\{\boldsymbol{a}_t^T(\theta_m)\}\sin\hat{\varphi}_l + \mathrm{Im}\{\boldsymbol{a}_t^T(\theta_m)\}\cos\hat{\varphi}_l\right]\Big),
\end{aligned}
\tag{28}
$$

$$
\begin{aligned}
y_{c,p}(l) =\beta\Big(&\mathrm{Re}\{\boldsymbol{a}_t^T(\theta_c)\}\cos\hat{\varphi}_l - \mathrm{Im}\{\boldsymbol{a}_t^T(\theta_c)\}\sin\hat{\varphi}_l \\
&+ j * \left[\mathrm{Re}\{\boldsymbol{a}_t^T(\theta_c)\}\sin\hat{\varphi}_l + \mathrm{Im}\{\boldsymbol{a}_t^T(\theta_c)\}\cos\hat{\varphi}_l\right]\Big).
\end{aligned}
\tag{29}
$$

Considering the delay of multipath returns and clutter relative to direct returns, the direct returns, multipath returns, and clutter of the pth pulse can be expressed as

$$
\boldsymbol{y}_{d,p} = \begin{bmatrix} y_{d,p}(1) \\ y_{d,p}(2) \\ \vdots \\ y_{d,p}(L) \\ \boldsymbol{0}_L \end{bmatrix},
\tag{30}
$$

$$
\boldsymbol{y}_{m,p} = \begin{bmatrix} \boldsymbol{0}_{l_m} \\ y_{m,p}(1) \\ y_{m,p}(2) \\ \vdots \\ y_{m,p}(L) \\ \boldsymbol{0}_{L-l_m} \end{bmatrix},
\tag{31}
$$

$$
\boldsymbol{y}_{c,p} = \begin{bmatrix} \mathbf{0}_{l_c} \\ y_{c,p}(1) \\ y_{c,p}(2) \\ \vdots \\ y_{c,p}(L) \\ \mathbf{0}_{L-l_c} \end{bmatrix}, \tag{32}
$$

where $\mathbf{0}_L$ denotes a column zero vector of length L. Taking into account the coherent pulse interval P, the direct returns, multipath returns, and clutter can be finally expressed as

$$
\boldsymbol{y}_d = \left[\boldsymbol{y}_{d,1}{}^T, \boldsymbol{y}_{d,2}{}^T, \dots, \boldsymbol{y}_{d,P}{}^T \right]^T, \tag{33}
$$

$$
\boldsymbol{y}_m = \left[\boldsymbol{y}_{m,1}{}^T, \boldsymbol{y}_{m,2}{}^T, \dots, \boldsymbol{y}_{m,P}{}^T \right]^T, \tag{34}
$$

$$
\boldsymbol{y}_c = \left[\boldsymbol{y}_{c,1}{}^T, \boldsymbol{y}_{c,2}{}^T, \dots, \boldsymbol{y}_{c,P}{}^T \right]^T. \tag{35}
$$

The signal power of the direct returns, multipath returns, clutter, and noise can be given by

$$
\begin{aligned}
P_d &= \frac{1}{L} \boldsymbol{y}_d{}^H \boldsymbol{y}_d \\
&= \frac{1}{L}\left(\mathrm{Re}\{\boldsymbol{y}_d\}^2 + \mathrm{Im}\{\boldsymbol{y}_d\}^2 \right),
\end{aligned} \tag{36}
$$

$$
\begin{aligned}
P_m &= \frac{1}{L} \boldsymbol{y}_m{}^H \boldsymbol{y}_m \\
&= \frac{1}{L}\left(\mathrm{Re}\{\boldsymbol{y}_m\}^2 + \mathrm{Im}\{\boldsymbol{y}_m\}^2 \right),
\end{aligned} \tag{37}
$$

$$
\begin{aligned}
P_c &= \frac{1}{L} \boldsymbol{y}_c{}^H \boldsymbol{y}_c \\
&= \frac{1}{L}\left(\mathrm{Re}\{\boldsymbol{y}_c\}^2 + \mathrm{Im}\{\boldsymbol{y}_c\}^2 \right),
\end{aligned} \tag{38}
$$

$$
P_n = P N_T L \sigma_n^2. \tag{39}
$$

The loss function of the network is set to be the inverse of the objective function in the above optimization problem and can be expressed as

$$
loss = \frac{1}{\mathrm{SINR}} = \frac{P_c + P_n}{P_d + P_m}. \tag{40}
$$

3.2.4. Adam Optimizer

The optimizer is a tool to guide the neural network for parameter updates. After the neural network achieves forward propagation once and calculates the loss function, the optimizer is needed to perform backward propagation to achieve the update of the network parameters. Adam optimizer is a classical deep learning optimizer based on gradient descent algorithm, combining the ideas of momentum method and adaptive learning rate, with the advantage of adaptively adjusting the learning rate to adapt to different data and models.

The Adam optimizer computes first-order moment estimates and second-order moment estimates of the gradient in each iteration, thus updating the learning rate and network parameters.

The gradient parameter of the loss function at the tth iteration with respect to the network weights w can be expressed as

$$g_t = \nabla loss = \frac{\partial loss}{\partial w_t}. \tag{41}$$

The first-order moments and second-order moments of the network weights in the tth iteration can be expressed as

$$m_t = \beta_1 \cdot m_{t-1} + (1 - \beta_1) \cdot g_t, \tag{42}$$

$$r_t = \beta_2 \cdot r_{t-1} + (1 - \beta_2) \cdot g_t \odot g_t, \tag{43}$$

where β_1 and β_2 are two hyperparameters of the Adam optimizer, usually set to $\beta_1 = 0.9$ and $\beta_2 = 0.999$.

At the beginning of the iteration, there is a deviation of m_t and r_t to the initial value, so it is necessary to correct the deviation of the first-order moments and second-order moments. The corrected first-order moments and second-order moments can be expressed as

$$\hat{m}_t = \frac{m_t}{1 - \beta_1^t}, \tag{44}$$

$$\hat{r}_t = \frac{r_t}{1 - \beta_2^t}. \tag{45}$$

We iterate over the network weights, and the network weight parameters can be updated and represented as follows

$$w_{t+1} = w_t - \frac{\hat{m}_t}{\sqrt{\hat{r}_t} + \delta}, \tag{46}$$

where δ refers to a small constant used for numerical stabilization, usually set to $\delta = 10^{-8}$. We summarize the proposed algorithm in Algorithm 1.

Algorithm 1: Proposed algorithm.

Input: Random normalized phase sequence x_I, learning rate of Adam γ, number of iterations E.

Output: Desired waveform phase sequence x_O.

Set $e = 0$, Adam learning rate set to $\gamma > 0$;
1: Construct forward propagation module according to Figure 2;
2: Input x_I to the forward propagation module to obtain output x_O;
3: Compute y_d, y_m and y_c according to (27)–(35);
4: Compute P_d, P_m, P_c and P_n according to (36)–(39), and the loss function is constructed by (40);
5: Optimizing loss function with Adam optimizer;
6: If $e = E$, stop and output the result. Otherwise, update e, i.e., $e = e + 1$, and back to the step 2.

3.2.5. Complexity Analysis

To evaluate the complexity of our proposed model, we measured its time complexity using floating-point arithmetic, FLOPs, and its spatial complexity using parameter quantities, Params. Assuming that the input of the network is a sequence with a length of 320, the complexity of the proposed model is shown in Table 1.

Table 1. The complexity of the proposed model.

Model	FLOPs	Params
The proposed model	786,432.0	396,416

4. Results and Discussion

In this section, a series of simulations were conducted to verify the effectiveness of the proposed algorithm and the superiority of the designed waveform. All simulations were analyzed on a PC with a i7-12700H CPU and a GTX3080 GPU and 16 GB RAM. The neural network in this article was processed with Python 3.7 and pytorch 1.12 and ran on GPU.

The MIMO radar employed a transceiver co-array that consisted of a uniform line array with half-wavelength spacing and a wavelength of λ. The simulation parameters shown in Table 2 were used in the subsequent simulations.

Table 2. Simulation Parameters.

Parameters	Value
Number of transmitting antennas N_T	20
Number of receiving antennas N_R	20
Number of snapshot L	16
Number of neurons J	128
CPI P	16
Carrier frequency f_c	3 GHz
Target azimuth θ_d	$20°$
Multipath azimuth θ_m	$-10°$
Clutter azimuth θ_c	$-5°$
θ_i	$45°$
θ_v	$10°$
Target velocity v	45 m/s
Relative delay for multipath returns l_m	5
Relative delay for clutter l_c	2
Specular reflection coefficient ρ	$0.8e^{j\pi/4}$
Signal-to-noise ratio SNR	20 dB
Interference-to-noise ratio INR	20 dB

4.1. Convergence

This section analyzes the convergence of the proposed algorithm. The learning rate of the network was $\gamma = 0.01$, and the number of iterations was $E = 1000$. The number of optimized signals was $N_T = 20$, and the number of snapshots for each signal was $L = 16$.

As illustrated in Figure 3, the curve of the loss function gradually decreased as the number of training iterations increased. Particularly, during the first 10 training sessions, the loss function decreased rapidly. After 20 training sessions, the loss function started to converge and eventually approached zero, thus indicating that the proposed algorithm can converge to the global optimal solution.

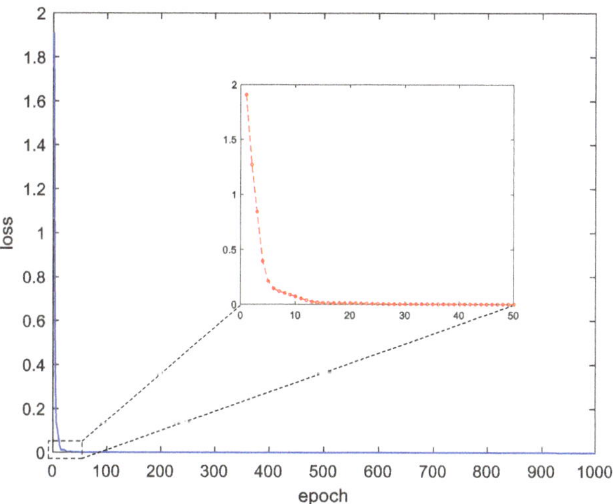

Figure 3. Loss function curve.

4.2. SINR Performance

In this section, the learning rate of the network and the number of iterations were $\gamma = 0.01$ and $E = 1000$. The number of optimized signals was $N_T = 20$, and the number of snapshots for each signal was $L = 16$. The interference-to-noise ratio was INR = 20 dB. The set of SNR were $[-20\,\text{dB}, -15\,\text{dB}, -10\,\text{dB}, -5\,\text{dB}, 0\,\text{dB}, 5\,\text{dB}, 10\,\text{dB}, 15\,\text{dB}, 20\,\text{dB}]$.

To evaluate the SINR performance of the waveform designed by the proposed algorithm, we compared it with the SINR performance of the waveforms designed using the multipath suppression algorithm [4] and the CAN algorithm [34] at different SNR conditions, as shown in Figure 4. It can be seen that the proposed algorithm achieved a higher SINR at both low SNR and high SNR. The SINR performance of the proposed algorithm was more than 2 dB higher than that of the multipath suppression algorithm at different SNR conditions. Because the purpose of the CAN algorithm is to generate orthogonal waveforms, the SINR performance of the generated waveform was much worse than that of the proposed algorithm and the multipath suppression algorithm.

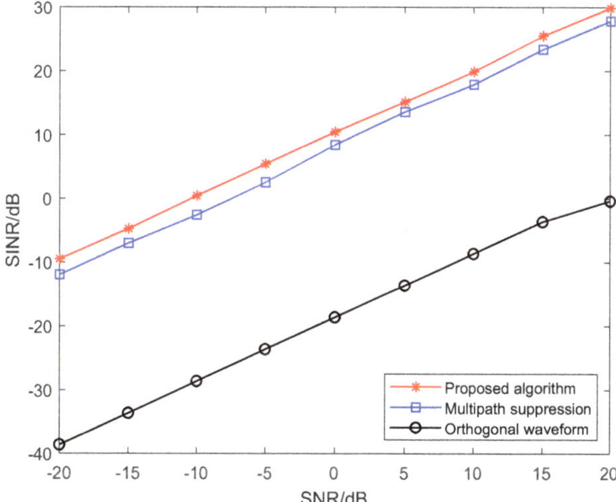

Figure 4. Output SINR versus the SNR.

To further highlight the superiority of the SINR performance of the waveform generated by the proposed algorithm, we conducted a comparative analysis assuming SNR = 20 dB and $N_T = 4$. Specifically, we compared the SINR performance of the proposed algorithm against an existing multipath exploitation algorithm [40] and a multipath suppression algorithm under these conditions. As shown in Table 3, the SINR performance of the proposed algorithm was higher than that of the multipath exploitation algorithm and multipath suppression algorithm.

Table 3. SINR performance of the proposed algorithm, existing multipath exploitation algorithm, and multipath suppression algorithm.

Methods	Proposed Algorithm	Multipath Exploitation	Multipath Suppression
SINR/dB	25.64	23.18	20.64

4.3. Transmit Beampattern

In this section, we set the learning rate of the network to $\gamma = 0.01$, and we set the number of iterations to $E = 1000$. The number of optimized signals was $N_T = 20$, and the number of snapshots for each signal was $L = 16$. The signal-to-noise ratio was SNR = 20 dB, and the interference-to-noise ratio was INR = 20 dB. The azimuth ofthe target, multipath, and clutter were $\theta_d = 20°$, $\theta_m = -10°$, and $\theta_c = -5°$. The azimuth ranged from $-90°$ to $90°$ with a grid point spacing of $1°$.

Figure 5 shows a comparison between the transmit beampattern generated by the proposed algorithm and the multipath suppression algorithm. As can be observed from the figure, the transmit beampattern generated by the proposed algorithm forms peaked in the target and multipath azimuths, while creating a deep notch in the clutter azimuth. In contrast, the multipath suppression algorithm treated the multipath returns as clutter, thus resulting in deep notches in both the multipath azimuth and clutter azimuth. It is evident from the transmit beampattern generated by the proposed algorithm and the multipath suppression algorithm that the multipath suppression algorithm achieved a suppression level of approximately -40 dB for both multipath returns and clutter. On the other hand, the proposed algorithm achieved a significantly superior suppression level of -70 dB for clutter energy.

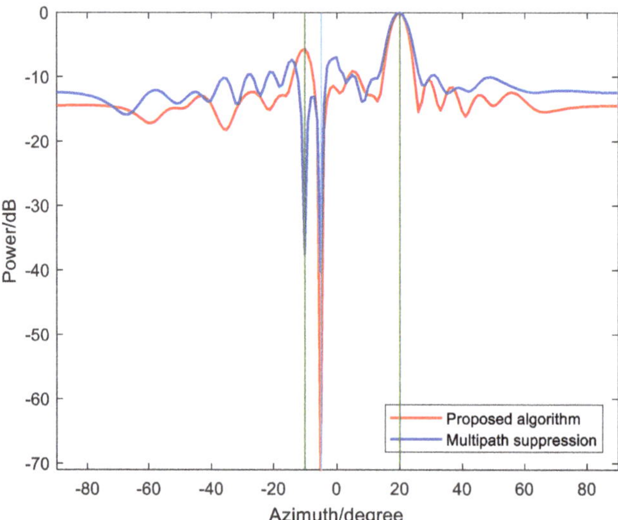

Figure 5. The transmit beampattern.

4.4. Detection Probability

In this section, we compared the detection performance of the waveforms generated by the proposed algorithm and multipath suppression algorithm under different specular reflection coefficients. The learning rate of the network was $\gamma = 0.01$, and the number of iterations was $E = 1000$. The number of optimized signals was $N_T = 20$, and the number of snapshots for each signal was $L = 16$. The set of SNR were [-20 dB, -15 dB, -10 dB, -5 dB, 0 dB, 5 dB, 10 dB, 15 dB, 20 dB], and the interference-to-noise ratio was INR = 20 dB. The set of specular reflection coefficients included the values of $0.7e^{j\pi/4}$, $0.8e^{j\pi/4}$, and $0.9e^{j\pi/4}$.

The corresponding detection probabilities can be calculated by

$$P_d = \mathrm{MarcQ}\left(\sqrt{2\mathrm{SINR}}, \sqrt{-2\log\left(P_{fa}\right)} \right), \tag{47}$$

where $\mathrm{MarcQ}(\cdot)$ denotes the Marcum-Q function [45], and P_{fa} denotes the false alarm probability.

Assuming the false alarm probability $P_{fa} = 10^{-6}$, the simulation results are shown in Figure 6. The proposed algorithm exhibited superior detection probability performance compared to the multipath suppression algorithm under the same input of SNR, as is evidenced by its higher P_d value.

Regarding the proposed algorithm for multipath exploitation, the detection probability increased with the increase in multipath reflection intensity. In contrast, for the multipath suppression algorithm, the detection probability was not significantly affected by the multipath reflection intensity, as the multipath returns were treated as clutter and were suppressed by the algorithm.

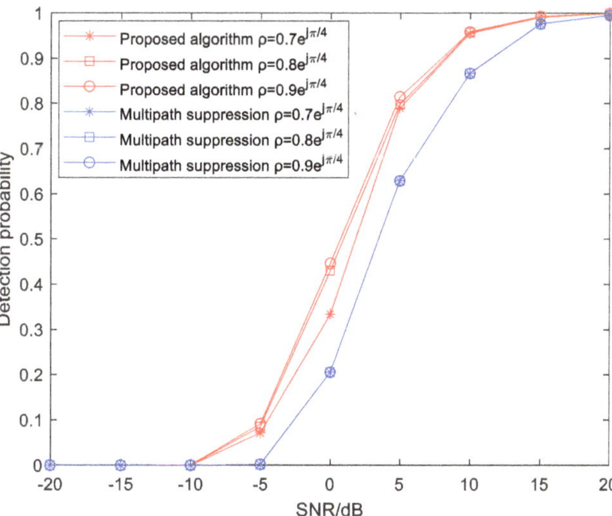

Figure 6. Comparisons of the detection probability.

To confirm the impact of the false alarm probability on the detection probability performance of the proposed algorithm, we evaluated and compared the detection probability performance of the proposed algorithm with existing and multipath exploitation algorithm and a multipath suppression algorithm under different false alarm probabilities. We set the signal-to-noise ratio to SNR = 20dB and the number of optimized signals to $N_T = 4$. As shown in Table 4, the detection probability of the proposed algorithm was higher than that of the multipath exploitation algorithm and multipath suppression algorithm under different false alarm probabilities.

Table 4. The detection probability performance under different false alarm probabilities.

Methods	Proposed Algorithm	Multipath Exploitation	Multipath Suppression
$P_{fa} = 10^{-7}$	0.9408	0.8872	0.7968
$P_{fa} = 10^{-6}$	0.9765	0.9491	0.8952
$P_{fa} = 10^{-5}$	0.9928	0.9820	0.9570

4.5. Effect of Initial Input

In this section, we set the learning rate of the network to $\gamma = 0.01$ and the number of iterations to $E = 1000$. The number of optimized signals was $N_T = 20$, and the number of snapshots for each signal was $L = 16$. The signal-to-noise ratio was SNR = 20 dB, and the interference-to-noise ratio was INR = 20 dB.

We compared the SINR performance of the proposed algorithm using two different initial inputs: a random normalized phase sequence and a sequence optimized by the CAN algorithm. As shown in Figure 7, the proposed algorithm was insensitive to the initial input. For both cases, where the initial input was a random normalized sequence and a sequence optimized by the CAN algorithm, the SINR performance of the proposed algorithm had an acceptable error at the beginning of training and tended to become consistent with the increase in training iterations.

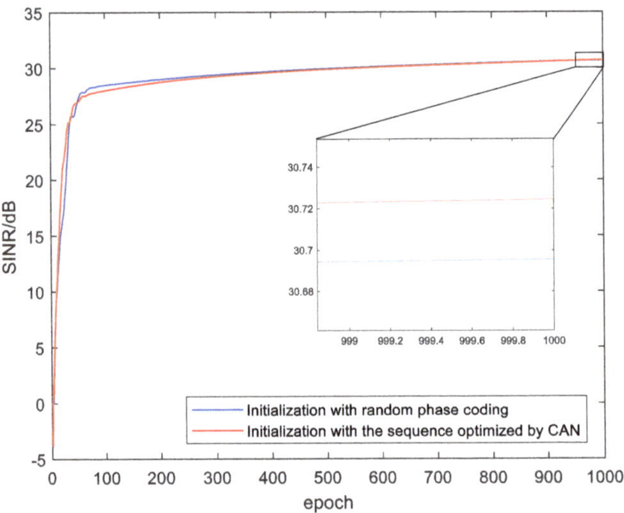

Figure 7. SINR of different initial inputs.

4.6. Effect of Number of Transmitting Antennas

In this section, we compared the SINR performance of the proposed algorithm with different numbers of transmit antennas. We set the learning rate of the network to $\gamma = 0.01$, and the number of iterations to $E = 1000$. The number of snapshots for each signal was $L = 16$. The signal-to-noise ratio was SNR = 20 dB, and the interference-to-noise ratio was INR = 20 dB. The set of the number of transmit antennas were [4, 8, 12, 16, 20, 24, 28].

As shown in Figure 8, the SINR performance of the proposed algorithm increased with the number of transmit array antennas N_T. This is because the increase in the number of transmit antennas provided higher degrees of freedom (DOFs), which enabled the algorithm to exploit the multipath reflections better and achieve a higher SINR. This implies that augmenting the number of transmit antennas in the proposed algorithm leads to a better detection performance of the MIMO radar. Nonetheless, the increase in the number of transmit antennas may also escalate the cost and computational complexity of MIMO radar systems, thereby necessitating a comprehensive consideration in choosing an appropriate number of transmit antennas for practical applications.

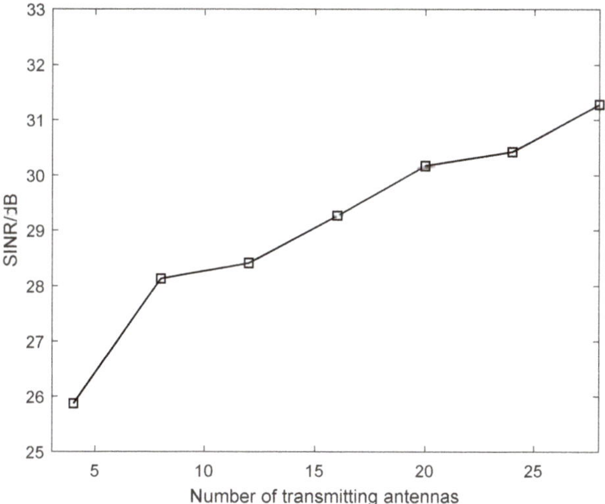

Figure 8. SINR with different numbers of transmitting antennas.

4.7. The Phase of the Waveform

In this section, we plotted the phase of the waveform optimized by the proposed algorithm under different numbers of transmit antennas. We set the learning rate of the network to $\gamma = 0.01$ and the number of iterations to $E = 1000$. The number of snapshots for each signal was $L = 16$. The signal-to-noise ratio was SNR $= 20$ dB, and the interference-to-noise ratio was INR $= 20$ dB.

As shown in Figure 9, the phase of the waveform optimized by the proposed algorithm varied between 0 and 2π for different numbers of transmit antennas. This result demonstrates the effectiveness of the algorithm in optimizing the transmission waveform.

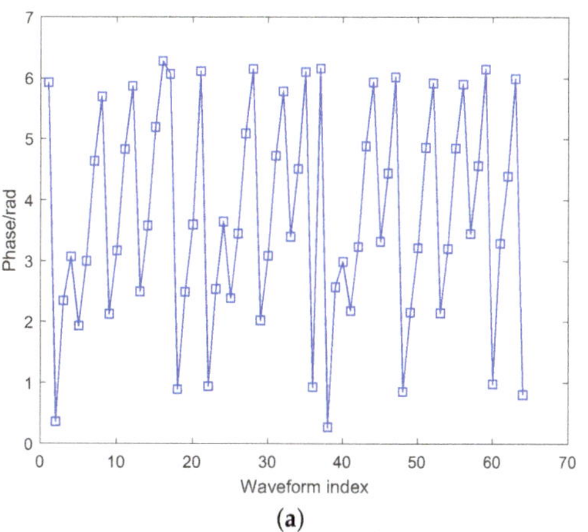

(a)

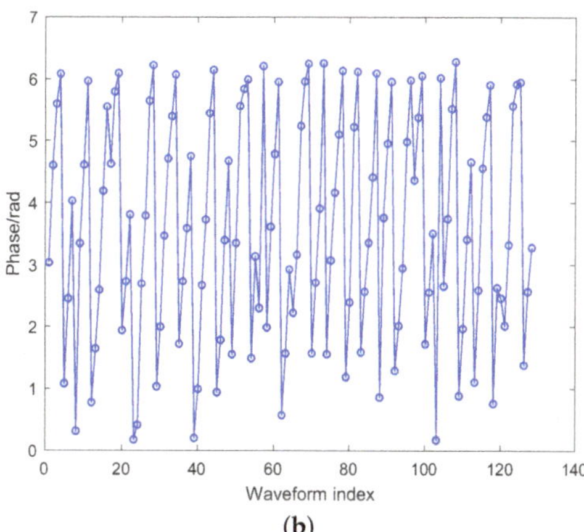

(b)

Figure 9. *Cont.*

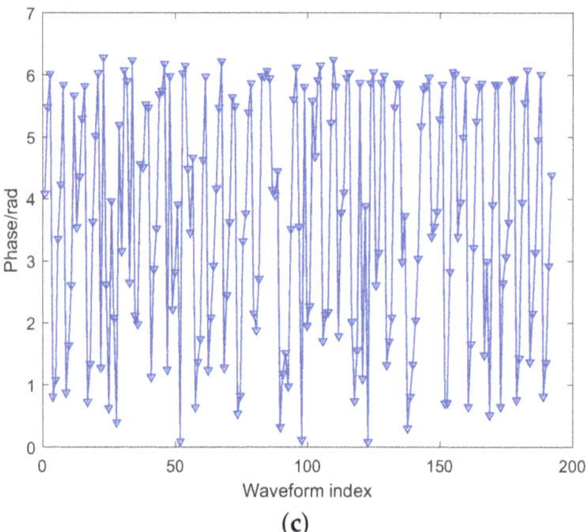
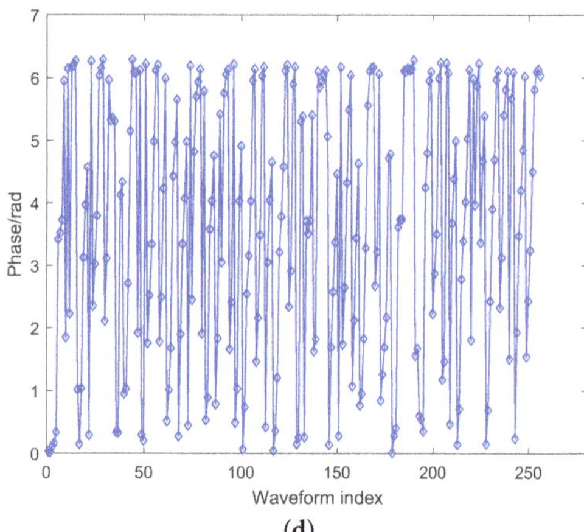

Figure 9. Phase of the optimized waveform with different numbers of transmit antennas. (**a**) $N_T = 4$, (**b**) $N_T = 8$, (**c**) $N_T = 12$, (**d**) $N_T = 16$.

5. Conclusions

In conclusion, this paper presents a novel approach for designing transmit waveforms for multipath exploitation using deep learning. We established the MIMO radar signal model for scenarios where a multipath exists and proposes an optimization objective that maximizes the receiver's SINR with a CM constraint. To solve this non-convex problem, we developed a deep residual optimization training network that directly and optimally solved the problem without relaxation.

Our simulation results demonstrate that the proposed algorithm effectively utilized the energy of multipath reflections, thus outperforming existing methods in terms of SINR performance and detection probability. We also showed that the proposed algorithm was robust to different initial inputs and benefited from the increased degrees of freedom provided by additional transmit antennas.

Overall, our research has important implications for MIMO radar applications, where optimizing waveform design can significantly impact the accuracy and reliability of target detection. Future work could explore the generalization of our approach to more complex scenarios, such as those with unknown multipath information or multi-target situations.

Author Contributions: Conceptualization, Z.Z. and Y.Z.; methodology, Y.Z.; software, Z.Z.; validation, Z.Z., Y.Z. and X.P.; formal analysis, Z.Z.; investigation, Z.Z.; resources, Y.Z.; data curation, Z.Z.; writing—original draft preparation, Z.Z.; writing—review and editing, Z.Z., X.P., H.X. and Y.Z.; visualization, J.M., J.C. and Y.S.; supervision, Y.Z.; project administration, Y.Z.; funding acquisition, Y.Z. All authors have read and agreed to the published version of the manuscript.

Funding: This research was funded by the National Natural Science Foundation of China under Grant U2133216.

Data Availability Statement: Not applicable.

Conflicts of Interest: The authors declare no conflict of interest.

References

1. Barton, D.K. Low-angle radar tracking. *Proc. IEEE* **1974**, *62*, 687–704. [CrossRef]
2. White, W.D. Low-Angle Radar Tracking in the Presence of Multipath. *IEEE Trans. Aerosp. Electron. Syst.* **1974**, *10*, 835–852. [CrossRef]
3. Bar-Shalom, Y.; Kumar, A.; Blair, W.D.; Groves, G.W. Tracking low elevation targets in the presence of multipath propagation. *IEEE Trans. Aerosp. Electron. Syst.* **1994**, *30*, 973–979. [CrossRef]

4. Hickman, G.; Krolik, J.L. MIMO GMTI radar with multipath clutter suppression. In Proceedings of the 2010 IEEE Sensor Array and Multichannel Signal Processing Workshop, Jerusalem, Israel, 4 October 2010; pp. 65–68.

5. Xin, J.; Sane, A. Linear prediction approach to direction estimation of cyclostationary signals in multipath environment. *IEEE Trans. Signal Process.* **2001**, *49*, 710–720. [CrossRef]

6. Xin, J.; Sane, A. Direction estimation of coherent narrowband signals using spatial signature. In Proceedings of the 2002 IEEE Sensor Array and Multichannel Signal Processing Workshop, Rosslyn, VA, USA, 6 August 2002; pp. 523–527.

7. Yu, J.; Krolik, J. MIMO adaptive beamforming for nonseparable multipath clutter mitigation. *IEEE Trans. Aerosp. Electron. Syst.* **2014**, *50*, 2604–2618. [CrossRef]

8. Aubry, A.; De Maio, A.; Foglia, G.; Orlando, D. Diffuse Multipath Exploitation for Adaptive Radar Detection. *IEEE Trans. Signal Process.* **2015**, *63*, 1268–1281. [CrossRef]

9. Hayvaci, H.T.; De Maio, A.; Erricolo, D. Improved detection probability of a radar target in the presence of multipath with prior knowledge of the environment. *IET Radar Sonar Navig.* **2013**, *7*, 36–46. [CrossRef]

10. Rong, Y.; Aubry, A.; De Maio, A.; Tang, M. Automatically tunable AMF for radar detection in diffuse multipath. In Proceedings of the 2020 IEEE Sensor Array and Multichannel Signal Processing Workshop, Hangzhou, China, 8 June 2020; pp. 1–5.

11. Yilmaz, S.H.G.; Hayvaci, H.T. Multipath exploitation radar with adaptive detection in partially homogeneous environments. *IET Radar Sonar Navig.* **2020**, *14*, 1475–1482. [CrossRef]

12. Li, J.; Stoica, P. MIMO Radar with Colocated Antennas. *IEEE Signal Process. Mag.* **2007**, *24*, 106–114. [CrossRef]

13. Fishler, E.; Haimovich, A.; Blum, R.; Chizhik, D.; Cimini, L.; Valenzuela, R. MIMO radar: An idea whose time has come. In Proceedings of the 2004 IEEE Radar Conference, Philadelphia, PA, USA, 29 April 2004; pp. 71–78.

14. Sun, H.; Brigui, F.; Lesturgie, M. Analysis and comparison of MIMO radar waveforms. In Proceedings of the 2014 International Radar Conference, Lille, France, 13 October 2014; pp. 1–6.

15. Stoica, P.; Li, J.; Xie, Y. On Probing Signal Design For MIMO Radar. *IEEE Trans. Signal Process.* **2007**, *55*, 4151–4161. [CrossRef]

16. Stoica, P.; Li, J.; Xu, L.; Roberts, W. On Parameter Identifiability of MIMO Radar. *IEEE Signal Process. Lett.* **2007**, *14*, 968–971.

17. Yu, X.; Qiu, H.; Yang, J.; Wei, W.; Cui, G.; Kong, L. Multi-spectrally constrained MIMO radar beampattern design via sequential convex approximation. *IEEE Trans. Aerosp. Electron. Syst.* **2022**, *58*, 2935–2949. [CrossRef]

18. Raei, E.; Alaee-Kerahroodi, M.; Shankar, M.B. Spatial-and range-ISLR trade-off in MIMO radar via waveform correlation optimization. *IEEE Trans. Signal Process.* **2021**, *69*, 3283–3298. [CrossRef]

19. Qian, J.Y.; Zheng, G.X.; Saleem, A. Channel modeling based on multilayer artificial neural network in metro tunnel environments. *ETRI J.* **2022**, 1–13. [CrossRef]

20. Jia, M.H.; Zheng, G.X.; Ji, W.L. A new model for predicting the characteristic of RF propagation in rectangular tunnel. In Proceedings of the 2008 IEEE International Conference on Personal Wireless Communications Conference Proceedings, Dalian, China, 10 September 2008; pp. 1–4.

21. Kermani, M.H.; Kamarei, M. A ray-tracing method for predicting delay spread in tunnel environments. In Proceedings of the 2000 IEEE International Conference on Personal Wireless Communications Conference Proceedings, Hyderabad, India, 17 December 2000; pp. 538–542.

22. Zhao, X.W.; Du, F.; Geng, S.Y.; Sun, N.Y.; Zhang, Y.; Fu, Z.H.; Wang, G.J. Neural network and GBSM based time-varying and stochastic channel modeling for 5G millimeter wave communications. *China Commun.* **2019**, *16*, 80–90. [CrossRef]

23. Sun, N.; Geng, S.; Li, S.; Zhao, X.; Wang, M.; Sun, S. Channel modeling by RBF neural networks for 5G Mm-wave communication. In Proceedings of the 2018 IEEE/CIC International Conference on Communications in China, Beijing, China, 12 August 2018; pp. 768–772.

24. Ertel, R.B.; Reed, J.H. Angle and time of arrival statistics for circular and elliptical scattering model. *IEEE J. Sel. Areas Commun.* **1999**, *17*, 1829–1840. [CrossRef]

25. Petrus, P.; Rappaport, T.S. Geometrical-based statistical macrocell channel model for mobile environments. *IEEE Trans. Commun.* **2002**, *50*, 495–502. [CrossRef]

26. Li, J.; Guerci, J.R.; Xu, L. Signal Waveform's Optimal Under Restriction Design for Active Sensing. In Proceedings of the 2006 IEEE Sensor Array and Multichannel Signal Processing Workshop, Waltham, MA, USA, 12 July 2006; pp. 382–386.

27. Tang, B.; Li, J.; Liang, J. Alternating direction method of multipliers for radar waveform design in spectrally crowded environments. *Signal Process.* **2018**, *142*, 398–402. [CrossRef]

28. Yu, X.; Cui, G.; Kong, L.; Li, J.; Gui, G. Constrained Waveform Design for Colocated MIMO Radar With Uncertain Steering Matrices. *IEEE Trans. Aerosp. Electron. Syst.* **2019**, *55*, 356–370. [CrossRef]

29. Aubry, A.; De Maio, A.; Piezzo, M.; Farina, A. Radar waveform design in a spectrally crowded environment via nonconvex quadratic optimization. *IEEE Trans. Aerosp. Electron. Syst.* **2014**, *50*, 1138–1152. [CrossRef]

30. Yang, Y.; Blum, R.S. MIMO radar waveform design based on mutual information and minimum mean-square error estimation. *IEEE Trans. Aerosp. Electron. Syst.* **2007**, *43*, 330–343. [CrossRef]

31. Naghsh, M.M.; Modarres-Hashemi, M.; Shahbazpanahi, S. Unified Optimization Framework for Multi-Static Radar Code Design Using Information-Theoretic Criteria. *IEEE Trans. Signal Process.* **2013**, *61*, 5401–5416. [CrossRef]

32. Yu, X.; Cui, G.; Zhang, T. Constrained Transmit Beampattern Design for Colocated MIMO Radar. *Signal Process.* **2017**, *144*, 145–154. [CrossRef]

33. Wang, Y.C.; Wang, X.; Liu, H. On the Design of Constant Modulus Probing Signals for MIMO Radar. *IEEE Trans. Signal Process.* **2012**, *60*, 4432–4438. [CrossRef]
34. Stoica, P.; He, H.; Li, J. New Algorithms for Designing Unimodular Sequences with Good Correlation Properties. *IEEE Trans. Signal Process.* **2009**, *57*, 1415–1425. [CrossRef]
35. He, H.; Stoica, P.; Li, J. Designing Unimodular Sequence Sets with Good Correlations—Including an Application to MIMO Radar. *IEEE Trans. Signal Process.* **2009**, *57*, 4391–4405. [CrossRef]
36. Li, Y.; Vorobyov, S.A. Fast Algorithms for Designing Unimodular Waveform(s) with Good Correlation Properties. *IEEE Trans. Signal Process.* **2018**, *66*, 1197–1212. [CrossRef]
37. Sen, S.; Hurtado, M.; Nehorai, A. Adaptive OFDM radar for detecting a moving target in urban scenarios. In Proceedings of the 2009 International Waveform Diversity and Design Conference, Kissimmee, FL, USA, 8 February 2009; pp. 268–272.
38. Sen, S.; Nehorai, A. OFDM MIMO radar with mutual-information waveform design for low-grazing angle tracking. *IEEE Trans. Signal Process.* **2010**, *58*, 3152–3162. [CrossRef]
39. Sen, S.; Nehorai, A. Adaptive OFDM radar for target detection in multipath scenarios. *IEEE Trans. Signal Process.* **2011**, *59*, 78–90. [CrossRef]
40. Xu, Z.; Fan, C.; Huang, X. MIMO Radar Waveform Design for Multipath Exploitation. *IEEE Trans. Signal Process.* **2021**, *69*, 5359–5371. [CrossRef]
41. Fan, C.; Xie, Z.; Wang, J.; Xu, Z.; Huang, X. Robust MIMO Waveform Design in the Presence of Unknown Mutipath Return. *Remote Sens.* **2023**, *14*, 4356. [CrossRef]
42. Imani, S.; Ghorashi, S.A. Sequential quasi-convex-based algorithm for waveform design in colocated multiple-input multiple-output radars. *IET Signal Process.* **2023**, *10*, 309–317. [CrossRef]
43. Hertz, J.A. *Introduction to the Theory of Neural Computation*; CRC Press: Boca Raton, FL, USA, 2018.
44. He, K.; Zhang, X.; Ren, S.; Sun, J. Deep Residual Learning for Image Recognition. In Proceedings of the 2016 IEEE Conference on Computer Vision and Pattern Recognition, Las Vegas, NV, USA, 27 June 2016; pp. 770–778.
45. Richards, M.A. *Fundamentals of Radar Signal Processing*; McGraw Hill: New York, NY, USA, 2005.

Technical Note

Robust MIMO Waveform Design in the Presence of Unknown Mutipath Return

Chongyi Fan, Zhuang Xie, Jian Wang *, Zhou Xu and Xiaotao Huang

College of Electronic Science and Technology, National University of Defense Technology, Changsha 410073, China
* Correspondence: jianwang_uwb@nudt.edu.cn

Abstract: Assuming uncertain multipath return, this paper considers the robust joint transmit waveform and the receiving filter bank design of a multiple-input–multiple-output (MIMO) radar for multipath exploitation. The actual multipath return is considered to belong to an uncertain set, and we focus on the worst-case optimization of the signal-to-interference-plus-noise ratio (SINR) in the output of the filter bank. The design is cast as a non-convex max–min problem, which is very hard to solve. To tackle it, an equivalent reformulation is utilized and a cyclic optimization paradigm is devised. At each iteration, the filter's optimization problem is equal to a set of separate solvable problems, the closed-form solution to which can be given directly. Moreover, we have shown that the max–min problem for the waveform optimization belongs to the area of generalized fractional programming, and it can be globally solved by resorting to Dinkelbach's algorithm. Through simulations, the superiority of the proposed algorithm is demonstrated via a number of examples.

Keywords: robust waveform design; multipath return; MIMO; max–min; worst-case SINR

Citation: Fan, C.; Xie, Z.; Wang, J.; Xu, Z.; Huang, X. Robust MIMO Waveform Design in the Presence of Unknown Mutipath Return. *Remote Sens.* **2022**, *14*, 4356. https://doi.org/10.3390/rs14174356

Academic Editor: Okan Yurduseven

Received: 7 July 2022
Accepted: 30 August 2022
Published: 2 September 2022

Publisher's Note: MDPI stays neutral with regard to jurisdictional claims in published maps and institutional affiliations.

1. Introduction

1.1. Background

Recent advances in radar waveform design have shown that, by transmitting flexible waveforms according to an operation scenario, a radar system can achieve significantly enhanced target detection abilities, improved resolution, and increased parameter estimation accuracy [1–7]. As a result, jointly designing the transmit waveform and receive filter of a radar system leveraging on the apriori knowledge about the operating environment has received considerable attention in the research field.

1.2. Related Work

Among the various types of new radar systems, the multiple-input–multiple-output (MIMO) radar has attracted significant interest from researchers in the field of waveform design. Compared with traditional phased-array radars that transmit single waveforms, MIMO radar is able to transmit totally different waveforms from its multiple channels [8–13]. In this context, its great potential in achieving better target detection, identification, and tracking performance via the design of suitable waveforms has been further explored. In the last decade, a number of studies on MIMO radar waveform optimization have been carried out with different objectives and constraints [14–20]. In [14], the MIMO radar waveform design problem under a constant modulus and similarity constraints was investigated, and two constrained sequential optimization algorithms were developed. In [15], the design of the transmitted waveforms and receiving filters was extended to an airborne MIMO radar for space-time processing. Unlike most studies, which constrained the waveform to have a constant modulus [14,16], in [17], the peak-to-average-power ratio (PAPR) constraint and similarity constraint were simultaneously imposed on the transmitted waveforms. While the aforementioned works optimized the waveforms and filters from the perspective of the signal-to-interference-plus-noise ratio (SINR), an information-theoretic approach was applied in [18], where information-theoretic criteria were employed as design metrics. Considering possible steering matrix mismatches, the worst-case SINR optimization was studied in [19] with the constant modulus constraint imposed on the waveform, where both the continuous and discrete probing waveform phase cases were considered. While most works have focused on the waveform design that maximizes the SINR for a single target, the waveform design for a multiple-target scenario was considered in [20].

In a practical radar operation, when there is a rough or glistening surface existing in a scene, the signal backscattered from the target will be received at the radar array through different propagation paths aside from the direct path [21,22]. In this case, the returns contain not only the line-of-sight (LOS) return, but also a multipath component [23,24]. Though there have been some studies devoted to the suppression of the multipath return, research has shown that the target detection performance can be further improved via proper multipath exploitation.

In this context, the radar waveform design technique offers a great potential advantage for the enhancement of the target detection performance for multipath exploitation. There have been some studies that have used the radar waveform design technique to improve the target detection performance in the presence of multipath effect. In [25], the constant-modulus orthogonal frequency division multiplexing (OFDM) waveform was applied in order to improve the moving-target detection performance to overcome the PAPR problem. In [26], based on the mutual information criterion, weighted OFDM waveforms were transmitted by an MIMO array to track the target for the exploitation of multipath reflections. In a nutshell, the radar for multipath exploitation in the existing researches is assumed to transmit fixed waveform which lacks the ability to adjust the waveform at different environment. The main exploration direction is to optimize the receive end for enhanced performance and the transmit waveform design in fast time domain for multipath exploitation, especially with MIMO radar, still remains an open issue in the current literature.

1.3. Motivation and Contributions of This Paper

In our previous work [27], an MIMO jointly transmitted waveform and receiving filter design for multipath exploitation was investigated. Therein, the propagation path of the multipath return that was treated was known in advance, and the transmitted waveform and receiving filter could be optimized by leveraging the semi-definite relaxation, followed by a randomization procedure. Nevertheless, it is worth pointing out that the algorithm proposed in [27] is relatively limited in terms of its capability in more practical cases when the knowledge of the environment is incomplete or imprecise, i.e., the exact a priori knowledge on the multipath configuration (reflective coefficient of the surface, arrival direction, fast time delay) is unavailable. In this sense, signal model mismatches may happen, and the performance of the waveform designed in [27] is not guaranteed. Therefore, the design of a waveform that works against potential model mismatches to ensure robust performance is of great significance, and this remains an open issue.

Aiming at filling this gap, in this paper, we consider a robust MIMO waveform design to account for a more practical situation in which the prior knowledge about the multipath return is imperfect. Specifically, we assume that there are several possible expressions for the multipath return, depending on the position and reflective coefficient of the reflective surface. In this context, we consider the worst-case SINR at the outputs of the filter banks, with each tuned to a specific multipath return case, as the figure of merit in order to guarantee a robust target detection ability regardless of the actual multipath return. It is shown that the resultant robust design is hard to solve, since it belongs to a class of non-convex max–min optimization problems. We reformulated the design into an equivalent form that is easier to tackle and showed that the worst-case SINR can, thereby, be sequentially optimized based on a cyclic optimization procedure. At each iteration, when the signal is fixed, each element of the filter bank can be globally optimized, and the closed-form solution can be given directly; for a given filter bank, the waveform can be optimized by resorting to the generalized Dinkelbach algorithm [28,29]. At the analysis stage, the effectiveness of the proposed algorithm and its superiority over the non-robust design are illustrated through a series of developed case studies.

The remaining content of this paper is organized as follows. In Section 2, the signal model for the robust MIMO waveform design under unknown multipath return is introduced. In Section 3, the non-convex max–min problem for the joint robust design of the MIMO waveform and receiving filter bank is formulated, and a sequential optimization-based algorithm is proposed to obtain the optimized waveform and filters. Section 4 provides several numerical experiments to demonstrate the effectiveness and superiority of the proposed algorithm. Finally, Section 5 draws the conclusions and outlines some possible future research tracks.

Notations: Throughout this paper, scalars are denoted by italic letters; vectors and matrices are denoted by bold italic lowercase and capital letters, respectively. The operator $\mathrm{tr}(\cdot)$ obtains the trace of a square matrix, and the operator $\mathrm{vec}(\cdot)$ performs the column-wise stacking of a matrix. $\mathbb{C}^N$ stands for the N-dimensional complex space. $\Re(x)$ and $\arg(x)$ obtain the the the real part and phase angle of the argument x, respectively. $\otimes$ denotes the Kronecker product, and $\mathrm{E}(\cdot)$ represents the expectation operator. The superscripts $(\cdot)^T$ and $(\cdot)^H$ denote the transpose and conjugate transpose, respectively. For a matrix $\mathbf{A}$, $\mathbf{A} \succ 0$ means that $\mathbf{A}$ is positive definite and $\mathbf{A} \succeq 0$ means that $\mathbf{A}$ is semi-definite.

2. Signal Model

Consider a colocated MIMO radar array consisting of N_T transmitters and N_R receivers that is deployed in a scene to detect a target in direction θ_t. We denote its signal-transmitting matrix by $\mathbf{S} = [\mathbf{s}_1, \mathbf{s}_2, ..., \mathbf{s}_{N_T}]^T \in \mathbb{C}^{N_T \times L}$, where $\mathbf{s}_n \in \mathbb{C}^L$ is the waveform transmitted at the nth channel, with L being the number of samples in fast time domain. As depicted in Figure 1, suppose that there is a reflective surface causing multipath return in the scenario. As a consequence, when the radar system transmits signals to illuminate the target, the echoes received by the array include not only the direct return, but also the backscattered target signal reflected from the reflective surface, which is defined as multipath return. There are also undesired returns from a cluttered area and other signal-independent disturbances. The echoes received by the radar are down-converted to the baseband, which can be expressed as

$$\mathbf{Y} = \mathbf{Y}_d + \mathbf{Y}_m + \mathbf{Y}_c + \mathbf{n} \in \mathbb{C}^{N_R \times 2L}, \tag{1}$$

where $\mathbf{Y}_d$, $\mathbf{Y}_m$, $\mathbf{Y}_c$, and $\mathbf{n}$ represent the direct return, multipath return, clutter, and noise, respectively. It is worth stressing that we focus on the received data, which consist of 2L-length range cells based on the model derived in [27], since longer propagation distances of the multipath return will lead to a longer fast time delay compared with the direct return.

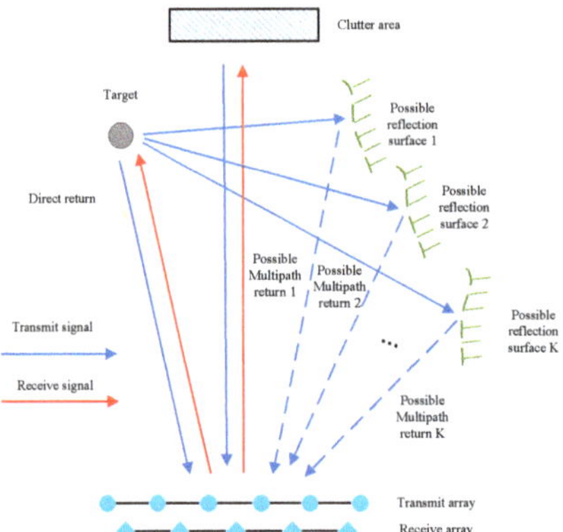

Figure 1. The considered scenario in the presence of multipath return.

2.1. Direct Return Model

The direct return signal from the target can be expressed as

$$\mathbf{Y}_d = \alpha_0 \mathbf{a}_r(\theta_d)\mathbf{a}_t^T(\theta_d)\mathbf{S}\mathbf{J}_0 \in \mathbb{C}^{N_R \times 2L}, \tag{2}$$

where α_0 is a complex parameter accounting for the two-way direct propagation and backscattering effects from the target, and $\mathbf{a}_t(\theta_d) \in \mathbb{C}^{N_T}$ and $\mathbf{a}_r(\theta_d) \in \mathbb{C}^{N_R}$ represent, respectively, the transmitting and receiving steering vector in the target direction θ_d. $\mathbf{J}_l$ is the $L \times 2L$ shift matrix, with its (m, n)th element given as

$$\mathbf{J}_l(m, n) = \begin{cases} 1, & \text{if } m - n + l = 0 \\ 0, & \text{if } m - n + l \neq 0 \end{cases}. \tag{3}$$

It is worth pointing out that the notation $\mathbf{J}_l$ is applied as a shift matrix for the rest of this paper. By utilizing the vectorization operator, we can obtain a $2N_R L \times 1$-dimensional direct return target vector that can be expressed as

$$\mathbf{y}_d = \text{vec}(\mathbf{Y}_d) = \alpha_0 \underbrace{\left(\mathbf{J}_0^T \otimes \left(\mathbf{a}_r(\theta_d)\mathbf{a}_t^T(\theta_d)\right)\right)}_{\mathbf{H}_d} \mathbf{s} \in \mathbb{C}^{2N_R L \times 1}$$

$$= \alpha_0 \mathbf{H}_d \mathbf{s}, \tag{4}$$

where $\mathbf{s} = \text{vec}(\mathbf{S}) \in \mathbb{C}^{N_T L \times 1}$.

2.2. Multipath Return Model

In practice, accurate knowledge about a reflective surface (including its position and reflective coefficient) is not reasonable, and accordingly, it is more rational to assume that the actual multipath return has many possible expressions. To be specific, we assume that we know that the physical set of the reflective surface belongs to an uncertain set consisting of K possible cases, as shown in Figure 1. Accordingly, there are K possible formulations for the multipath return from the target in the direction θ_d. Based on above analysis, we can get the baseband signal of the kth possible target multipath target return expression through slight modification of (2), which is expressed as

$$\mathbf{Y}_{m,k} = \rho_k \alpha_0 \mathbf{a}_r(\theta_{m,k})\mathbf{a}_t^T(\theta_d)\mathbf{S}\mathbf{J}_{r_{m,k}} \in \mathbb{C}^{N_R \times 2L}, \tag{5}$$

where ρ_k and $\theta_{m,k}$ are, respectively, the corresponding reflective coefficient of the surface and the arrival direction of the multipath return. $r_{m,k}$ is the relative fast time delay of the kth possible path compared with the direct return.

Similarly to the derivation in (4), we can rewrite $\mathbf{Y}_{m,k}$ as a $2N_R L \times 1$ vector by resorting to the vectorization operator:

$$\mathbf{y}_{m,k} = \mathrm{vec}(\mathbf{Y}_{m,k}) = \rho_k \alpha_0 \underbrace{\left(\mathbf{J}_{r_{m,k}}^T \otimes \mathbf{a}_r(\theta_{m,k}) \mathbf{a}_t^T(\theta_d) \right)}_{\mathbf{H}_{m,k}} \mathbf{s} \in \mathbb{C}^{2N_R L \times 1}$$

$$= \rho_k \alpha_0 \mathbf{H}_{m,k} \mathbf{s}. \tag{6}$$

2.3. Disturbance Model

In a radar operation, there are inevitable undesired reflections from other objects in the scene, e.g., the ground, water, and trees, which will have an obvious impact on the received echo. Herein, aside from the clutter within the same range cell as the target signal, we also take into consideration the impact of the clutter from the neighboring range cells. Therefore, the signal-dependent clutter is modeled as the superposition of signals from different directions and range cells. Suppose that there are P scatters in the scene interfering with the received echoes; then, based on the previous analysis, the clutter caused by the pth considered scatter can be expressed as

$$\mathbf{Y}_{c,p} = \beta_p \mathbf{a}_r(\theta_p) \mathbf{a}_t^T(\theta_p) \mathbf{S} \mathbf{J}_{l_{c,p}} \in \mathbb{C}^{N_R \times 2L}, \tag{7}$$

where β_p represents the complex backscattering coefficient of the pth interfering object, θ_p is its relative direction with respect to (w.r.t) the radar, and $l_{c,p}$ is the relative fast time delay compared with the direct return. It should be pointed out that, here, we assume that no multipath return is involved in the clutter item. Hence, the total clutter item can be expressed as

$$\mathbf{Y}_c = \sum_{p=1}^{P} \beta_p \mathbf{a}_r(\theta_p) \mathbf{a}_t^T(\theta_p) \mathbf{S} \mathbf{J}_{l_{c,p}} \in \mathbb{C}^{N_R \times 2L}. \tag{8}$$

Similarly to the derivations in the previous subsections, we write $\mathbf{Y}_c$ in vector form through column-wise stacking as follows:

$$\mathbf{y}_c = \mathrm{vec}(\mathbf{Y}_c) = \sum_{p=1}^{P} \beta_p \underbrace{\left(\mathbf{J}_{l_{c,p}}^T \otimes \left(\mathbf{a}_r(\theta_p) \mathbf{a}_t^T(\theta_p) \right) \right)}_{\mathbf{H}_{c,p}} \mathbf{s}$$

$$= \sum_{p=1}^{P} \beta_p \mathbf{H}_{c,p} \mathbf{s} \in \mathbb{C}^{2N_R L \times 1}. \tag{9}$$

On the other hand, there are also signal-independent disturbances in the received data, which result from the system noise. Without loss of generality, we consider the signal-independent noise to be Gaussian, with its covariance matrix given as $\mathbf{R}_n$.

3. Problem Formulation and Proposed Algorithm

In addition to the various waveform design criteria, which include mutual information (MI) maximization [18,30,31], ambiguity function shaping [32–35], and Cramer–Rao-bound (CRB) minimization [36,37], the SINR is widely utilized as the optimization criterion due to its close relationship with the target detection and parameter estimation performance [38–40]. In this context, we deal with the joint design of the transmitted waveform and the corresponding receiving filter bank by optimizing the worst-case SINR over the unknown multipath return for enhanced target detection performance. Unlike in the single-filter case, herein, our goal is to enhance the worst-case SINR, namely, the minimum SINR among the available branches, at the output of the filter bank. By doing this, we are able to improve the output SINR performance of the radar system in comparison with that of the conventional design regardless of the actual multipath return expression. Specifically, we propose the processing of the received data through a filter array consisting of K

filters, with each filter being tuned to a specific multipath return case. As for the constraint imposed on the transmitted waveform, we restrict the waveform to have finite energy, i.e., $\|\mathbf{s}\|_2^2 = 1$. We denote by $\mathbf{w}_k$ the kth filter; its output SINR under the kth multipath return case can be calculated with

$$\text{SINR}_k = \frac{|\alpha_0|^2 |\mathbf{w}_k(\mathbf{H}_d + \rho_k \mathbf{H}_{m,k})\mathbf{s}|^2}{\mathbf{w}_k^H \left(\sum\limits_{p=1}^{P} \sigma_{\bar{c},p}^2 \mathbf{H}_{c,p} \mathbf{s}\mathbf{s}^H \mathbf{H}_{c,p}^H + \mathbf{R}_n \right) \mathbf{w}_k} . \tag{10}$$

where $\sigma_{\bar{c},p}^2 = \mathrm{E}\left[|\beta_p|^2\right]$ is the interfering power of the pth scatter. Herein, the numerator of SINR_k denotes the useful energy at the output of the filter, and the denominator of SINR_k represents the remaining energy of the clutter and signal-independent noise. The worst-case SINR at the output of the filter bank is defined as

$$\widetilde{\text{SINR}} = \min_{k=1,\dots,K} \text{SINR}_k . \tag{11}$$

Based on the above analysis, the robust joint transmitted waveform and receiving filter bank design problem can be cast as

$$\mathcal{P} \begin{cases} \max\limits_{\mathbf{s},\{\mathbf{w}_k\}_{k=1}^{K}} \min\limits_{k=1,\dots,K} \dfrac{|\alpha_0|^2 |\mathbf{w}_k^H (\mathbf{H}_d + \rho_k \mathbf{H}_{m,k})\mathbf{s}|^2}{\mathbf{w}_k^H \left(\sum\limits_{p=1}^{P} \sigma_{\bar{c},p}^2 \mathbf{H}_{c,p} \mathbf{s}\mathbf{s}^H \mathbf{H}_{c,p}^H + \mathbf{R}_n \right) \mathbf{w}_k} . \\[20pt] s.t. \ \|\mathbf{s}\|_2^2 = 1 \end{cases} \tag{12}$$

It is seen that $\mathcal{P}$ is a non-convex problem (max–min objective and the energy constraint) and it NP-hard, which is difficult to solve. Nevertheless, as will be shown in the remainder of this part, a novel algorithm can be devised in order to effectively obtain a sub-optimal solution to $\mathcal{P}$. To begin with, the following proposition provides an equivalent formulation of $\mathcal{P}$, which lays the foundation for the subsequent derivations.

Proposition 1. *In terms of the optimal solution, Problem $\mathcal{P}$ is equal to the following problem:*

$$\widetilde{\mathcal{P}} \begin{cases} \max\limits_{\mathbf{s},\{\mathbf{w}_k\}_{k=1}^{K}} \min\limits_{k=1,\dots,K} \dfrac{|\alpha_0| \Re\left(\mathbf{w}_k^H (\mathbf{H}_d + \rho_k \mathbf{H}_{m,k})\mathbf{s}\right)}{\sqrt{\mathbf{w}_k^H \left(\sum\limits_{p=1}^{P} \sigma_{\bar{c},p}^2 \mathbf{H}_{c,p} \mathbf{s}\mathbf{s}^H \mathbf{H}_{c,p}^H + \mathbf{R}_n \|\mathbf{s}\|_2^2 \right) \mathbf{w}_k}} . \\[24pt] s.t. \quad \Re\left(\mathbf{w}_k^H (\mathbf{H}_d + \rho_k \mathbf{H}_{m,k})\mathbf{s}\right) \geq 0, \ k = 1,2,\dots K \end{cases} \tag{13}$$

This means that, once an optimal solution to $\widetilde{\mathcal{P}}$ is obtained, it is possible to build an optimal solution to $\mathcal{P}$, and vice versa.

Proof. To begin with, it is obvious that the problem $\mathcal{P}$ shares the same optimal objective value with the following unconstrained max–min problem:

$$\max_{\mathbf{s},\{\mathbf{w}_k\}_{k=1}^{K}} \min_{k=1,\dots,K} \frac{|\alpha_0|^2 \left|\mathbf{w}_k^H (\mathbf{H}_d + \rho_k \mathbf{H}_{m,k})\frac{\mathbf{s}}{\|\mathbf{s}\|_2}\right|^2}{\mathbf{w}_k^H \left(\sum\limits_{p=1}^{P} \sigma_{\bar{c},p}^2 \mathbf{H}_{c,p} \frac{\mathbf{s}\mathbf{s}^H}{\|\mathbf{s}\|_2^2} \mathbf{H}_{c,p}^H + \mathbf{R}_n \right) \mathbf{w}_k} . \tag{14}$$

Moreover, since the value of the objective function in (14) is invariant by multiplying a scalar for the numerator and denominator at the same time, we can further recast (14) as

$$\max_{\mathbf{s},\{\mathbf{w}_k\}_{k=1}^{K}} \ \min_{k=1,\ldots,K} \ \frac{|\alpha_0|^2 \left| \mathbf{w}_k^H (\mathbf{H}_d + \rho_k \mathbf{H}_{m,k}) \mathbf{s} \right|^2}{\mathbf{w}_k^H \left(\sum\limits_{p=1}^{P} \sigma_{c,p}^2 \mathbf{H}_{c,p} \mathbf{s} \mathbf{s}^H \mathbf{H}_{c,p}^H + \mathbf{R}_n \|\mathbf{s}\|_2^2 \right) \mathbf{w}_k}. \tag{15}$$

It is immediately ascertainable that (15) is equivalent to the following problem in the sense of optimal solutions:

$$\max_{\mathbf{s},\{\mathbf{w}_k\}_{k=1}^{K}} \ \min_{k=1,\ldots,K} \ \frac{|\alpha_0| \left| \mathbf{w}_k^H (\mathbf{H}_d + \rho_k \mathbf{H}_{m,k}) \mathbf{s} \right|}{\sqrt{\mathbf{w}_k^H \left(\sum\limits_{p=1}^{P} \sigma_{c,p}^2 \mathbf{H}_{c,p} \mathbf{s} \mathbf{s}^H \mathbf{H}_{c,p}^H + \mathbf{R}_n \|\mathbf{s}\|_2^2 \right) \mathbf{w}_k}}. \tag{16}$$

We denote the feasible points of (13) and (16) by Ω_1 and Ω_2, respectively. Then, it is obvious that $\Omega_1 \subseteq \Omega_2$. Moreover, we note that, for any given point in Ω_1, the objective value of (16) is larger than that of (13), which leads to the result that the optimal value of (16) is higher than that of (13). On the other hand, $\forall \bar{\mathbf{x}} = (\bar{\mathbf{s}}, \{\bar{\mathbf{w}}_k\}_{k=1}^{K}) \in \Omega_2$, so one can verify that $\hat{\mathbf{x}} = \left(\bar{\mathbf{s}}, \left\{ \bar{\mathbf{w}}_k \cdot e^{j \arg\{\bar{\mathbf{w}}_k^H (\mathbf{H}_d + \rho_k \mathbf{H}_{m,k}) \bar{\mathbf{s}}\}} \right\}_{k=1}^{K} \right)$ is in Ω_1, and the objective in (13) achieves the same value at $\hat{\mathbf{x}}$ as the objective in (16) attains at $\bar{\mathbf{x}}$. That is to say, the optimal value of (13) is not lower than that of (16). In conclusion, (13) and (16) share the same optimal value, and any optimal solution for (13) is also optimal for (16). It is sufficient for us to deal with (13). Therefore, the proof is completed. $\square$

Based on Proposition 1, we can focus on dealing with $\widetilde{\mathcal{P}}$ in the following content. Though the reformulated form $\widetilde{\mathcal{P}}$ is still non-convex and NP-hard due to the max–min fractional objective function, in what follows, we develop an effective algorithm based on alternative optimization to obtain a sub-optimal solution to $\widetilde{\mathcal{P}}$. In particular, at the nth iteration, we focus on tackling the problems of $\widetilde{\mathcal{P}}_1^{(n)}$ and $\widetilde{\mathcal{P}}_2^{(n)}$ w.r.t. the waveform and filters, respectively:

$$\widetilde{\mathcal{P}}_1^{(n)} \begin{cases} \max\limits_{\{\mathbf{w}_k\}_{k=1}^{K}} \ \min\limits_{k=1,\ldots,K} \ \dfrac{|\alpha_0| \Re\left(\mathbf{w}_k^H (\mathbf{H}_d + \rho_k \mathbf{H}_{m,k}) \mathbf{s}^{(n-1)} \right)}{\sqrt{\mathbf{w}_k^H \left(\sum\limits_{p=1}^{P} \sigma_{c,p}^2 \mathbf{H}_{c,p} \mathbf{s}^{(n-1)} \mathbf{s}^{(n-1)H} \mathbf{H}_{c,p}^H + \mathbf{R}_n \|\mathbf{s}\|_2^2 \right) \mathbf{w}_k}} \\[2em] s.t. \quad \Re\left(\mathbf{w}_k^H (\mathbf{H}_d + \rho_k \mathbf{H}_{m,k}) \mathbf{s}^{(n-1)} \right) \geq 0, \ k = 1, 2, \ldots K \end{cases} \tag{17}$$

and

$$\widetilde{\mathcal{P}}_2^{(n)} \begin{cases} \max\limits_{\mathbf{s}} \ \min\limits_{k=1,\ldots,K} \ \dfrac{|\alpha_0| \Re\left(\mathbf{w}_k^{(n)H} (\mathbf{H}_d + \rho_k \mathbf{H}_{m,k}) \mathbf{s} \right)}{\sqrt{\mathbf{w}_k^{(n)H} \left(\sum\limits_{p=1}^{P} \sigma_{c,p}^2 \mathbf{H}_{c,p} \mathbf{s} \mathbf{s}^H \mathbf{H}_{c,p}^H + \mathbf{R}_n \|\mathbf{s}\|_2^2 \right) \mathbf{w}_k^{(n)}}}, \\[2em] s.t. \quad \Re\left(\mathbf{w}_k^{(n)H} (\mathbf{H}_d + \rho_k \mathbf{H}_{m,k}) \mathbf{s} \right) \geq 0, \ k = 1, 2, \ldots K \end{cases} \tag{18}$$

where $\mathbf{s}^{(n)}$ and $\left\{ \mathbf{w}_k^{(n)} \right\}_{k=1}^{K}$ denote the optimized waveform and filters at the nth iteration, respectively.

3.1. Optimizing Filters with a Fixed Waveform

Since $\widetilde{\mathcal{P}}_1^{(n)}$ consists of K objective functions that are separable, we can proceed to dealing with the problem w.r.t. $\mathbf{w}_k$, which is formulated as

$$\widetilde{\mathcal{P}}_{1,k}^{(n)} \begin{cases} \max\limits_{\mathbf{w}_k} \dfrac{|\alpha_0|\Re\left(\mathbf{w}_k^H(\mathbf{H}_d + \rho_k\mathbf{H}_{m,k})\mathbf{s}^{(n-1)}\right)}{\sqrt{\mathbf{w}_k^H\left(\sum\limits_{p=1}^{P}\sigma_{c,p}^2\mathbf{H}_{c,p}\mathbf{s}^{(n-1)}\mathbf{s}^{(n-1)H}\mathbf{H}_{c,p}^H + \mathbf{R}_n\|\mathbf{s}^{(n-1)}\|_2^2\right)\mathbf{w}_k}} \\[3ex] s.t. \quad \Re\left(\mathbf{w}_k^H(\mathbf{H}_d + \rho_k\mathbf{H}_{m,k})\mathbf{s}^{(n-1)}\right) \geq 0 \end{cases} \tag{19}$$

We denote by

$$\begin{cases} \mathbf{d}_k(\mathbf{s}^{(n-1)}) = |\alpha_0|(\mathbf{H}_d + \rho_k\mathbf{H}_{m,k})\mathbf{s}^{(n-1)} \\[1ex] \mathbf{Q}(\mathbf{s}^{(n-1)}) = \sum\limits_{p=1}^{P}\sigma_{c,p}^2\mathbf{H}_{c,p}^H\mathbf{s}^{(n-1)}\mathbf{s}^{(n-1)H}\mathbf{H}_{c,p} + \mathbf{R}_n\|\mathbf{s}^{(n-1)}\|_2^2 \end{cases} \tag{20}$$

and recast the filter design problem as

$$\widetilde{\mathcal{P}}_{1,k}^{(n)} \begin{cases} \max\limits_{\mathbf{w}_k} \dfrac{\Re(\mathbf{w}_k^H\mathbf{d}_k(\mathbf{s}^{(n-1)}))}{\mathbf{w}_k^H\mathbf{Q}(\mathbf{s}^{(n-1)})\mathbf{w}_k} \\[2ex] s.t. \quad \Re(\mathbf{w}_k^H\mathbf{d}_k(\mathbf{s}^{(n-1)})) \geq 0 \end{cases} \tag{21}$$

One can easily verify that $\widetilde{\mathcal{P}}_{1,k}^{(n)}$ is equivalent to the following problem:

$$\widetilde{\mathcal{P}}_{1,k}^{(n)\prime} \begin{cases} \max\limits_{\mathbf{w}_k} \mathbf{w}_k^H\mathbf{Q}(\mathbf{s}^{(n-1)})\mathbf{w}_k \\[1ex] s.t. \quad \Re(\mathbf{w}_k^H\mathbf{d}_k(\mathbf{s}^{(n-1)})) = 1 \end{cases} \tag{22}$$

The optimal solution of the kth filter at the nth iteration can be immediately obtained [41], and it is given as

$$\mathbf{w}_k^{(n)} = \mathbf{Q}^{-1}(\mathbf{s}^{(n-1)}) \cdot \mathbf{d}_k(\mathbf{s}^{(n-1)}). \tag{23}$$

3.2. Optimizing the Waveform with Fixed Filters

For fixed filters $\left\{\mathbf{w}_k^{(n)}\right\}_{k=1}^{K}$, the optimization problem of $\mathbf{s}^{(n)}$ is cast as

$$\widetilde{\mathcal{P}}_2^{(n)} \begin{cases} \max\limits_{\mathbf{s}} \min\limits_{k=1,\dots,K} \dfrac{|\alpha_0|\Re\left(\mathbf{w}_k^{(n)H}(\mathbf{H}_d + \rho_k\mathbf{H}_{m,k})\mathbf{s}\right)}{\sqrt{\mathbf{w}_k^{(n)H}\left(\sum\limits_{p=1}^{P}\sigma_{c,p}^2\mathbf{H}_{c,p}\mathbf{s}\mathbf{s}^H\mathbf{H}_{c,p}^H + \mathbf{R}_n\|\mathbf{s}\|_2^2\right)\mathbf{w}_k^{(n)}}} \\[3ex] s.t. \quad \Re\left(\mathbf{w}_k^{(n)H}(\mathbf{H}_d + \rho_k\mathbf{H}_{m,k})\mathbf{s}\right) \geq 0,\, k = 1, 2, \dots K \end{cases} \tag{24}$$

Note that the denominator of the objective in $\widetilde{\mathcal{P}}_2^{(n)}$ can be equivalently formulated as

$$\sqrt{\mathbf{w}_k^{(n)H}\left(\sum_{p=1}^{P}\sigma_{c,p}^2\mathbf{H}_{c,p}\mathbf{s}\mathbf{s}^H\mathbf{H}_{c,p}^H + \mathbf{R}_n\|\mathbf{s}\|_2^2\right)\mathbf{w}_k^{(n)}} = \sqrt{\mathbf{s}^H\left(\sum_{p=1}^{P}\sigma_{c,p}^2\mathbf{H}_{c,p}^H\mathbf{w}_k^{(n)}\mathbf{w}_k^{(n)H}\mathbf{H}_{c,p} + \mathbf{w}_k^{(n)H}\mathbf{R}_n\mathbf{w}_k^{(n)}\right)\mathbf{s}}$$
$$= \left\|\left(\sum_{p=1}^{P}\sigma_{c,p}^2\mathbf{H}_{c,p}^H\mathbf{w}_k^{(n)}\mathbf{w}_k^{(n)H}\mathbf{H}_{c,p} + \mathbf{w}_k^{(n)H}\mathbf{R}_n\mathbf{w}_k^{(n)}\right)^{\frac{1}{2}}\mathbf{s}\right\|_2 \tag{25}$$

Therefore, we can reformulate $\widetilde{\mathcal{P}}_2^{(n)}$ as

$$
\widetilde{\mathcal{P}}_2^{(n)'}
\begin{cases}
\max\limits_{\mathbf{s}}\ \min\limits_{k=1,\dots,K}\ \dfrac{|\alpha_0|\Re\left(\mathbf{w}_k^{(n)H}(\mathbf{H}_d+\rho_k\mathbf{H}_{m,k})\mathbf{s}\right)}{\left\|\left(\sum\limits_{p=1}^{P}\sigma_{c,p}^2\mathbf{H}_{c,p}^{H}\mathbf{w}_k^{(n)}\mathbf{w}_k^{(n)H}\mathbf{H}_{c,p}+\mathbf{w}_k^{(n)H}\mathbf{R}_n\mathbf{w}_k^{(n)}\right)^{\frac{1}{2}}\mathbf{s}\right\|_2}\ . \\[2em]
s.t.\quad \Re\left(\mathbf{w}_k^{(n)H}(\mathbf{H}_d+\rho_k\mathbf{H}_{m,k})\mathbf{s}\right)\ge 0,\ k=1,2,\dots K
\end{cases}
\tag{26}
$$

It is obvious that $\widetilde{\mathcal{P}}_2^{(n)'}$ is non-convex. By defining the functions

$$
\begin{cases}
u_k^{(n)}(\mathbf{s})=|\alpha_0|\Re\left(\mathbf{w}_k^{(n)H}(\mathbf{H}_d+\rho_k\mathbf{H}_{m,k})\mathbf{s}\right) \\[1em]
v_k^{(n)}(\mathbf{s})=\left\|\left(\sum\limits_{p=1}^{P}\sigma_{c,p}^2\mathbf{H}_{c,p}^{H}\mathbf{w}_k^{(n)}\mathbf{w}_k^{(n)H}\mathbf{H}_{c,p}+\mathbf{w}_k^{(n)H}\mathbf{R}_n\mathbf{w}_k^{(n)}\right)^{\frac{1}{2}}\mathbf{s}\right\|_2
\end{cases},\quad k=1,2,\dots,K,
\tag{27}
$$

we can re-express $\widetilde{\mathcal{P}}_2^{(n)'}$ as

$$
\widetilde{\mathcal{P}}_2^{(n)'}
\begin{cases}
\max\limits_{\mathbf{s}}\ \min\limits_{k=1,\dots,K}\ \dfrac{u_k^{(n)}(\mathbf{s})}{v_k^{(n)}(\mathbf{s})} \\[1.5em]
s.t.\quad u_k^{(n)}(\mathbf{s})\ge 0,\ k=1,2,\dots K
\end{cases}
\tag{28}
$$

It is worth noting that $\widetilde{\mathcal{P}}_2^{(n)'}$ is a typical generalized fractional programming problem $\left\{u_k^{(n)}(\mathbf{s})\right\}_{k=1}^{K}$—are all non-negative concave functions; $\left\{v_k^{(n)}(\mathbf{s})\right\}_{k=1}^{K}$—are all positive convex). Therefore, we propose that we tackle $\widetilde{\mathcal{P}}_2^{(n)'}$ by exploiting the idea of generalized Dinkelbach fractional programming. In detail, the approach involves an iteration process. At the qth iteration, the following problem is considered:

$$
\mathcal{P}_2^{(n,q)}
\begin{cases}
\max\limits_{\mathbf{s}}\ \min\limits_{k=1,\dots,K}\ u_k^{(n)}(\mathbf{s})-\varepsilon^{(n,q-1)}v_k^{(n)}(\mathbf{s}) \\[1em]
s.t.\quad u_k^{(n)}(\mathbf{s})\ge 0,\ k=1,2,\dots K
\end{cases}
\tag{29}
$$

where $\varepsilon^{(n,q-1)}=\min\limits_{1\le k\le K}\dfrac{u_k^{(n)}(\mathbf{s}^{(n,q-1)})}{v_k^{(n)}(\mathbf{s})^{(n,q-1)}}$, with $\mathbf{s}^{(n,q-1)}$ being the optimized waveform at the $(q-1)$th iteration. Subsequently, we will show that, by resorting to the epigraph form, $P_2^{(n,q)}$ can be recast into a maximization as

$$
\mathcal{P}_2^{(n,q)'}
\begin{cases}
\max\limits_{\mathbf{s},t}\ t \\[0.8em]
s.t.\quad u_k^{(n)}(\mathbf{s})\ge 0, \\[0.6em]
\qquad\ \ u_k^{(n)}(\mathbf{s})-\varepsilon^{(n,q-1)}v_k^{(n)}(\mathbf{s})\ge t, \\[0.6em]
\qquad\ \ k=1,2,\dots,K
\end{cases}
\tag{30}
$$

where t is an auxiliary variable. It is seen that $\mathcal{P}_2^{(n,q)'}$ is actually a second-order cone programming (SOCP) problem that can be solved in polynomial time via the interior point method. Based on the generalized Dinkelbach fractional programming paradigm, we can obtain a global optimal solution to $\widetilde{\mathcal{P}}_2^{(n)}$ by solving a series of problems $\left\{\mathcal{P}_2^{(n,q)'}\right\}_{q=1}^{\infty}$. For clearer illustration, we summarize the developed approach in Algorithm 1. It is worth stressing that $\mathbf{H}_d$, $\left\{\mathbf{H}_{m,k}\right\}_{k=1}^{K}$, $\left\{\rho_k\right\}_{k=1}^{K}$, $\left\{\mathbf{H}_{c,p}\right\}_{p=1}^{P}$, $\left\{\sigma_{c,p}^2\right\}_{p=1}^{P}$, and $\mathbf{R}_n$ are assumed to be known in advance.

Algorithm 1 Developed approach.

Input : $\mathbf{H}_d$, $\left\{\mathbf{H}_{m,k}\right\}_{k=1}^{K}$, $\left\{\rho_k\right\}_{k=1}^{K}$, $\left\{\mathbf{H}_{c,p}\right\}_{p=1}^{P}$, $\left\{\sigma_{c,p}^2\right\}_{p=1}^{P}$, $\mathbf{R}_n$

Step 1. $n = 0$, initialize the waveform $\mathbf{s}^{(0)}$.

Step 2. $n = n + 1$. Compute $\mathbf{Q}(\mathbf{s}^{(n-1)})$ and get the optimized kth filter as $\mathbf{w}_k^{(n)} = \mathbf{Q}^{-1}(\mathbf{s}^{(n-1)}) \cdot \mathbf{d}_k(\mathbf{s}^{(n-1)})$.

Step 3. Obtain the optimized waveform $\mathbf{s}^{(n)}$ by solving a series of SOCP problems $\left\{\mathcal{P}_2^{(n,q)'}\right\}_{q=1}^{\infty}$

Step 4. Repeat steps 2 and 3 until convergence.

Output Optimized waveform $\mathbf{s}^{\star} = \mathbf{s}^{(n)}$ and filters $\left\{\mathbf{w}_k^{\star}\right\}_{k=1}^{K} = \left\{\mathbf{w}_k^{(n)}\right\}_{k=1}^{K}$.

3.3. Complexity Analysis

As to the computational complexity of the proposed algorithm, it is dependent on the number of iterations and the complexity in each inner iteration. In particular, the overall complexity is linear w.r.t. the number of iterations, and it includes the computation of the matrix inversion and the solution of a series of SOCP problems in each iteration. While the optimization of the filters involves $\mathcal{O}((2N_R L)^3)$ floating-point operations, the optimization of the waveform is related to the number of inner iterations, and $\mathcal{O}((2N_T L)^{3.5})$ floating-point operations are required for solving the SOCP problem $\mathcal{P}_2^{(n,q)'}$ in each inner iteration [42].

4. Simulation Results

In this section, a series of examples are given to demonstrate the effectiveness of the proposed design. Specifically, we consider an MIMO radar with an inter-element spacing of $2d$ for both the transmitting and receiving array, where d is a half-wavelength. Throughout this part, the considered elements for the MIMO radar are fixed at $N_R = 4$. The carrier frequency of the probing signals is set to 1 GHz, and the length of the code to be designed in each channel is set to $L = 20$. As for the target, it is placed in the direction $\theta_d = -20°$. With reference to the two-way direct propagation and backscattering effects from the target, $\alpha_0 = 1$ is set. We assume that there are $K = 20$ possible reflective surface configurations in the scenario, accordingly resulting in 20 possible multipath return parameter pairs. Specifically, for the kth multipath return, $\theta_{m,k}$, $|\rho_k|$, and $\arg(\rho_k)$ are randomly drawn from the sectors $[-30°, 0°]$, $[0.92, 0.94]$, and $[-\frac{\pi}{4}, -\frac{\pi}{12}]$, and $r_{m,k}$ is drawn as a random integer from the region $[L - 5, L]$. We consider the signal-independent disturbance to be white Gaussian noise with a uniting power of $\sigma_n^2 = 1$. For the signal-dependent item, we assume that there are $P = 30$ scatters in the clutter area with azimuth directions $\{\theta_p\}_{p=1}^{30}$ that are uniformly drawn from the region $[-90°, 90°]$. The relative delay $\{l_{c,p}\}_{p=1}^{30}$ is uniformly drawn from integers in the region $[0, 2L]$. The scatters are set to have the same mean interfering power $\sigma_{c,p}^2 = \sigma^2$, which is defined by the interference-to-noise ratio (INR), which is calculated as $\mathrm{INR} = 10\log\frac{\sigma^2}{\sigma_n^2}$. Throughout this part, the INR is set to 20 dB.

4.1. Convergence and Computation Time Analysis

In this example, we check the effectiveness of the developed algorithm on the sequentially improvement of the output worst-case SINR, as well as its convergence and computation time. As depicted in Figure 2, the worst-case SINR curves are plotted against the number of iterations for different transmitting array sizes N_T. As expected, the developed algorithm monotonically enhances the worst-case SINR, which significantly improves the target detection ability w.r.t. unknown multipath return. Furthermore, it is observed that a higher transmitting array size leads to a higher worst-case SINR due to the higher number of degrees of freedom (DOFs) that it introduces. Finally, the results shed light on the effectiveness of the developed algorithm in ensuring an optimized worst-case (w.r.t. unknown multipath return) target detectability. Indeed, significant performance gains through iterations can be explicitly observed. The corresponding computation times are reported in Table 1. One can clearly observe that, though higher worst-case SINR values can be achieved as the transmitting array size becomes larger, the computational burden accordingly becomes heavier, which coincides with the complexity analysis in Section 3.3. Moreover, to illustrate the effectiveness of the proposed algorithm in optimizing the transmitted waveform more directly, we draw the modulus and the phase of the optimized waveform with different transmitting array sizes. As depicted in Figure 3, the waveforms optimized under different N_T values exhibit different behaviors in the fast time domain, but all are constrained by the energy constraint.

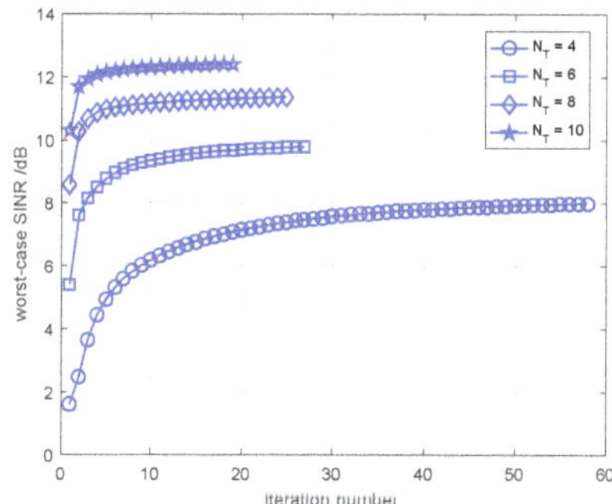

Figure 2. Worst-case SINR values of the waveform and filter with the proposed design versus the iteration number.

Table 1. Computational times of the proposed algorithm with different transmitting array sizes.

N_T	4	6	8	10
Computational time/s	627.077466	770.360833	1007.908756	1303.744407

4.2. Robustness of the Developed Design against Different Multipath Returns

In this example, we further check the robustness of the waveforms and filters designed with the developed algorithm against the uncertainty of multipath return. In particular, we compute the actual achieved output SINR over various multipath returns with different transmitting array sizes. Moreover, for better illustration of the superiority of the proposed design, a non-robust design is added for comparison [27], which is defined as the solution to the following problem:

$$\begin{cases} \max_{\mathbf{s},\mathbf{w}} \dfrac{|\alpha_0|^2 |\mathbf{w}(\mathbf{H}_d + \rho_5 \mathbf{H}_{m,5})\mathbf{s}|^2}{\mathbf{w}^H \left(\sum_{p=1}^{P} \sigma_{c,p}^2 \mathbf{H}_{c,p} \mathbf{s}\mathbf{s}^H \mathbf{H}_{c,p}^H + \mathbf{R}_n \right) \mathbf{w}} \\ s.t. \ \|\mathbf{s}\|_2^2 = 1 \end{cases} \tag{31}$$

This means that we utilize the fifth multipath return case for the non-robust design. As shown in Figure 4, the worst-case output SINR provided by the proposed design significantly outperforms that of the non-robust design. Moreover, in these four situations, it can be observed that the proposed design provides a robust output SINR over various multipath propagation scenarios. The superiority of the proposed design is highlighted by observing that, in almost all multipath return cases, the SINR provided by the proposed design stayed around 3.5 dB higher than that of the non-robust design, except in the fifth multipath propagation case. This can be explained by the proposed design's improvement of the worst-case SINR by averaging the performance over all possible cases, while the non-robust design was only effective in the fifth case. In practices, if there are mismatches between the models of the assumed and the actual scenarios, the performance of the non-robust design will be significantly degraded, but our design enjoys quite stable performance as the scenario changes. The results also show that a larger transmitting array size leads to a higher SINR value.

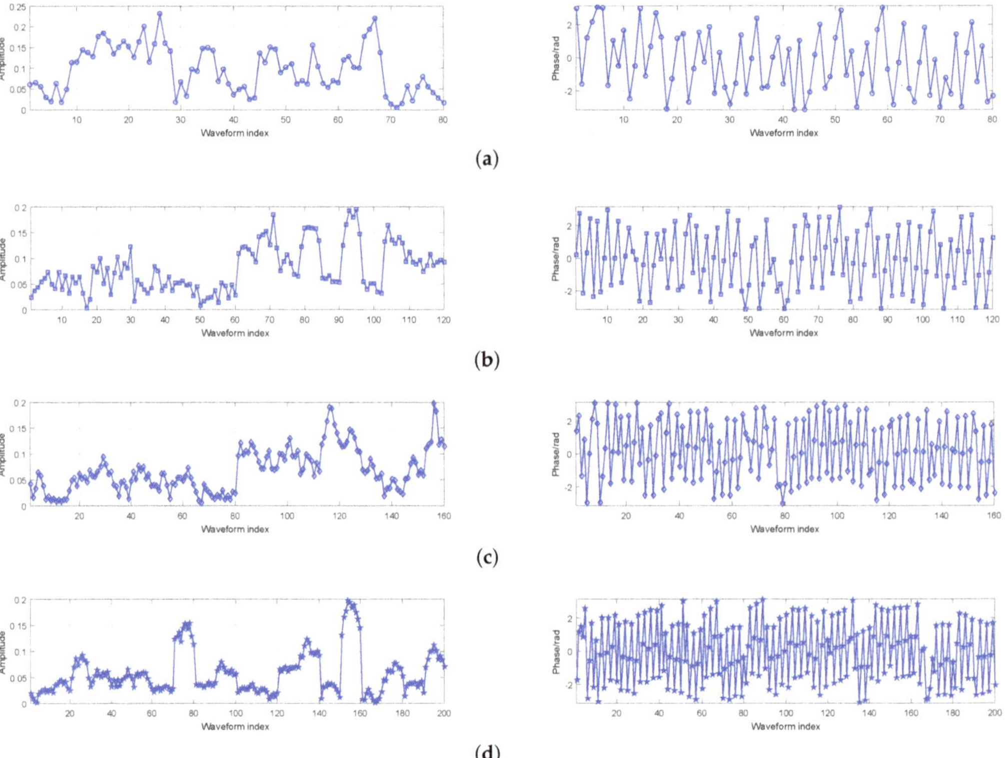

Figure 3. Modulus and phase of the optimized waveform with different transmitting array sizes. (**a**) $N_T = 4$, (**b**) $N_T = 6$, (**c**) $N_T = 8$, and (**d**) $N_T = 10$.

4.3. The Impact of the Transmitting Array Size on the Worst-Case SINR

In the final example, we check the impact of the transmitting array size N_T on the worst-case SINR provided by the proposed algorithm. In particular, we plot the worst-case SINR curve with N_T values varying from 4 to 10 in Figure 5, with the other parameters fixed. Unsurprisingly, the result clearly shows an increasing SINR trend w.r.t N_T, as a larger number of antennas can bring more DOFs in the waveform design process for disturbance suppression; consequently, higher worst-case SINR values can be obtained.

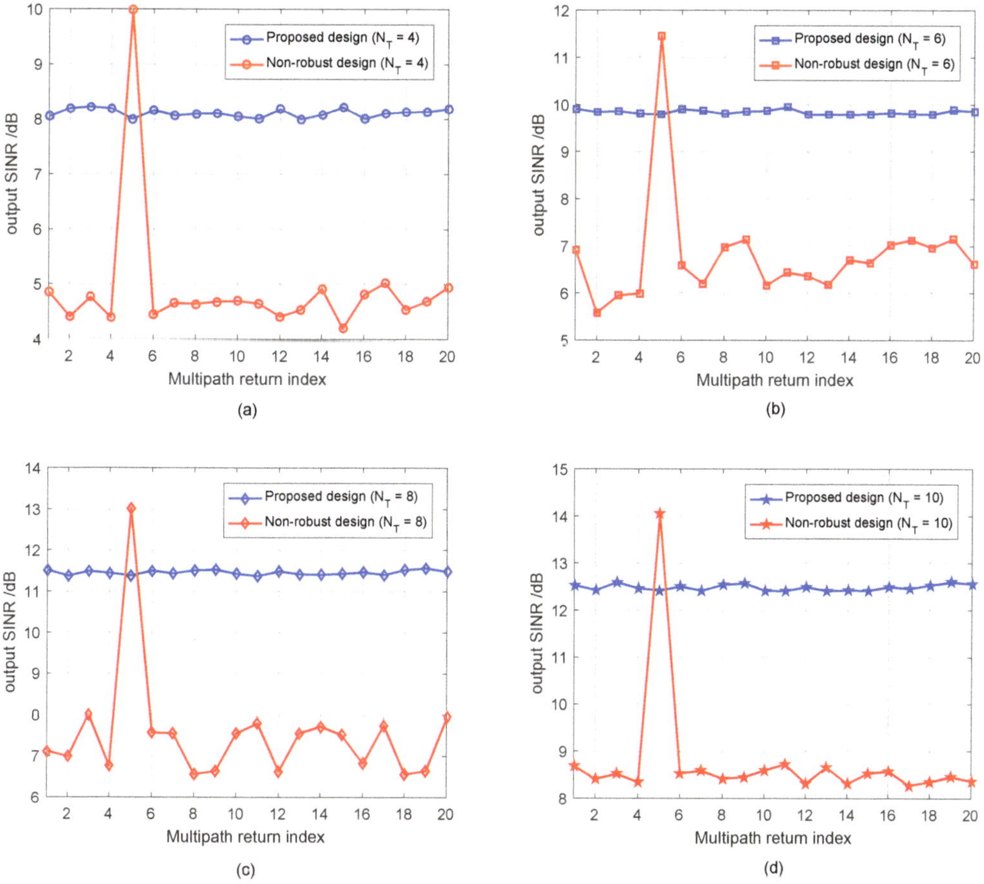

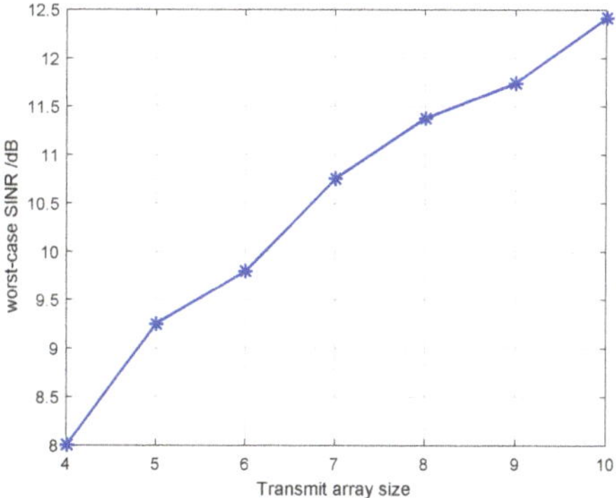

Figure 4. SINR values achieved with the proposed design and the non-robust design over different multipath propagation cases with various transmitting array sizes. (**a**) $N_T = 4$, (**b**) $N_T = 6$, (**c**) $N_T = 8$, and (**d**) $N_T = 10$.

Figure 5. The worst-case output SINR provided by the proposed design versus the transmitting array size.

5. Conclusions

In this paper, a robust joint MIMO design for a transmitted waveform and receiving filter bank under uncertain multipath return was investigated. The goal of the robust design was to make the designed waveform and filters provide relatively stable output SINR values regardless of the actual multipath return expression, which was cast as a max–min problem and was hard to solve. A corresponding algorithm was developed based on a cyclic optimization paradigm. The simulation results show that the proposed design can achieve robustness against different multipath scenarios; this is of great significance in real applications, since actual information about an operating scene may be imprecise. Future research may include this robust design under more complex constraints on the transmitted waveform, such as spectral compatibility constraints.

Author Contributions: Conceptualization, C.F. and J.W.; methodology, J.W.; software, C.F. and J.W.; validation, C.F., J.W. and X.H.; formal analysis, C.F.; investigation, C.F.; resources, J.W.; writing—original draft preparation, C.F.; writing—review and editing, C.F., Z.X. (Zhuang Xi), Z.X. (Zhou Xu) and J.W.; visualization, C.F.; supervision, X.H. All authors have read and agreed to the published version of the manuscript.

Funding: This research received no external funding.

Conflicts of Interest: The authors declare no conflict of interest.

References

1. Zhou, K.; Quan, S.; Li, D.; Liu, T.; He, F.; Su, Y. Waveform and filter joint design method for pulse compression sidelobe reduction. *IEEE Trans. Geosci. Remote Sens.* **2021**, *60*, 1–15. [CrossRef]
2. Yu, X.; Qiu, H.; Yang, J.; Wei, W.; Cui, G.; Kong, L. Multi-spectrally constrained MIMO radar beampattern design via sequential convex approximation. *IEEE Trans. Aerosp. Electron. Syst.* **2022**, *58*, 2935–2949. [CrossRef]
3. Argenti, F.; Facheris, L. Radar pulse compression methods based on nonlinear and quadratic optimization. *IEEE Trans. Geosci. Remote Sens.* **2020**, *59*, 3904–3916. [CrossRef]
4. Yang, J.; Aubry, A.; De Maio, A.; Yu, X.; Cui, G. Design of constant modulus discrete phase radar waveforms subject to multi-spectral constraints. *IEEE Signal Process. Lett.* **2020**, *27*, 875–879. [CrossRef]
5. Raei, E.; Alaee-Kerahroodi, M.; Shankar, M.B. Spatial-and range-ISLR trade-off in MIMO radar via waveform correlation optimization. *IEEE Trans. Signal Process.* **2021**, *69*, 3283–3298. [CrossRef]
6. Jiang, W.; Haimovich, A.M.; Simeone, O. Joint design of radar waveform and detector via end-to-end learning with waveform constraints. *IEEE Trans. Aerosp. Electron. Syst.* **2021**, *58*, 552–567. [CrossRef]
7. Liu, Q.; Ren, W.; Hou, K.; Long, T.; Fathy, A.E. Design of Polyphase Sequences with Low Integrated Sidelobe Level for Radars With Spectral Distortion via Majorization-Minimization Framework. *IEEE Trans. Aerosp. Electron. Syst.* **2021**, *57*, 4110–4126. [CrossRef]
8. Li, J.; Stoica, P. MIMO radar with colocated antennas. *IEEE Signal Process. Mag.* **2007**, *24*, 106–114. [CrossRef]
9. Li, J.; Stoica, P.; Xu, L.; Roberts, W. On parameter identifiability of MIMO radar. *IEEE Signal Process. Lett.* **2007**, *14*, 968–971.
10. Fishler, E.; Haimovich, A.; Blum, R.; Chizhik, D.; Cimini, L.; Valenzuela, R. MIMO radar: An idea whose time has come. In Proceedings of the 2004 IEEE Radar Conference (IEEE Cat. No. 04CH37509), Philadelphia, PA, USA, 29 April 2004; IEEE: Toulouse, France, 2004; pp. 71–78.
11. Yang, Y.; Blum, R.S. MIMO radar waveform design based on mutual information and minimum mean-square error estimation. *IEEE Trans. Aerosp. Electron. Syst.* **2007**, *43*, 330–343. [CrossRef]
12. He, H.; Stoica, P.; Li, J. Designing unimodular sequence sets with good correlations—Including an application to MIMO radar. *IEEE Trans. Signal Process.* **2009**, *57*, 4391–4405. [CrossRef]
13. Cheng, Z.; He, Z.; Zhang, S.; Li, J. Constant modulus waveform design for MIMO radar transmit beampattern. *IEEE Trans. Signal Process.* **2017**, *65*, 4912–4923. [CrossRef]
14. Cui, G.; Li, H.; Rangaswamy, M. MIMO radar waveform design with constant modulus and similarity constraints. *IEEE Trans. Signal Process.* **2013**, *62*, 343–353. [CrossRef]
15. Tang, B.; Tang, J. Joint design of transmit waveforms and receive filters for MIMO radar space-time adaptive processing. *IEEE Trans. Signal Process.* **2016**, *64*, 4707–4722. [CrossRef]
16. Aldayel, O.; Monga, V.; Rangaswamy, M. Successive QCQP refinement for MIMO radar waveform design under practical constraints. *IEEE Trans. Signal Process.* **2016**, *64*, 3760–3774. [CrossRef]
17. Cheng, Z.; He, Z.; Liao, B.; Fang, M. MIMO radar waveform design with PAPR and similarity constraints. *IEEE Trans. Signal Process.* **2017**, *66*, 968–981. [CrossRef]
18. Naghsh, M.M.; Modarres-Hashemi, M.; Kerahroodi, M.A.; Alian, E.H.M. An information theoretic approach to robust constrained code design for MIMO radars. *IEEE Trans. Signal Process.* **2017**, *65*, 3647–3661. [CrossRef]

19. Yu, X.; Cui, G.; Kong, L.; Li, J.; Gui, G. Constrained waveform design for colocated MIMO radar with uncertain steering matrices. *IEEE Trans. Aerosp. Electron. Syst.* **2018**, *55*, 356–370. [CrossRef]
20. Yu, X.; Alhujaili, K.; Cui, G.; Monga, V. MIMO radar waveform design in the presence of multiple targets and practical constraints. *IEEE Trans. Signal Process.* **2020**, *68*, 1974–1989. [CrossRef]
21. Aubry, A.; De Maio, A.; Foglia, G.; Orlando, D. Diffuse multipath exploitation for adaptive radar detection. *IEEE Trans. Signal Process.* **2015**, *63*, 1268–1281. [CrossRef]
22. Rong, Y.; Aubry, A.; De Maio, A.; Tang, M. Diffuse multipath exploitation for adaptive detection of range distributed targets. *IEEE Trans. Signal Process.* **2020**, *68*, 1197–1212. [CrossRef]
23. Richards, M.A. *Principles of Modern Radar. Volume I, Basic Principles [Electronic Resource]*; Richards, M.A., Scheer, J.A., Holm, W.A., Eds.; SciTech Pub.: Raleigh, NC, USA, 2010.
24. Fante, R.L.; Torres, J.A. Cancellation of diffuse jammer multipath by an airborne adaptive radar. *IEEE Trans. Aerosp. Electron. Syst.* **1995**, *31*, 805–820. [CrossRef]
25. Sen, S.; Nehorai, A. Adaptive OFDM radar for target detection in multipath scenarios. *IEEE Trans. Signal Process.* **2010**, *59*, 78–90. [CrossRef]
26. Sen, S.; Nehorai, A. OFDM MIMO radar with mutual-information waveform design for low-grazing angle tracking. *IEEE Trans. Signal Process.* **2010**, *58*, 3152–3162. [CrossRef]
27. Xu, Z.; Fan, C.; Huang, X. MIMO radar waveform design for multipath exploitation. *IEEE Trans. Signal Process.* **2021**, *69*, 5359–5371. [CrossRef]
28. Crouzeix, J.P.; Ferland, J.A. Algorithms for generalized fractional programming. *Math. Program.* **1991**, *52*, 191–207. [CrossRef]
29. Aubry, A.; De Maio, A.; Naghsh, M.M. Optimizing radar waveform and Doppler filter bank via generalized fractional programming. *IEEE J. Sel. Top. Signal Process.* **2015**, *9*, 1387–1399. [CrossRef]
30. Tang, B.; Liu, J.; Wang, H.; Hu, Y. Constrained radar waveform design for range profiling. *IEEE Trans. Signal Process.* **2021**, *69*, 1924–1937. [CrossRef]
31. Tang, B.; Li, J. Spectrally constrained MIMO radar waveform design based on mutual information. *IEEE Trans. Signal Process.* **2018**, *67*, 821–834. [CrossRef]
32. Cui, G.; Fu, Y.; Yu, X.; Li, J. Local ambiguity function shaping via unimodular sequence design. *IEEE Signal Process. Lett.* **2017**, *24*, 977–981. [CrossRef]
33. Li, Y.; Vorobyov, S.A.; Koivunen, V. Ambiguity function of the transmit beamspace-based MIMO radar. *IEEE Trans. Signal Process.* **2015**, *63*, 4445–4457. [CrossRef]
34. Arlery, F.; Kassab, R.; Tan, U.; Lehmann, F. Efficient gradient method for locally optimizing the periodic/aperiodic ambiguity function. In Proceedings of the 2016 IEEE Radar Conference (RadarConf), Philadelphia, PA, USA, 2–6 May 2016; IEEE: Toulouse, France, 2016; pp. 1–6.
35. Aubry, A.; De Maio, A.; Jiang, B.; Zhang, S. Ambiguity function shaping for cognitive radar via complex quartic optimization. *IEEE Trans. Signal Process.* **2013**, *61*, 5603–5619. [CrossRef]
36. Sen, S. PAPR-constrained pareto-optimal waveform design for OFDM-STAP radar. *IEEE Trans. Geosci. Remote Sens.* **2013**, *52*, 3658–3669. [CrossRef]
37. Li, J.; Xu, L.; Stoica, P.; Forsythe, K.W.; Bliss, D.W. Range compression and waveform optimization for MIMO radar: A Cramér–Rao bound based study. *IEEE Trans. Signal Process.* **2007**, *56*, 218–232.
38. Li, Z.; Shi, J.; Liu, W.; Pan, J.; Li, B. Robust joint design of transmit waveform and receive filter for MIMO-STAP radar under target and clutter uncertainties. *IEEE Trans. Veh. Technol.* **2021**, *71*, 1156–1171. [CrossRef]
39. Cui, G.; Fu, Y.; Yu, X.; Li, J.; Gui, G. Robust transmitter–receiver design for extended target in signal-dependent interference. *Signal Process.* **2018**, *147*, 60–67. [CrossRef]
40. Yang, J.; Aubry, A.; De Maio, A.; Yu, X.; Cui, G. Multi-spectrally constrained transceiver design against signal-dependent interference. *IEEE Trans. Signal Process.* **2022**, *70*, 1320–1332. [CrossRef]
41. Van Trees, H.L. *Optimum Array Processing: Part IV of Detection, Estimation, and Modulation Theory*; John Wiley & Sons: Hoboken, NJ, USA, 2004.
42. Ben-Tal, A.; Nemirovski, A. *Lectures on Modern Monvex Mptimization (2012)*; SIAM: Philadelphia, PA, USA, 2011.

Article

MIMO DFRC Signal Design in Signal-Dependent Clutter

Xue Yao [1,*], Bunian Pan [2], Tao Fan [2], Xianxiang Yu [2], Guolong Cui [2] and Xiangfei Nie [3]

[1] School of Information Science and Engineering, Southeast University, Nanjing 210096, China
[2] School of Information and Communication Engineering, University of Electronic Science and Technology of China, Chengdu 611731, China; bunianpan@std.uestc.edu.cn (B.P.); taofan@std.uestc.edu.cn (T.F.); xianxiangyu@uestc.edu.cn (X.Y.); cuiguolong@uestc.edu.cn (G.C.)
[3] School of Electronic and Information Engineering, Chongqing Three Gorges University, Chongqing 404020, China; niexf@sanxiau.edu.cn
* Correspondence: xueyao@seu.edu.cn; Tel.: +86-152-2364-3806

Abstract: This paper deals with the Dual-Function Radar and Communication (DFRC) signal design for a Multiple-Input–Multiple-Output (MIMO) system, considering the presence of signal-dependent clutter. A modulation methodology called Spectral Position Index and Amplitude (SPIA) modulation is proposed, which involves selecting passband and stopband positions and applying amplitude modulation. Signal to Interference plus Noise Ratio (SINR) is maximized to enhance radar detectability. Meanwhile, variable modulus and communication modulation constraints are enforced to ensure compatibility with the current hardware techniques and communication demand, respectively. In addition, the mainlobe width and sidelobe level constraints used to concentrate energy in a specific area of space are enforced. To tackle the resulting nonconvex and NP-hard optimization problem, an Iterative Block Enhancement (IBE) framework that alternately updates each signal in each emitting antenna is exploited to monotonically increase SINR. Each block involves the Dinkelbach's Iterative Procedure (DIP), Sequential Convex Approximation (SCA) and Alternating Direction Method of Multipliers (ADMM) to obtain a single signal. The computational complexity and convergence of the algorithm are analyzed. Finally, the numerical results highlight the effectiveness of the proposed dual-function scheme in sidelobe signal-dependent clutter.

Keywords: dual-function radar and communication system; DFRC signal design; clutter suppression

Citation: Yao, X.; Pan, B.; Fan, T.; Yu, X.; Cui, G.; Nie, X. MIMO DFRC Signal Design in Signal-Dependent Clutter. *Remote Sens.* **2023**, *15*, 3256. https://doi.org/10.3390/rs15133256

Academic Editor: Andrzej Stateczny

Received: 22 May 2023
Revised: 19 June 2023
Accepted: 22 June 2023
Published: 24 June 2023

1. Introduction

The Dual-Function Radar and Communication (DFRC) system has attracted significant attention due to its advantages of a small size, low cost, and strong mobility. This system can be classified into two categories: communication-centric and radar-centric. The radar-centric DFRC system primarily focuses on detection and emits a dual-function signal for both detection and communication using the radar system's available resources. However, designing a dual-function signal poses a considerable challenge. Existing works primarily concentrate on utilizing Degrees of Freedom (DoFs) of different dimensions to achieve this dual function.

Earlier studies explored the combination of a Linear Frequency Modulation (LFM) signal and Phase Shift Keying (PSK) modulation [1,2]. However, LFM-PSK signals often exhibit higher autocorrelation sidelobe levels compared to standard LFM. Another approach involves embedding information in the transform domain of the dual-function signal. Various modulation methods have been proposed, such as frequency nulling modulation [3], Ambiguity Function (AF) sidelobe nulling modulation [4], and watermarking demodulation [5]. The data rates of the mentioned methods [1–5] are limited as they only consider modulation in the time and frequency domains.

The Multiple-Input–Multiple-Output (MIMO) system has garnered attention due to its potential for spatial domain modulation. In [6–8], researchers have embedded information in an orthogonal frequency-hopping signal cluster, capitalizing on the increased possibilities offered by MIMO systems. Alternatively, [9–11] form different sidelobe levels or/and different phases for the communication users by designing the weight vector and exploiting signal diversity. In [12], a spatio-spectral method is proposed by shaping the energy spectral density of synthetic signals in users' directions. Ref. [13] maximizes the Signal to Interference plus Noise Ratio (SINR) under power and communication modulation constraints. To improve the data rate, related studies [14–19] have introduced the concept of index modulation in dual-function signal design by drawing inspiration from index modulation techniques employed in the communication literature [20–22]. It is important to note that the distinction between these studies lies in the specific application scenario. The former focuses on DFRC systems, while the latter primarily pertains to communication systems. For instance, the communication scheme of the DFRC system in [17] exploits the agile profile of the radar signals to convey its message via frequency and spatial index modulation.

However, the radar-centric DFRC system always operates in a complex electromagnetic countermeasure environment. The existing non-uniform, changeable and strong independent clutter degrades the radar detection performance. Many works have been employed in MIMO radar signal design with uniform or non-uniform clutter [23–31]. Recently, a radar-centric dual-function signal design in a signal-dependent clutter environment ias pursued to maximize the SINR [32–34]. In [32], a constant-envelope DFRC signal was designed with constraints regarding the synthesis error associated with each communication user, which is different from using an indirect method to control the overall synthesis error in [33]. Ref. [34] proposes a novel Spacetime-Adaptive Processing (STAP)-Symbol-level Precoding (SLP)-based DFRC signal design method that enjoys the advantages of both techniques. In general, the works in [32–34] complete the dual-function signal design based on a conventional communication sequence (e.g., PSK) in which communication information is priori embedded.

In this paper, we still focus on the dual-function signal design in signal-dependent clutter. Different from the existing approaches [32–34], a new Spectral Position Index and Amplitude (SPIA) modulation methodology is proposed instead of using a conventional communication sequence. The main contributions of this paper are as follows:

(1) DFRC Signal Design Scheme via SPIA Modulation: A novel information modulation method, namely Spectral Position Index and Amplitude (SPIA) modulation via spectral passband and stopband position selection and amplitude modulation, is proposed. To realize radar detection and multi-user communication simultaneously, a nonconvex and NP-hard optimization problem is formed. More specifically, SINR is maximized to enhance radar detectability. Variable modulus and communication modulation constraints are enforced to ensure compatibility with the current hardware techniques and communication demand, respectively and the mainlobe width and sidelobe level constraints used to concentrate energy in a specific area of space are enforced.

(2) Convergence and Computational Complexity: With the help of the Iterative Block Enhancement (IBE) framework, we split the original nonconvex NP-hard problem into smaller subproblems. Dinkelbach's Iterative Procedure, Sequential Convex Approximation and the Alternating Direction Method Of Multipliers (DSADMM) were used to solve the subproblems. Finally, DSADMM is incorporated into the IBE framework to form the IBE-DSADMM algorithm for implementing the dual-function signal design. The IBE-DSADMM algorithm guarantees that the SINR value monotonically increases and converges to a finite value. Furthermore, the computational complexity is also analyzed.

Note that numerous studies have extensively addressed the optimization of beampatterns [35–38]. For instance, in [35], the perturbation of the zeros in the radiation pattern enabled the achievement of individually adjustable sidelobe levels for linear and planar antenna arrays. Another interesting study [38] employed the Coordinate Descent (CD) framework for bi-objective Pareto optimization, focusing on minimizing the Integrated Sidelobe Level Ratio (ISLR) both spatially and across the range while considering various practical constraints. It is important to highlight that there are notable distinctions between our work and the aforementioned studies. (1) Dual-Function Background: Our work emphasizes the dual-function background, wherein the optimized beampattern in this paper facilitates simultaneous multi-user communication and detection by designing a dual-function signal. (2) Novel modulation method: In this paper, the proposed SPIA modulation focuses on index modulation via inter-pulse modulation, where the baseband signal is modified. (3) Novel Optimization Problem: Using SINR as the optimization criterion, a new optimization problem is formed by considering the mainlobe width and sidelobe level constraints, variable modulus and communication modulation constraints.

Notation:

We use boldface for vectors $\mathbf{a}$ (lower case), and matrices $\mathbf{A}$ (upper case). The transpose, the complex conjugate, the conjugate transpose and the factorial operators are denoted by the symbols $(\cdot)^T$, $(\cdot)^*$, $(\cdot)^H$ and $(\cdot)!$, respectively. $\mathbb{C}^{N \times M}$ and $\mathbb{C}^N$ are, respectively, the sets of $N \times M$-dimensional matrices and N-dimensional vectors of complex numbers. The letter j represents the imaginary unit (i.e., $j = \sqrt{-1}$). $\Re\{\}$ means the real part of a complex valued scalar. We use $|\cdot|$ for the magnitude or cardinality of a scalar value or a set, respectively. In addition, $\lfloor \rfloor$ denotes the floor function. $\|\cdot\|^2$ and $\otimes$ refer to the Euclidean norm of a vector and the Kronecker product.

2. Methods

2.1. System Model

Assume that a colocated narrow band MIMO-DFRC system with N_t transmitting N_R receiving antennas is detecting targets while transmitting information to C communication users in signal-dependent clutter, as shown in Figure 1. The discrete baseband signal transmitted by the n-th antenna is $\mathbf{s}_n = [\mathbf{s}_n(1), \cdots, \mathbf{s}_n(M)]^T$, where $n = 1, \cdots; N_t$, M denotes fast time sampling number. Thus, the m-th signal sample received by the N_R antennas is

$$\mathbf{x}_m = a_0 \mathbf{A}(\theta_0)\bar{\mathbf{s}}_m + \mathbf{d}_m + \mathbf{v}_m, \tag{1}$$

where $\bar{\mathbf{s}}_m = [\mathbf{s}_1(m), \cdots, \mathbf{s}_{N_t}(m)]^T$ denotes the m-th sample of N_t transmitted signals. a_0 and θ_0 are the complex amplitude and direction of the target, respectively. $\mathbf{A}(\theta) = \mathbf{a}_r^*(\theta)\mathbf{a}_t^H(\theta)$ is the spacial steering matrix in direction of θ, therein $\mathbf{a}_r(\theta)$ and $\mathbf{a}_t(\theta)$, which represent the normalized receiving and transmitting spacial steering vectors, respectively. Uniform Linear Arrays (ULAs), are given by

$$\mathbf{a}_r(\theta) = \frac{1}{\sqrt{N_R}}[1, e^{j2\pi d_r \sin\theta/\lambda}, \cdots, e^{j2\pi d_r(N_R-1)\sin\theta/\lambda}]^T, \tag{2}$$

$$\mathbf{a}_t(\theta) = \frac{1}{\sqrt{N_t}}[1, e^{j2\pi d_t \sin\theta/\lambda}, \cdots, e^{j2\pi d_t(N_t-1)\sin\theta/\lambda}]^T, \tag{3}$$

where d_t and d_r are the array element spacings of the transmitting and receiving arrays, respectively. λ is the wavelength.

$\mathbf{d}_m$ denotes the signal-dependent clutter. Specifically, considering p, uncorrelated interfering sources from the same range bin with the target, each located at a specific azimuth φ_p, $\mathbf{d}_m$ are written as

$$\mathbf{d}_m = \sum_{p=1}^{P} a_p \mathbf{A}(\varphi_p)\bar{\mathbf{s}}_m, \tag{4}$$

therein, a_p denotes the the complex amplitude of the p-th clutter. For the P-interfering sources, $a_p, p = 1, \cdots, P$ are independent complex Gaussian variables with zero mean and variance $\delta_p^2 = \mathbb{E}[|a_p|^2]$.

$\mathbf{v}_m$ denotes the additive noise matrix, whose entries modelled as zero mean independent random complex Gaussian variables with variance σ_v^2.

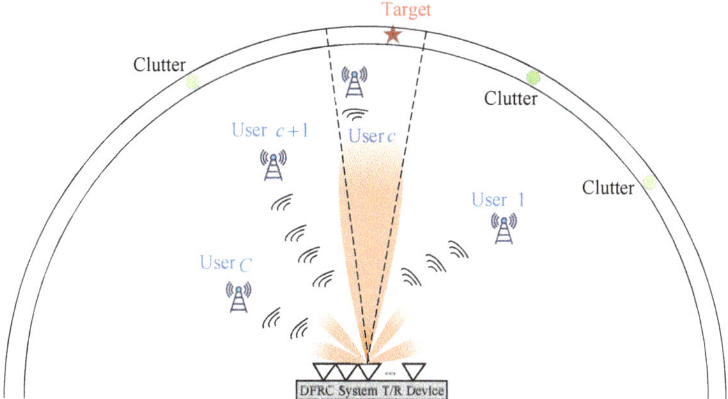

Figure 1. DFRC work scenario for detection and communication in signal-dependent clutter.

According to Equation (1), the power of the target echo is given by

$$\frac{1}{M}\sum_{m=1}^{M}\mathbb{E}\left[\|a_0\mathbf{A}(\theta_0)\bar{\mathbf{s}}_m\|^2\right]$$

$$=\mathbb{E}[\|a_0\|^2]\frac{1}{M}\sum_{m=1}^{M}\bar{\mathbf{s}}_m^H\mathbf{a}_\mathrm{t}(\theta_0)\mathbf{a}_\mathrm{r}^T(\theta_0)\mathbf{a}_\mathrm{r}^*(\theta_0)\mathbf{a}_\mathrm{t}^H(\theta_0)\bar{\mathbf{s}}_m$$

$$=\mathbb{E}[\|a_0\|^2]\mathrm{tr}\left(\mathbf{a}_\mathrm{t}^H(\theta_0)\frac{1}{M}\sum_{m=1}^{M}\bar{\mathbf{s}}_m\bar{\mathbf{s}}_m^H\mathbf{a}_\mathrm{t}(\theta_0)\right)$$

$$=\delta_0^2\mathbf{a}_\mathrm{t}^H(\theta_0)\mathbf{R}\mathbf{a}_\mathrm{t}(\theta_0), \tag{5}$$

where $\delta_0^2=\mathbb{E}[\|a_0\|^2]$, and $\mathbf{R}$ represents the covariance matrix of the signal, expressed as

$$\mathbf{R}=\frac{1}{M}\sum_{m=1}^{M}\bar{\mathbf{s}}_m\bar{\mathbf{s}}_m^H, \tag{6}$$

Similarly, the power of the clutter is given by

$$\frac{1}{M}\sum_{m=1}^{M}\mathbb{E}\left[\left\|\sum_{p=1}^{P}a_p\mathbf{A}(\varphi_p)\bar{\mathbf{s}}_m\right\|^2\right]=\sum_{p=1}^{P}\delta_p^2\mathbf{a}_\mathrm{t}^H(\varphi_p)\mathbf{R}\mathbf{a}_\mathrm{t}(\varphi_p). \tag{7}$$

2.1.1. Spectral Position Index and Amplitude Modulation

In this subsection, a modulation method is proposed by spectral passband and stopband position selection and amplitude modulation to embed information. Specifically, we shaped the Energy Spectral Densities (ESDs) of the spatially synthetic signal arriving at the users' directions to form the passbands and stopbands. Assume that the C users are located at azimuths $\theta_c, c=1,\cdots,C$. The spatially synthetic signal $\mathbf{x}_c\in\mathbb{C}^M$ in the direction θ_c is given by [12]

$$\mathbf{x}_c=\mathbf{S}\mathbf{a}^*(\theta_c), \tag{8}$$

where $\mathbf{S}=[\mathbf{s}_1,\mathbf{s}_2,\cdots,\mathbf{s}_{N_t}]\in\mathbb{C}^{M\times N_t}$.

The normalized frequency band of $\mathbf{x}_c, c = 1, \cdots, C$ is $\Omega_c \in [0,1)$ with L_c frequency subbands $\Omega_{l_c} = (f_{l_c,1}, f_{l_c,2}) \in \Omega, l_c = 1_c, 2_c, \cdots, L_c$ where $f_{l_c,1}$ and $f_{l_c,2}$ denote the lower and upper normalized frequencies associated with Ω_{l_c}, respectively. $\mathcal{L}_c$ is the set with L_c available frequency subbands elements, given by

$$\mathcal{L}_c := \{\Omega_{1_c}, \Omega_{2_c}, \cdots, \Omega_{L_c}\}. \tag{9}$$

Spectral position index modulation selects different subband positions to embed information. As for user c, according to the communication requirements, Q_c subbands are chosen from $\mathcal{L}_c$ for embedding information. The set of possible subband position selections is denoted by

$$\mathcal{T}_c := \{\mathcal{L}_c^{(k)} \,||\, \mathcal{L}_c^{(k)}| = Q_c, \mathcal{L}_c^{(k)} \subset \mathcal{L}_c\}, \tag{10}$$

where (k) stands for the k-th selection pattern in the set $\mathcal{T}_c$. The number of possible position selections is

$$|\mathcal{T}_c| = C_{L_c}^{Q_c} = \frac{(L_c)!}{(Q_c)!(L_c - Q_c)!}. \tag{11}$$

According to Equation (11), the maximum number of bits transmitted via spectral position index modulation is

$$D_{c,1} = \left\lfloor \log_2\left(C_{L_c}^{Q_c}\right) \right\rfloor. \tag{12}$$

Furthermore, the amplitudes of selected subbands can also be used to embed information. Therein, a stopband that corresponds to data is "1" and a passband that corresponds to data is "0". Therefore, the number of bits transmitted by amplitude modulation is $D_{c,2} = Q_c$. As a result, the data rate towards the user c is

$$D_c = (D_{c,1} + D_{c,2})/T_r, \tag{13}$$

where T_r is Pulse Repetition Time (PRT).

An example is given in the following. Suppose that the MIMO-DFRC system transmits information to two users located in θ_1, θ_2. According to Equation (8), the spatially synthetic far-field baseband discrete signals can be obtained as $\mathbf{x}_1 = \mathbf{Sa}^*(\theta_1), \mathbf{x}_2 = \mathbf{Sa}^*(\theta_2)$. Then, frequency subband sets used to transmit information are (Without loss of generality, we assume $\mathcal{L}_1 = \mathcal{L}_2$)

$$\begin{aligned}
\mathcal{L}_1 &:= \{\Omega_{1_1}, \Omega_{2_1}, \cdots, \Omega_{L_1}\} = \{(0.1, 0.13), (0.3, 0.33), (0.6, 0.63), (0.8, 0.83)\} \\
\mathcal{L}_2 &:= \{\Omega_{1_2}, \Omega_{2_2}, \cdots, \Omega_{L_2}\} = \{(0.1, 0.13), (0.3, 0.33), (0.6, 0.63), (0.8, 0.83)\}
\end{aligned} \tag{14}$$

Assuming $Q_1 = 2$ positions are selected from L_1 to deliver information to user 1, six possible selection patterns are

$$\begin{aligned}
\mathcal{L}_1^{(1)} &= \{(0.1, 0.13), (0.3, 0.33)\}, \mathcal{L}_1^{(2)} = \{(0.1, 0.13), (0.6, 0.63)\}, \mathcal{L}_1^{(3)} = \{(0.1, 0.13), (0.8, 0.83)\}, \\
\mathcal{L}_1^{(4)} &= \{(0.3, 0.33), (0.6, 0.63)\}, \mathcal{L}_1^{(5)} = \{(0.3, 0.33), (0.8, 0.83)\}, \mathcal{L}_1^{(6)} = \{(0.6, 0.63), (0.8, 0.83)\}.
\end{aligned} \tag{15}$$

According to Equation (12), the maximum number of transmitted bits is $D_{1,1} = 2$. Without losing generality, $\mathcal{L}_1^{(1)}, \mathcal{L}_1^{(2)}, \mathcal{L}_1^{(3)}, \mathcal{L}_1^{(6)}$ are selected to convey binary sequence "00", "01", "10" and "11", respectively. Furthermore, the amplitudes of selected subbands can also be used to embed information. Specifically, the conveyed binary data are "1" when the subband is a stopband; otherwise, the data are "0" for a passband. Figure 2 shows the subband selection and the energy control of selected subbands when the binary sequence "0101" is transmitted to user 1. In more detail, $\mathcal{L}_1^{(2)} = \{(0.1, 0.13), (0.6, 0.63)\}$ is used to transmit the first two bits of "0101". $\Omega_{1_1} = (0.1, 0.13)$ is enforced as a passband to transmit the third bit "0" and $\Omega_{1_1} = (0.6, 0.63)$ is enforced as a stopband to transmit the third bit "1".

Similarly, assuming $Q_2 = 1$ position is selected from L_2 to deliver information to user 2, four possible selection patterns are

$$\mathcal{L}_2^{(1)} = \{(0.1, 0.13)\}, \mathcal{L}_2^{(2)} = \{(0.3, 0.33)\}, \mathcal{L}_2^{(3)} = \{(0.6, 0.63)\}, \mathcal{L}_2^{(4)} = \{(0.8, 0.83)\}. \quad (16)$$

Generally, $\mathcal{L}_2^{(1)}, \mathcal{L}_2^{(2)}, \mathcal{L}_2^{(3)}, \mathcal{L}_2^{(4)}$ represent binary sequence "00", "01", "10" and "11", respectively. Figure 2 also shows the subband selection and the energy control of selected subbands when the binary sequence "010" is transmitted to user 2. Specifically, $\mathcal{L}_2^{(2)} = \{(0.3, 0.33)\}$ is used to transmit the first two bits of "010". $\Omega_{2_2} = (0.3, 0.33)$ is enforced as a passband to transmit the third bit "0".

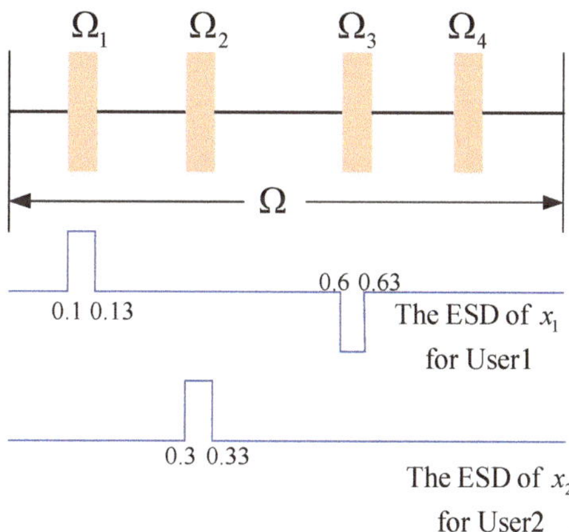

Figure 2. An example of SPIA modulation.

2.1.2. Spectral Position Index and Amplitude Demodulation

The communication demodulation method is discussed in this subsection. Assuming that the channel state information is known, the baseband signal received by the user c located in θ_c is

$$\tilde{\mathbf{x}}_c = \beta_c \mathbf{S}\mathbf{a}^*(\theta_c) + \mathbf{n}_c, \quad (17)$$

where β_c is the channel coefficient. $\mathbf{n}_c \in \mathbb{C}^{M \times 1}$ denotes the additive Gaussian white noise vector with zero mean and variance δ_c^2.

The frequency subband energy of the received signal $\tilde{\mathbf{x}}_c$ is utilized to demodulate the information. In more detail, we performed K points Discrete Fourier Transformation (DFT) of the signal $\tilde{\mathbf{x}}_c$ and then calculated the energy e_{l_c} (in dB) corresponding to each subband $\Omega_{l_c} = (f_{l_c,1}, f_{l_c,2}), l_c = 1, \cdots, L_c$, given by [12]

$$e_{l_c} = 10 \lg(\hat{\mathbf{x}}_c^H \mathbf{F}_{l_c} \mathbf{F}_{l_c}^H \hat{\mathbf{x}}_c / N_{l_c}), \quad (18)$$

where $\hat{\mathbf{x}}_c = [\hat{\mathbf{x}}_c^T, \mathbf{0}_{K-M}^T]^T \in \mathbb{C}^K$. $\mathbf{F}_{l_c} \in \mathbb{C}^{K \times (N_{l_c}+1)}$ represents the DFT matrix, the n_{l_c}-th column of $\mathbf{F}_{l_c}$ is $[1, e^{j2\pi f_{l_c}^{n_l}}, \cdots, e^{j2\pi f_{l_c}^{n_l}(K-1)}]^T, f_{l_c}^{n_l} = f_{l_c,1} + (n_{l_c} - 1)2\pi/K, n_{l_c} = 1, \ldots, N_{l_c} + 1, N_{l_c} = \lfloor \frac{f_{l_c,2} - f_{l_c,1}}{2\pi/K} \rfloor$.

The average energy of all selected sub-bands is

$$\bar{e}_c = \sum_{l=1}^{L_c} e_{l_c}/L_c.$$ (19)

Then, the double threshold is set as $\bar{e}_c + \gamma$ and $\bar{e}_c - \gamma$, where γ is a small positive constant. If $e_{l_c} > \bar{e}_c + \gamma$, the subband is detected as a passband. Similarly, if $e_{l_c} < \bar{e}_c - \gamma$, the subband is detected as a stopband. As a result, the positions and amplitudes of subbands used to embed information are detected. The information demodulation procedure is shown in Figure 3.

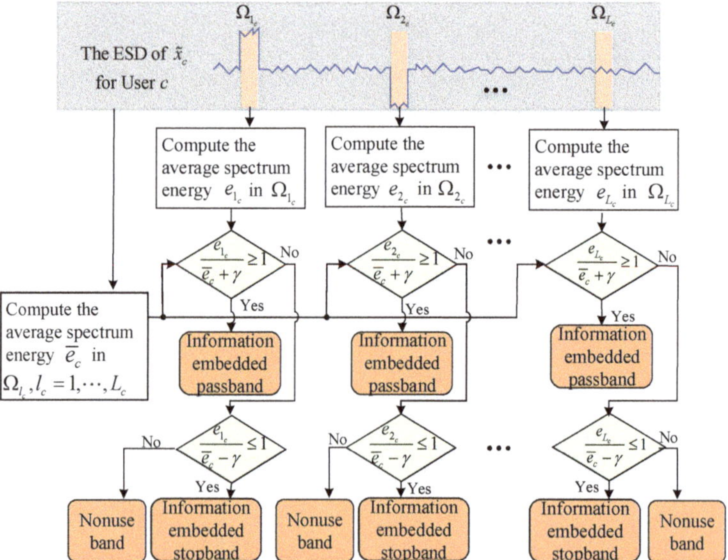

Figure 3. Information demodulation procedure.

2.2. Problem Formulation

In this section, we formalize the dual-function signal design optimization problem via the maximization of the SINR with communication modulation, mainlobe level, sidelobe level and variable modulus constraints.

2.2.1. Waveform Design Metric

Since a higher SINR provides a better radar detection performance, the DFRC signal design is pursued to maximize SINR in this paper. According to Equations (5) and (7), SINR at the output of the receiver is written as

$$\text{SINR} = \frac{\delta_0^2 \mathbf{a}_t^H(\theta_0)\mathbf{R}\mathbf{a}_t(\theta_0)}{\sum\limits_{p=1}^{P} \delta_p^2 \mathbf{a}_t^H(\varphi_p)\mathbf{R}\mathbf{a}_t(\varphi_p) + \delta^2},$$ (20)

where $\delta^2 = MN_t\delta_v^2$ is the noise power. It can clearly be seen that the calculation of SINR needs the priori information of target location, target echo power, clutter location, clutter power, and noise power. These parameters can be acquired by exploiting a cognitive paradigm [39,40].

2.2.2. Waveform Constraints

(1) Communication Modulation Constraint: Assume that Q_c subbands are selected to transmit information to user c. Specifically, let $\Omega_{q_c} = (f_{q_c,1}, f_{q_c,2}) \in \mathcal{L}_c, q_c = 1, \cdots, Q_c$, the subband energy for the user c can be expressed as

$$E_c = \sum_{q_c=1}^{Q_c} \alpha_{q_c} \mathbf{x}_c^H \mathbf{R}_{q_c} \mathbf{x}_c = \sum_{q_c=1}^{Q_c} \alpha_{q_c} \mathbf{a}^T(\theta_c) \mathbf{S}^H \mathbf{R}_{q_c} \mathbf{Sa}^*(\theta_c), \tag{21}$$

therein, the (m, n)-th element of $\mathbf{R}_{q_c}$ is [40–43]

$$\mathbf{R}_{q_c}(m, n) = \begin{cases} \dfrac{e^{j2\pi f_{q_c,2}(m-n)} - e^{j2\pi f_{q_c,1}(m-n)}}{j2\pi(m-n)}, & m \neq n \\ (f_{q_c,2} - f_{q_c,1}), & m = n \end{cases}, \tag{22}$$

where $\alpha_{q_c} \in \{0, 1\}$ is the weighted factor of the q_c-th subband.

As $\text{tr}[\mathbf{AXBX}^T\mathbf{C}] = \text{vect}(\mathbf{X})^T(\mathbf{B}^T \otimes \mathbf{CA})\text{vect}(\mathbf{X})$ [44], we can obtain

$$E_c = \sum_{q_c=1}^{Q_c} \alpha_{q_c} \mathbf{s}^H [\mathbf{R}_{q_c}^H \otimes (\mathbf{a}(\theta_c)\mathbf{a}^H(\theta_c))]\mathbf{s}, \tag{23}$$

where $\mathbf{s} = \text{vec}\{[\mathbf{s}_1, \mathbf{s}_2, \cdots, \mathbf{s}_{N_t}]^T\} \in \mathbb{C}^{N_t M}$.

Further, the total spectral stopband energy of C users is

$$E_s = \sum_{c=1}^{C} E_c = \mathbf{s}^H \mathbf{R}_s \mathbf{s}, \tag{24}$$

where the value of α_{q_c} obeys the following rule if the q_c-th subband is stopband, $\alpha_{q_c} = 1$; otherwise, $\alpha_{q_c} = 0$ and

$$\mathbf{R}_s = \sum_{c=1}^{C} \sum_{q_c=1}^{Q_c} \alpha_{q_c} [\mathbf{R}_{q_c}^H \otimes (\mathbf{a}(\theta_c)\mathbf{a}^H(\theta_c))]. \tag{25}$$

To communicate effectively, the stopband constraint is enforced, as given by [12]

$$\mathbf{s}^H \mathbf{R}_s \mathbf{s} \leq \eta_s, \tag{26}$$

where η_s is the upper bound of energy.

Similarly, the spectral passband energy for the user c is

$$E_{pc} = \mathbf{s}^H \mathbf{R}_c \mathbf{s}, \tag{27}$$

where the value of α_{q_c} obeys the following rule if the q_c-th subband is passband, $\alpha_{q_c} = 1$; otherwise, $\alpha_{q_c} = 0$ and

$$\mathbf{R}_c = \sum_{q_c=1}^{Q_c} \alpha_{q_c} [\mathbf{R}_{q_c}^H \otimes (\mathbf{a}(\theta_c)\mathbf{a}^H(\theta_c))]. \tag{28}$$

For reliable communication, the spectral passband energy constraint for each user is enforced, as given by [12]

$$\eta_{pc} \leq \mathbf{s}^H \mathbf{R}_c \mathbf{s}, c = 1, \cdots, C, \tag{29}$$

where η_{pc} is the lower bound of transmitting energy for the user c.

(2) Integrated Sidelobe Level (ISL) Constraint: To reduce the echo intensity in non-target directions, the beampattern ISL constraint is enforced. $\vartheta_k, k = 1, \cdots, \tilde{K}$ and $\phi_k, k = 1, \cdots, \tilde{K}$ denote the mainlobe region and the sidelobe region, respectively. Therefore, the beampattern ISL is formulated as [12]

$$\frac{\mathbf{s}^H \mathbf{A}_s \mathbf{s}}{\mathbf{s}^H \mathbf{A}_m \mathbf{s}} \leq \varepsilon, \tag{30}$$

where $\mathbf{A}_s = \frac{1}{M} \sum_{k=1}^{\tilde{K}} \mathbf{I}_M \otimes [\mathbf{a}(\phi_k)\mathbf{a}^H(\phi_k)]$, $\mathbf{A}_m = \frac{1}{M} \sum_{k=1}^{\tilde{K}} \mathbf{I}_M \otimes [\mathbf{a}(\vartheta_k)\mathbf{a}^H(\vartheta_k)]$, and ε is the upper bound.

(3) Mainlobe Width Constraint: We can enforce the mainlobe width constraint to concentrate energy in a specific area of space, as [12]

$$P_L - \delta \leq \frac{\mathbf{s}^H \mathbf{A}_k \mathbf{s}}{\mathbf{s}^H \mathbf{A}_0 \mathbf{s}} \leq P_L + \delta, k = 1, 2, \tag{31}$$

where $\mathbf{A}_k = \left(\mathbf{I}_M \otimes [\mathbf{a}(\theta_{mk})\mathbf{a}^H(\theta_{mk})] \right) / M$. $\Phi(\theta_{m2} \geq \theta_{m0} \geq \theta_{m1})$ denotes the main-beam width, $P_L \in (0, 1)$, and δ is a small positive value.

(4) Variable Modulus Constraint: To mitigate the effects of signal nonlinear distortion, the variable modulus constraint [45,46] is enforced as follows

$$\frac{1}{N_t} - \kappa \leq |\mathbf{s}_n(m)|^2 \leq \frac{1}{N_t} + \kappa, \tag{32}$$

where $\frac{1}{N_t} - \kappa$ and $\frac{1}{N_t} + \kappa$ are the upper and lower bounds, respectively, and κ is a small positive constant. The above constraint degrades into a constant-envelop constraint when $\kappa = 0$.

2.2.3. Waveform Design Problem

To ensure a good detection performance while enabling multi-user communication, we constructed the following optimization problem, which maximizes SINR with several practical constraints,

$$\mathcal{P}_1 \begin{cases} \max_{s} & \frac{\mathbf{s}^H \mathbf{X}_0 \mathbf{s}}{\mathbf{s}^H \mathbf{Y}_0 \mathbf{s}} \\ \text{s.t.} & ① \frac{\mathbf{s}^H \mathbf{A}_s \mathbf{s}}{\mathbf{s}^H \mathbf{A}_m \mathbf{s}} \leq \varepsilon, \\ & ② \mathbf{s}^H \mathbf{R}_s \mathbf{s} \leq \eta_s, \\ & ③ \eta_{pc} \leq \mathbf{s}^H \mathbf{R}_c \mathbf{s}, c = 1, \cdots, C, \\ & ④ P_L - \delta \leq \frac{\mathbf{s}^H \mathbf{A}_k \mathbf{s}}{\mathbf{s}^H \mathbf{A}_0 \mathbf{s}} \leq P_L + \delta, k = 1, 2, \\ & ⑤ \frac{1}{N_t} - \kappa \leq |\mathbf{s}_n(m)|^2 \leq \frac{1}{N_t} + \kappa. \end{cases} \tag{33}$$

Therein, $\mathbf{A}_0 = \left(\mathbf{I}_M \otimes [\mathbf{a}(\theta_0)\mathbf{a}^H(\theta_0)] \right) / M$, $\mathbf{A}_p = \left(\mathbf{I}_M \otimes [\mathbf{a}(\varphi_p)\mathbf{a}^H(\varphi_p)] \right) / M$, $\mathbf{X}_0 = \delta_0^2 \mathbf{A}_0$, $\mathbf{Y}_0 = \sum_{p=1}^{P} \delta_p^2 \mathbf{A}_p + \mathbf{I}_{MN_t} \delta_v^2 N_t$.

Since both the objective function and the feasible domain are non-convex, $\mathcal{P}_1$ is a non-convex NP-hard problem. Next, we introduce an iterative algorithm.

2.3. Optimization Technique via IBE-DSADMM for Solving $\mathcal{P}_1$

To solve the NP-hard optimization problem $\mathcal{P}_1$, firstly, we split the original non-convex NP-hard problem into the smaller subproblems using via the Iterative Block Enhancement (IBE) framework. Then, the DSADMM algorithm, via jointing Dinkelbach's Iterative Procedure (DIP), Sequential Convex Approximation (SCA) and Alternating Direction Method Of Multipliers (ADMM), is used to solve the subproblems. Finally, DSADMM is incorporated into the IBE framework to form the IBE-DSADMM algorithm to monotonically increase SINR. The convergence and computational complexity are analyzed.

2.3.1. IBE-DSADMM Algorithm

This subsection presents the IBE-DSADMM algorithm to solve $\mathcal{P}_1$. Firstly, the IBM framework orderly optimizes $f(\mathbf{s}_1, \cdots, \mathbf{s}_{N_t})$ over one block variable (e.g., $\mathbf{s}_n$) in $(\mathbf{s}_1, \cdots, \mathbf{s}_{N_t})$, keeping the others fixed. Let $\mathbf{s}_n^{(i)}$ be the i-th optimized signal transmitted via n-th antenna, $n = 1, \cdots, N_t$. Therefore, in the i-th iteration, the following non-convex subproblems need to be solved:

$$\mathcal{P}_{\mathbf{s}_n^{(i)}} \begin{cases} \max_{\mathbf{s}_n} & f(\mathbf{s}_1^{(i)}, \cdots, \mathbf{s}_{n-1}^{(i)}, \mathbf{s}_n, \mathbf{s}_{n+1}^{(i-1)}, \cdots, \mathbf{s}_N^{(i-1)}) \\ \text{s.t.} & \mathbf{s}_n \in \mathcal{S}_n^{(i)}, \end{cases} \tag{34}$$

where $\mathcal{S}_n^{(i)}$ is the feasible set of $\mathbf{s}_n$ at the i-th iteration.

Then, the DSADMM algorithm is utilized to handle Problem $\mathbf{s}_n^{(i)}$. Before proceeding, Problem (34) is equivalently transformed into

$$\mathcal{P}_{\mathbf{s}_n^{(i)}} \begin{cases} \max_{\mathbf{s}_n} & f(\mathbf{s}_n; \bar{\mathbf{s}}_{-n}^{(i)}) \\ \text{s.t.} & ① \mathbf{s}_n^H \mathbf{A}_{sn} \mathbf{s}_n + \Re\{\mathbf{a}_{sn}^H \mathbf{s}_n\} + a_{sn} \leq \varepsilon(\mathbf{s}_n^H \mathbf{A}_{mn} \mathbf{s}_n + \Re\{\mathbf{a}_{mn}^H \mathbf{s}_n\} + a_{mn}) \\ & ② \mathbf{s}_n^H \mathbf{R}_{sn} \mathbf{s}_n + \Re\{\mathbf{s}_{sn}^H \mathbf{s}_n\} + r_{sn} \leq 0, \\ & ③ \eta_{pc} \leq \mathbf{s}_n^H \mathbf{R}_{cn} \mathbf{s}_n + \Re\{\mathbf{r}_{cn}^H \mathbf{s}_n\} + r_{cn}, c = 1, \cdots, C, \\ & ④ (P_L - \delta)(\mathbf{s}_n^H \mathbf{A}_{0n} \mathbf{s}_n + \Re\{\mathbf{a}_{0n}^H \mathbf{s}_n\} + a_{0n}) - (\mathbf{s}_n^H \mathbf{A}_{kn} \mathbf{s}_n + \Re\{\mathbf{a}_{kn}^H \mathbf{s}_n\} + a_{kn}) \leq 0, k = 1, 2, \\ & ⑤ (\mathbf{s}_n^H \mathbf{A}_{kn} \mathbf{s}_n + \Re\{\mathbf{a}_{kn}^H \mathbf{s}_n\} + a_{kn}) - (P_L + \delta) \times (\mathbf{s}_n^H \mathbf{A}_{0n} \mathbf{s}_n + \Re\{\mathbf{a}_{0n}^H \mathbf{s}_n\} + a_{0n}) \leq 0, k = 1, 2, \\ & ⑥ |\mathbf{s}_n(m)|^2 \leq \frac{1}{N_t} + \kappa, \\ & ⑦ \frac{1}{N_t} - \kappa \leq |\mathbf{s}_n(m)|^2, \end{cases}$$

where $f(\mathbf{s}_n; \bar{\mathbf{s}}_{-n}^{(i)}) = \frac{\mathbf{s}_n^H \mathbf{X}_n \mathbf{s}_n + \Re\{\mathbf{x}_n^H \mathbf{s}_n\} + x_n}{\mathbf{s}_n^H \mathbf{Y}_n \mathbf{s}_n + \Re\{\mathbf{y}_n^H \mathbf{s}_n\} + y_n}$.

Proof. See Appendix A for the derivation. $\square$

Nevertheless, $\mathcal{P}_{\mathbf{s}_n^{(i)}}$ is still a non-convex problem. A parameter w is introduced to transform the objective function $f(\mathbf{s}_n; \bar{\mathbf{s}}_{-n}^{(i)})$ into the following representation,

$$\chi(y, \mathbf{s}_n) = f_0(\mathbf{s}_n) - w f_1(\mathbf{s}_n),$$

where $f_0(\mathbf{s}_n) = \mathbf{s}_n^H \mathbf{X}_n \mathbf{s}_n + \Re\{\mathbf{x}_n^H \mathbf{s}_n\} + x_n$ and $f_1(\mathbf{s}_n) = \mathbf{s}_n^H \mathbf{Y}_n \mathbf{s}_n + \Re\{\mathbf{y}_n^H \mathbf{s}_n\} + y_n$. Therefore, the problem is rewritten as

$$\begin{cases} \max_{w \; \mathbf{s}_n} & \chi(w, \mathbf{s}_n) \\ \text{s.t.} & \{\mathbf{s}_n\} \in \mathcal{S}_n^{(i)}, \end{cases} \tag{35}$$

$w_{(t)}$ and $\mathbf{s}_{n(t)}$ denote the t-th iteration solutions. Invoking DIP method, Problem (35) is equivalent to

- Given $\mathbf{s}_{n(t-1)}$ and $w_{(t)} = f_0(\mathbf{s}_{n(t-1)})/f_1(\mathbf{s}_{n(t-1)})$.
- Given $w_{(t)}$, $\mathbf{s}_{n(t)}$ is updated by solving

$$\begin{cases} \max_{\mathbf{s}_n} & \chi(w_{(t)}, \mathbf{s}_n) \\ \text{s.t.} & \{\mathbf{s}_n\} \in \mathcal{S}_n^{(i)}. \end{cases} \tag{36}$$

- Repeat the above steps until convergence.

Unfortunately, Problem (36) is still non-convex. Leveraging the SCA algorithm, we can approximately recast this to a convex problem

$$
\mathcal{P}_{\mathbf{s}_{n(t)}}
\begin{cases}
\max\limits_{\mathbf{s}_n} & \mathbf{s}_n^H \mathbf{D}_n \mathbf{s}_n + \Re\{\mathbf{d}_n^H \mathbf{s}_n\} + d_n \\[4pt]
\text{s.t.} & \mathbf{s}_n^H \mathbf{B}_n \mathbf{s}_n + \Re\{\mathbf{b}_n^H \mathbf{s}_n\} + b_n \le 0, \\[4pt]
& \mathbf{s}_n^H \mathbf{R}_{sn} \mathbf{s}_n + \Re\{\mathbf{r}_{sn}^H \mathbf{s}_n\} + r_{sn} \le 0, \\[4pt]
& \mathbf{s}_n^H \bar{\mathbf{A}}_{kn} \mathbf{s}_n + \Re\{\bar{\mathbf{a}}_{kn}^H \mathbf{s}_n\} + \bar{a}_{kn} \le 0, k = 1, 2, \\[4pt]
& \mathbf{s}_n^H \tilde{\mathbf{A}}_{kn} \mathbf{s}_n + \Re\{\tilde{\mathbf{a}}_{kn}^H \mathbf{s}_n\} + \tilde{a}_{kn} \le 0, k = 1, 2, \\[4pt]
& \Re\{\bar{\mathbf{r}}_{cn}^H \mathbf{s}_n\} + \bar{r}_{cn} \le 0, c = 1, \ldots, C, \\[4pt]
& \mathbf{s}_n(m)^H \mathbf{s}_n(m) - 1/N_t - \kappa \le 0, m = 1, \cdots, M \\[4pt]
& \Re\{\bar{p}_1 \mathbf{s}_n(m)\} + \bar{p}_2 \le 0, m = 1, \cdots, M
\end{cases}
\tag{37}
$$

Proof. See Appendix B for the derivation. $\square$

Finally, the ADMM algorithm is used to solve $\mathcal{P}_{\mathbf{s}_{n(t)}}$. By introducing auxiliary variables $\{\mathbf{h}_{\bar{c}}\}, \{\mathbf{v}_c\}, \mathbf{z}$, the problem $\mathcal{P}_{\mathbf{s}_{n(t)}}$ is transformed into

$$
\begin{cases}
\min\limits_{\mathbf{h}, \{\mathbf{h}_{\bar{c}}\}, \{\mathbf{v}_c\}, \mathbf{z}} & -\mathbf{h}^H \mathbf{D}_n \mathbf{h} - \Re\{\mathbf{d}_n^H \mathbf{h}\} - d_n \\[4pt]
\text{s.t.} & \mathbf{h}_1 = \mathbf{h}, \mathbf{h}_1^H \mathbf{R}_{sn} \mathbf{h}_1 + \Re\{\mathbf{r}_{sn}^H \mathbf{h}_1\} + r_{sn} \le 0, \\[4pt]
& \mathbf{h}_2 = \mathbf{h}, \mathbf{h}_2^H \mathbf{B}_n \mathbf{h}_2 + \Re\{\mathbf{b}_n^H \mathbf{h}_2\} + b_n \le 0, \\[4pt]
& \mathbf{h}_k = \mathbf{h}, k = 3, 4, \mathbf{h}_k^H \bar{\mathbf{A}}_{k_0 n} \mathbf{h}_k + \Re\{\bar{\mathbf{a}}_{k_0 n}^H \mathbf{h}_k\} + \bar{a}_{k_0 n} \le 0 \\[4pt]
& \mathbf{h}_{\bar{k}} = \mathbf{h}, \bar{k} = 5, 6, \mathbf{h}_{\bar{k}}^H \tilde{\mathbf{A}}_{k_1 n} \mathbf{h}_{\bar{k}} + \Re\{\tilde{\mathbf{a}}_{k_1 n}^H \mathbf{h}_{\bar{k}}\} + \tilde{a}_{k_1 n} \le 0 \\[4pt]
& \mathbf{v}_c = \mathbf{h}, c = 1, \cdots, C, \Re\{\bar{\mathbf{r}}_{cn}^H \mathbf{v}_c\} + \bar{r}_{cn} \le 0, \\[4pt]
& \mathbf{z} = \mathbf{h}, \mathbf{z}(m)^H \mathbf{z}(m) - 1/N_t - \kappa \le 0, m = 1, \cdots, M \\[4pt]
& \Re\{\bar{p}_1 \mathbf{z}(m)\} + \bar{p}_2 \le 0, m = 1, \cdots, M
\end{cases}
\tag{38}
$$

where $k_0 = k - 2, k_1 = \bar{k} - 4$. The augmented Lagrangian of the above problem is constructed as follows [47]:

$$
L_\varrho(\mathbf{h}, \{\mathbf{h}_{\bar{c}}\}, \{\mathbf{v}_c\}, \mathbf{z}, \{\mu_{\bar{c}}\}, \{\mu_c\}, \mu_z) = -\mathbf{h}^H \mathbf{D}_n \mathbf{h} - \Re\{\mathbf{d}_n^H \mathbf{h}\} - d_n + \sum_{\bar{c}=1}^{6} \varrho/2 \|\mathbf{h}_{\bar{c}} - \mathbf{h} + \mu_{\bar{c}}/\varrho\|^2
$$

$$
+ \sum_{c=1}^{C} \varrho/2 \|\mathbf{v}_c - \mathbf{h} + \mu_c/\varrho\|^2 + \varrho/2 \|\mathbf{z} - \mathbf{h} + \mu_z/\varrho\|^2
$$

where $\{\mu_{\bar{c}}\}, \{\mu_c\}, \mu_z$ are dimensional multiplier vectors and $\varrho > 0$ is the penalty parameter.

We minimize $L_\varrho(\mathbf{h}, \{\mathbf{h}_{\bar{c}}\}, \{\mathbf{v}_c\}, \mathbf{z}, \{\mu_{\bar{c}}\}, \{\mu_c\}, \mu_z)$ via the ADMM algorithm. See Appendix C for details of the algorithm.

Finally, the procedure of the IBE-DSADMM algorithm is reported in Algorithm 1.

Algorithm 1: IBE-DSADMM for $\mathcal{P}_1$

Input: Feasible starting point $\mathbf{s}_0$;
Output: An optimized solution $\mathbf{s}^{(*)}$ to $\mathcal{P}_1$;

1: For $i = 0$, initialize $\mathbf{s}^{(i)} = \mathbf{s}_0$ and calculate $f(\mathbf{s}_1^{(i)}, \cdots, \mathbf{s}_N^{(i)})$;
2: $i := i + 1, n = 0$;
3: $n := n + 1$;
4: $t = 0, \mathbf{s}_{n(t)} = \mathbf{s}_n^{(i-1)}$;
5: $t = t + 1$;
6: Calculate $w_{(t)} = f_0(\mathbf{s}_{n(t-1)})/f_1(\mathbf{s}_{n(t-1)})$, $\mathbf{D}_n$ and $\mathbf{d}_n$;
7: Find an optimal solution $\mathbf{s}_{n(t)}^*$ to $\mathcal{P}_{\mathbf{s}_{n(t)}}$ by using ADMM;
8: If $|y_{(t)} - y_{(t-1)}| \le \kappa_1, \mathbf{s}_n^{(i)} = \mathbf{s}_{n(t)}^*$; Otherwise, go to Step 5;
9: If $n < N$, go to Step 3;
10: If $|f(\mathbf{s}_1^{(i-1)}, \cdots, \mathbf{s}_N^{(i-1)}) - f(\mathbf{s}_1^{(i)}, \cdots, \mathbf{s}_N^{(i)})| \le \kappa_2$, output $\mathbf{s}^{(*)} = \mathbf{s}^{(i)}$; Otherwise,
    go to Step 2.

2.3.2. Algorithm Initialization

An initial feasible point $\mathbf{s}_0$ is necessary to start Algorithm 1. Thus, the following problem is introduced to find $\mathbf{s}_0$

$$\mathcal{P}_\mathbf{s}\begin{cases} \text{find} & \mathbf{s} \\ \text{s.t.} & \textcircled{1}\dfrac{\mathbf{s}^H\mathbf{A}_s\mathbf{s}}{\mathbf{s}^H\mathbf{A}_m\mathbf{s}} \leq \varepsilon, \\ & \textcircled{2}\mathbf{s}^H\mathbf{R}_s\mathbf{s} \leq \eta_s, \\ & \textcircled{3}\eta_{pc} \leq \mathbf{s}^H\mathbf{R}_c\mathbf{s}, c = 1,\cdots,C, \\ & \textcircled{4}P_L - \delta \leq \dfrac{\mathbf{s}^H\mathbf{A}_k\mathbf{s}}{\mathbf{s}^H\mathbf{A}_0\mathbf{s}} \leq P_L + \delta, k = 1,2, \\ & \textcircled{5}\frac{1}{N_t} - \kappa \leq |\mathbf{s}(n)|^2 \leq \frac{1}{N_t} + \kappa,, n = 1,\cdots,MN_t. \end{cases}\tag{39}$$

Similarly, the non-convex constraints in Problem (39) are replaced with their first-order conditions. Meanwhile, some slack variables are introduced to ensure the feasibility [48]. Specifically, $\textcircled{1}$ is a convex constraint if $\mathbf{A}_s \succeq \varepsilon\mathbf{A}_m$. Otherwise, it is approximated as $\mathbf{s}^H\mathbf{A}_s\mathbf{s} - \varepsilon\left(\bar{\mathbf{s}}^H\mathbf{A}_m\bar{\mathbf{s}} + 2\Re\{\bar{\mathbf{s}}^H\mathbf{A}_m(\mathbf{s} - \bar{\mathbf{s}})\}\right) \leq 0$, where $\bar{\mathbf{s}}$ is the previous iteration solution. Follow the same method for the other constraints. Thus, Problem (39) is approximated as a convex problem as follows

$$\begin{aligned} \min_{\mathbf{s},\bar{b},\{b_c\},\{b_{\tilde{c}}\}} \quad & \bar{\rho}\Big[\sum_{c=1}^{C} b_c + \sum_{\tilde{c}=1}^{5} b_{\tilde{c}} + \bar{b}\Big] \\ \text{s.t.} \quad & \mathbf{s}^H\mathbf{R}_s\mathbf{s} \leq \eta_s, \\ & \Re\{\bar{\mathbf{r}}_c^H\mathbf{s}\} + \bar{r}_c - b_c \leq 0, b_c \geq 0, c = 1,\cdots,C, \\ & \mathbf{s}^H\tilde{\mathbf{A}}_s\mathbf{s} + \Re\{\tilde{\mathbf{a}}_s^H\mathbf{s}\} + \tilde{a}_s - b_{\tilde{c}} \leq 0, b_{\tilde{c}} \geq 0, \tilde{c} = 1, \\ & \mathbf{s}^H\tilde{\mathbf{A}}_k\mathbf{s} + \Re\{\tilde{\mathbf{a}}_k^H\mathbf{s}\} + \tilde{a}_k - b_{\tilde{c}} \leq 0, b_{\tilde{c}} \geq 0, \tilde{c} = 2,3, \\ & \mathbf{s}^H\tilde{\mathbf{A}}_k\mathbf{s} + \Re\{\tilde{\mathbf{a}}_k^H\mathbf{s}\} + \tilde{a}_k - b_{\tilde{c}} \leq 0, b_{\tilde{c}} \geq 0, \tilde{c} = 4,5, \\ & \mathbf{s}(n)^H\mathbf{s}(n) - 1/N_t - \kappa \leq 0, n = 1,\cdots,MN_t \\ & \Re\{\bar{q}_1\mathbf{s}(n)\} + \bar{q}_2 - \bar{b} \leq 0, \bar{b} \geq 0, \end{aligned}\tag{40}$$

where $\bar{b}, \{b_c\}, \{b_{\tilde{c}}\}$ are slack variables. $\bar{\rho}$ is a large enough positive number to penalize slack variables approaching zero. $\bar{\mathbf{r}}_c = -2\mathbf{R}_c\bar{\mathbf{s}}, \bar{r}_c = \eta_{pc} + \bar{\mathbf{s}}\mathbf{R}_c\bar{\mathbf{s}}, \bar{q}_1 = -2\mathbf{s}(n), \bar{q}_2 = 1/N_t - \kappa + \bar{\mathbf{s}}(n)^H\bar{\mathbf{s}}(n)$. If constraints $\textcircled{1}$ and $\textcircled{4}$ are convex, $\tilde{\mathbf{A}}_s = \mathbf{A}_s - \varepsilon\mathbf{A}_m, \tilde{\mathbf{A}}_k = (P_L - \delta)\mathbf{A}_0 - \mathbf{A}_k, \tilde{\mathbf{A}}_k = \mathbf{A}_k - (P_L + \delta)\mathbf{A}_0, \tilde{\mathbf{a}}_s = 0, \tilde{\mathbf{a}}_k = 0, \tilde{\mathbf{a}}_k = 0, \tilde{a}_s = 0, \tilde{a}_k = 0, \tilde{a}_k = 0$. Otherwise, $\tilde{\mathbf{A}}_s = \mathbf{A}_s, \tilde{\mathbf{A}}_k = (P_L - \delta)\mathbf{A}_0, \tilde{\mathbf{A}}_k = \mathbf{A}_k, \tilde{\mathbf{a}}_s = -2\varepsilon\mathbf{A}_m\bar{\mathbf{s}}, \tilde{\mathbf{a}}_k = -2\mathbf{A}_k\bar{\mathbf{s}}, \tilde{\mathbf{a}}_k = -2(P_L + \delta)\mathbf{A}_0\bar{\mathbf{s}}, \tilde{a}_s = \varepsilon(\bar{\mathbf{s}}^H\mathbf{A}_m\bar{\mathbf{s}}), \tilde{a}_k = \bar{\mathbf{s}}^H\mathbf{A}_k\bar{\mathbf{s}}, \tilde{a}_k = (P_L + \delta)(\bar{\mathbf{s}}^H\mathbf{A}_0\bar{\mathbf{s}})$. We solve the above convex problem using the CVX toolbox.

2.3.3. Computational Complexity

The computational complexity of IBE-DSADMM mainly depends on two factors. One is the iteration of variables $x_n, y_n, a_{sn}, a_{mn}, r_{sn}, r_{cn}, a_{kn}, a_{0n}$, with a computational complexity of $O((N_tM)^2)$. Another is the complexity of ADMM for solving $\mathcal{P}_{\mathbf{s}_{n(t)}}$ (e.t. $O(L_AT_DM^2)$). Therein, L_A and T_D are the iteration times of ADMM and DIP algorithms, respectively. Thus, the total computational complexity is $O(I_BN_t^3M^2) + O(I_BL_AT_DN_tM^2) + O(N_tM^3)$ for IBE-DSADMM, where I_B is the iteration time of the whole algorithm. $O(N_tM^3)$ is the cost of the matrix inversion of $\mathbf{D}_n$.

2.3.4. Convergence Analysis

We analyze the convergence of the IBE-DSADMM algorithm in this subsection.

(1) The ADMM algorithm decomposes $\mathcal{P}_{\mathbf{s}_{n(t)}}$ into multiple convex subproblems with closed-form solutions.

(2) The DSADMM algorithm ensures that the $w_{(t)}$ sequence is monotonically increasing to convergence [12].

(3) According to the IBE framework, it follows that

$$
\begin{aligned}
f(\mathbf{s}_1^{(i-1)}, \cdots, \mathbf{s}_N^{(i-1)}) &\leq f(\mathbf{s}_1^{(i)}, \mathbf{s}_2^{(i-1)}, \cdots, \mathbf{s}_N^{(i-1)}) \leq \cdots \leq \\
f(\mathbf{s}_1^{(i)}, \cdots, \mathbf{s}_{N-1}^{(i)}, \mathbf{s}_N^{(i-1)}) &\leq f(\mathbf{s}_1^{(i)}, \cdots, \mathbf{s}_{N-1}^{(i)}, \mathbf{s}_N^{(i)}),
\end{aligned}
\tag{41}
$$

which implies that the objective function value increases monotonically with iterations. In addition, the upper bound of $f(\mathbf{s}_1, \cdots, \mathbf{s}_N)$ is the maximum eigenvalue of matrix $\mathbf{Y}_0^{-1}\mathbf{X}_0$. Thus, the objective function increases monotonically to convergence.

3. Results

This section evaluates the performance of the proposed SPIA modulation method in terms of detection performance and communication performance. To highlight the superiority of the IBE-DSADMM algorithm, the IBE-DSIPM algorithm [49] is introduced as a benchmarker, which uses the Interior Point Method (IPM) algorithm instead of the ADMM algorithm to solve optimization problems.

Unless otherwise stated, Table 1 shows the parameter settings.

Table 1. The setting of simulational parameters.

Variable Name	Variable Setting
Transmitting array	Uniform linear array
Antenna spacing	half-wave
Antenna number	$N_t = 8$
Signal sample number	$M = 32$
Parameters of mainlobe width constraints	$\theta_0 = 15°, \theta_1 = 5°, \theta_2 = 25°, P_L = 0.5$ and $\delta = 0.05$
Sidelobe region	$[-90°, 5°] \cup [25°, 90°]$
Upper limit of antenna pattern ISL	$\varepsilon = 1.5$
Upper limit of stopband energy	$\eta_s = 5 \times 10^{-5} \times n_s$
Number of stopband	n_s
Exit condition value of IBE	$\kappa_2 = 10^{-5}$
Exit condition value of DIP	$\kappa_1 = 10^{-2}$
The location of communication user	$-70°$
The location of target	$\theta_0 = 15°$
The power of target	$\delta_0^2 = 20$ dB
The location of two reference sources	$30°$ and $60°$
The power of two reference sources	25 dB, 30 dB
Normalized available frequency subbands	$\Omega_{1_1} = (0.1, 0.13)$, $\Omega_{2_1} = (0.2, 0.23)$, $\Omega_{3_1} = (0.3, 0.33)$, $\Omega_{4_1} = (0.4, 0.43)$, $\Omega_{5_1} = (0.5, 0.53)$, $\Omega_{6_1} = (0.6, 0.63)$, $\Omega_{7_1} = (0.7, 0.73)$, $\Omega_{8_1} = (0.8, 0.83)$, $\Omega_{9_1} = (0.9, 0.93)$

3.1. Beampattern Performance

We aim to analyze the beampattern performance of the IBE-DSADMM algorithm considering different η_{p1} and κ for communication passband modulation and variable modulus constraints in this subsection. The influence of information embedding subband number Q_1 is also considered.

$Q_1 = 4$ subbands were selected for information embedding in Figure 4. In moredetail, the $Q_1 = 4$ subbands were set as $\Omega_{5_1} = (0.5, 0.53)$, $\Omega_{6_1} = (0.6, 0.63)$, $\Omega_{7_1} = (0.7, 0.73)$, $\Omega_{9_1} = (0.9, 0.93)$, respectively, and the bit sequence "0101" was transmitted by amplitude modulation. Figure 4a shows SINRs (in dB) versus iteration number for $\eta_{p1} = 2$, $\kappa = 0.0001, 0.05$. It can be observed that the SINRs (in dB) increase monotonously along with the iteration for both IBE-DSADMM and IBE-DSIPM algorithms. Moreover, the two algorithms share a converged SINR value. This is reasonable because the ADMM algorithm can obtain the optimal solution to the convex problem $\mathcal{P}_{\mathbf{s}_{n(t)}}$. The corresponding SINRs (in dB) versus iteration time are also depicted in Figure 4b. It is worth mentioning that the IBE-DSADMM algorithm converges faster than the IBE-DSIPM algorithm. Finally, it is obvious that the larger κ, the larger the obtained converged SINR value owning to the enlarged feasible set on $\mathcal{P}_1$. Figure 4c reports SINRs' (in dB) value against iteration number for $\eta_{p1} = 2, 5, \kappa = 0.05$. The results again exhibit that IBE-DSADMM and IBE-DSIPM obtain near-converged SINR values. Figure 4d also depicts the corresponding SINRs value versus CPU time. The results again demonstrate the advantage of the IBE-DSADMM algorithm in terms of convergence speed. Finally, the curves also show that the smaller the η_{p1}, the higher the converged SINR value will be owing to there being more DoFs on $\mathcal{P}_1$.

In Figure 4e, the beampatterns vesus angle are depicted for $\eta_{p1} = 2, \kappa = 0.0001, 0.05$. The results show all optimized beampatterns have a high sidelobe level around the communication direction $-70°$. The reason for this is that more energy is required to transmit to communication users, which is located in the sidelobe domain. Furthermore, the optimized SINR guarantees the maximum radiation power of the target direction (i.e., $\theta_0 = 15°$) and the minimum radiation energy of clutter directions (i.e., $-30°$ and $60°$), which enhances the target detection and clutter suppression capability. The results in Figure 4e are consistent with the previous Figure 4a,b. Specifically, the larger the κ, the better the beampattern performance will be. Figure 4f illustrates the normalized beampatterns obtained for $\eta_{p1} = 2, 5, \kappa = 0.05$. it is clearly observed that beampattern nullings are formed in clutter directions (i.e., $-30°$ and $60°$). In addition, these curves highlight again that IBE-DSADMM and IBE-DSIPM have an almost identical beampattern and the smaller the η_{p1} value, the better the obtained beampattern performance.

The impact of information embedding subbands number Q_1 is analyzed in Figure 5. The parameters are set as $\eta_{p1} = 2, \kappa = 0.05$ with $Q_1 = 2, 3, 4$. We chose 250 information embedding subband position selection patterns. Figure 5a shows the specific converged SINR values of the aforementioned 250 trails. It is clearly seen that the converged SINR values fluctuate within a very small range. The normalized beampatterns versus angle with the maximum converged SINR values are depicted in Figure 5b. The results again exhibit that different Q_1 share closely converged SINR values, which is consistent with Figure 5a.

3.2. Communication Performance

The communication performance of the proposed SPIA modulation method is discussed in this subsection. Figure 6a illustrates Symbol Error Ratio (SER) versus Power Noise Ratio (PNR) with $\eta_{p1} = 2, 5, \kappa = 0.05, 0.0001, Q_1 = 2, 4$. According to the based band echo $\tilde{\mathbf{x}}_c$ in (17), the PNR received by the c-th user is defined as $|\beta_c|^2 / \delta_c^2$. The curves show that the larger the η_{p1}, the smaller the SER will be due to more energy being transmitted to communication users. In addition, a smaller κ leads to a higher sidelobe level in communication direction, which also results in a lower SER. Nevertheless, an increase in η_{p1} and a decrease in κ raise the sidelobe level, which may degrade the detection performance. The results indicate that careful selection of η_{p1} and κ is needed to balance radar detection with SER performance. Figure 6a also reveals that a smaller Q_1 leads to a better SER performance as preciser demodulation is performed. Note that although Figure 5 shows that a different Q_1 almost cannot affect SINR performance, a trade-off between data rate and SER performance also exists. Figure 6b–d shows the optimized frequency band energy (in dB) towards communication users with selected $Q = 2, 3, 4$ subbands for information embedding, respectively. The curves highlight that the passbands and stopbands are formed in the accurate position corresponding to the delivered bit sequence.

3.3. Performance Analysis for Different Application Scenarios

This subsection evaluates the performance of the proposed SPIA modulation method for different application scenarios. Specifically, we consider different numbers of communication users and interfering sources. The signal-dependent interfering sources are assumed to be in the same range bin with the target. Figure 7a,b show optimized beampatterns with two users and three interfering sources and one user and four interfering sources, respectively. It can be seen that the energy could be focused near the target direction and beampattern nullings are formed in clutter directions. In addition, a higher sidelobe level is achieved with increasing η_{p1}.

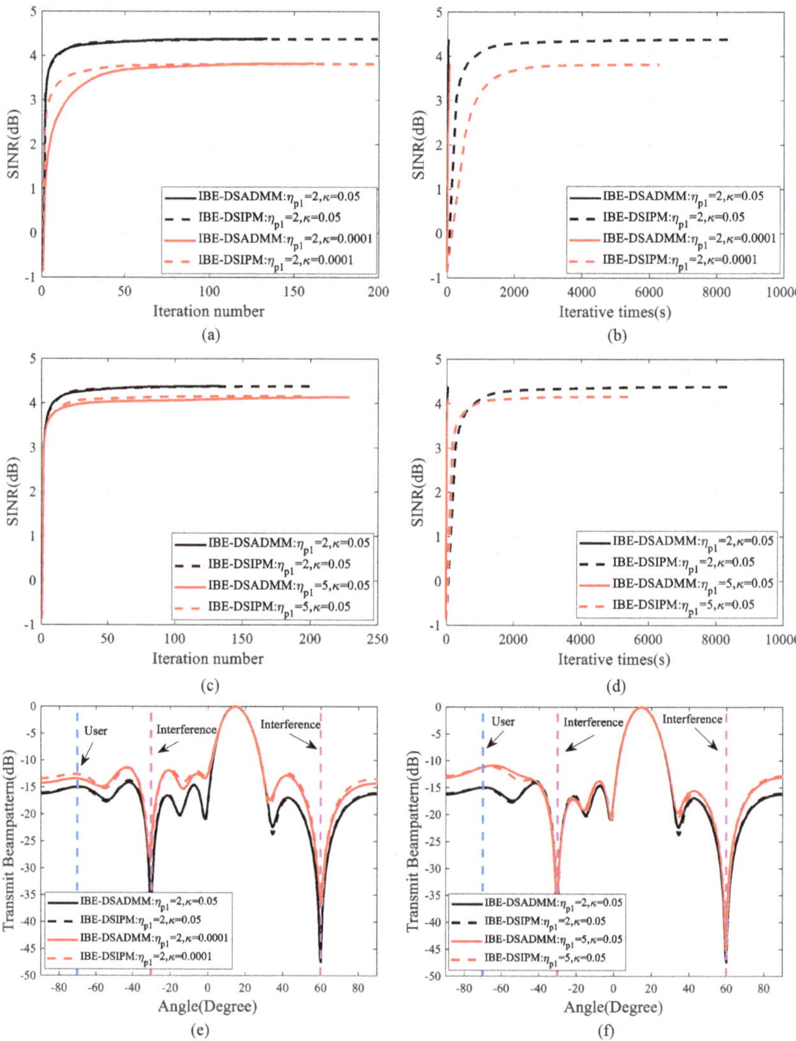

Figure 4. Beampattern performance under different parameters (**a**) SINR versus iteration number for $\eta_{p1} = 2, \kappa = 0.0001, 0.05$, (**b**) SINR versus iteration time (in seconds) $\eta_{p1} = 2, \kappa = 0.0001, 0.05$, (**c**) SINR versus iteration number for $\eta_{p1} = 2, 5, \kappa = 0.05$, (**d**) SINR versus iteration time (in seconds) $\eta_{p1} = 2, 5, \kappa = 0.05$, (**e**) beampattern versus angle for $\eta_{p1} = 2, \kappa = 0.0001, 0.05$, (**f**) beampattern versus angle for $\eta_{p1} = 2, 5, \kappa = 0.05$.

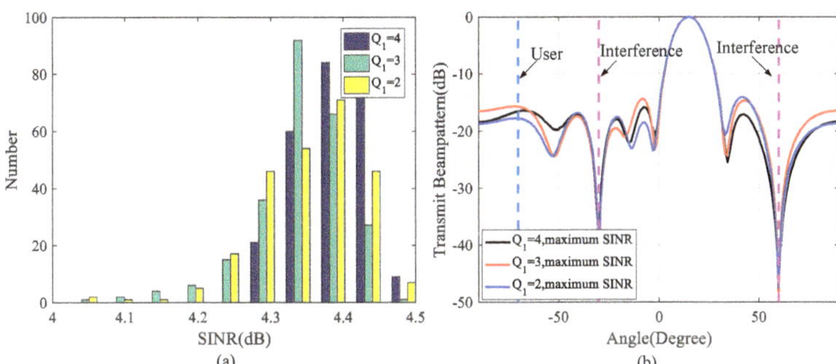

Figure 5. Beampattern performance for $Q_1 = 2, 3, 4$ (**a**) the histogram of converged SINR values, (**b**) beampattern versus angle with the maximum converged SINR values.

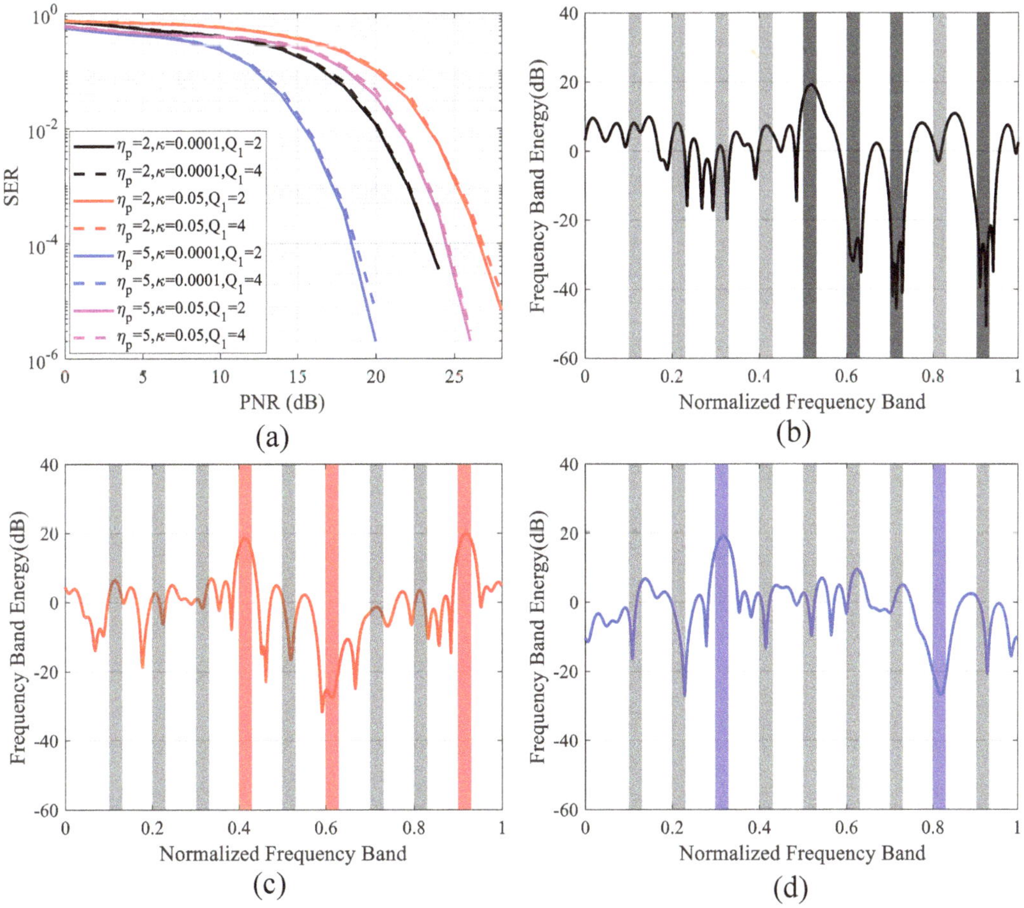

Figure 6. Communication performance (**a**) SER versus PNR for $\eta_p = 2, 5, \kappa = 0.0001, 0.05, Q_1 = 2, 4$, (**b**) optimized frequency band energy for communication user assuming $Q_1 = 4$, (**c**) optimized frequency band energy for communication user assuming $Q_1 = 3$, (**d**) optimized frequency band energy for communication user assuming $Q_1 = 2$.

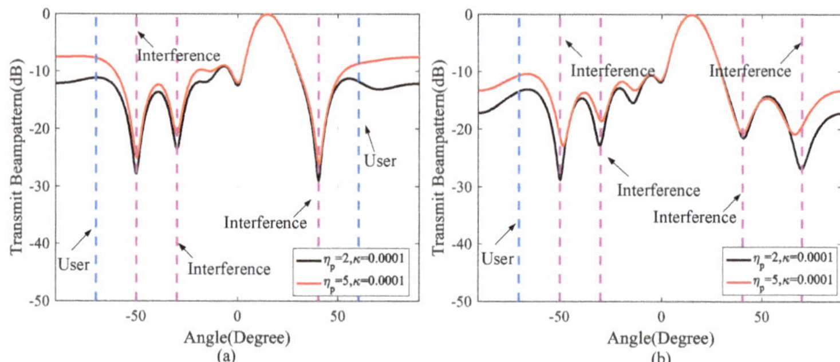

Figure 7. Beampattern performance for different application scenarios (**a**) beampattern versus angle (two users and three interfering sources), (**b**) beampattern versus angle (one user FL and four interfering sources).

3.4. Comparison with Related Information Embedding Methods

In this section, we compare the proposed SPIA modulation method with the sidelobe modulation [9] and our previous work, named spatio-spectral modulation [12].

In [9], the focus is on the design of joint transmit weight vectors and the utilization of signal diversity to introduce variations in the sidelobe levels (SLLs) for communication. The number of orthogonal signals used is equal to the number of bits being transmitted. By controlling the sidelobes in the communication directions, two distinct levels can be achieved, requiring the use of two transmit beamforming weight vectors. The communication SLLs are enforced to be either 0.1 (-20 dB) or 0.01 (-40 dB). Parameters such as array type, antenna spacing, antenna number, communication user direction, mainlobe region, and sidelobe region are the same as specified in Table 1. Figure 8a,b illustrate the beampatterns achieved through sidelobe modulation [9] and the proposed SPIA modulation method, respectively. The number of information embedding subbands in the SPIA modulation method, denoted as Q1, is set to 4. The results indicate that the beampattern performance of the proposed SPIA modulation method is superior to the sidelobe modulation method. The former method achieves its beampattern by matching it with an ideal mask pattern, while the latter maximizes the SINR. It is important to note that the proposed SPIA modulation method creates nullings in the direction of interference sources, which aids in clutter suppression. However, the former method does not take into account the presence of clutter. Additionally, the former method requires designing a specific orthogonal signal set size to meet the data rate requirements.

To highlight the superiority of the proposed SPIA modulation method, the spatio-spectral modulation [12] is taken as a benchmarker. The parameter settings are shown in Table 2. Figure 9a depicts the optimized beampatterns obtained by the proposed SPIA modulation method and spatio-spectral modulation [12], which delineates that the proposed SPIA modulation method can form the beampattern nullings in clutter directions. Figure 9b presents the number of bits transmitted in a PRT versus the number of selected information embedding subbands. Since the spatio-spectral modulation in [12] uses the energy of subbands to embed information. Thereby, the data rate is fixed in 20 bits per PRT when the available frequency subband number is 20. According to Equation(13), the data rate in the proposed SPIA method increases along with the number of selected information embedding subbands. Note that the proposed SPIA method provides a higher data rate than the spatio-spectral modulation in [12] when more than six subbands are selected. Figure 9c shows the optimized frequency band energy (in dB) towards communication user 1 using the proposed SPIA modulation method. Ten information subbands selected to embed information are formed by the corresponding passbands and stopbands. Figure 9d shows the optimized frequency band energy (in dB) towards communication user 1 via spatio-spectral modulation [12]. The curve highlights that all 20 available subbands are used to embed information.

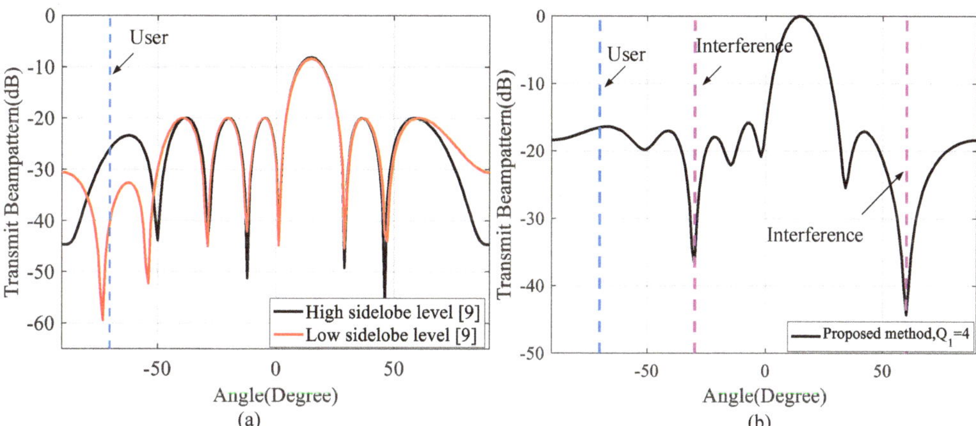

Figure 8. The comparison with sidelobe modulation in [9] (**a**) beampatterns versus angle using sidelobe modulation [9], (**b**) beampatterns versus angle using proposed SPIA modulation.

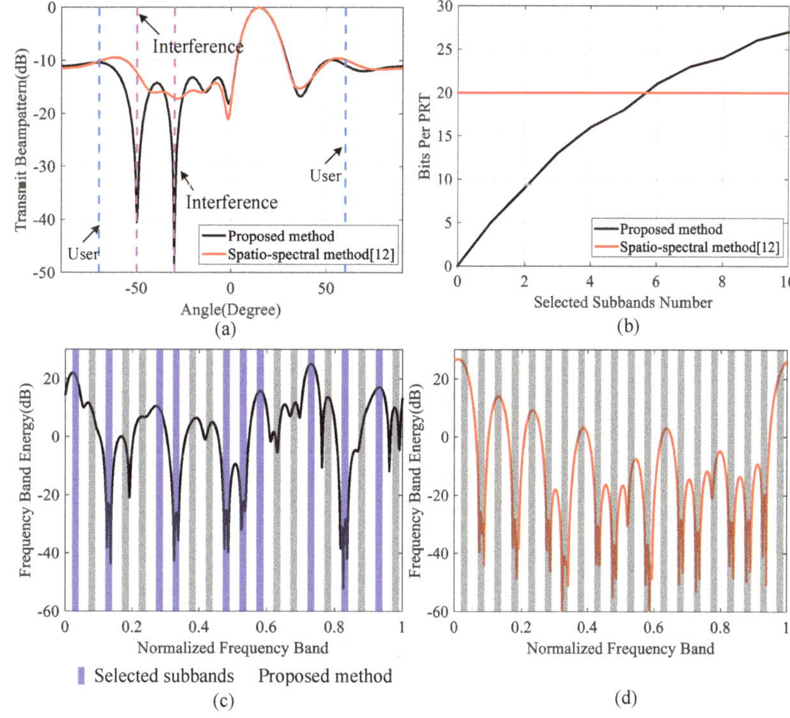

Figure 9. The comparison with spatio-spectral modulation scheme in [12] (**a**) beampatterns versus angle, (**b**) data rate versus selected subband number, (**c**) optimized frequency band energy using proposed SPIA method (**d**) optimized frequency band energy using spatio-spectral modulation method [12].

Table 2. The setting of simulational parameters.

Number	SPIA Method	Number	Spatio-Spectral Method [12]
1	Two users positions $-70°, 60°$	1	Two users positions $-70°, 60°$
2	Lower bound of passband energies $\eta_{p1} = \eta_{p2} = 2$	2	Lower bound of passband energies $\eta_{p1} = \eta_{p2} = 2$
3	Upper bound of stopband energy $\eta_s = 5 \times 10^{-5} \times n_s$	3	Upper bound of stopband energy $\eta_s = 2.5 \times 10^{-5} \times n_s$
4	Two interference sources' positions $-30°, -50°$	4	Without this parameter
5	Power of two interference sources $25\,\text{dB}, 30\,\text{dB}$	5	Without this parameter
6	Power of target $\delta_0^2 = 0\,\text{dB}$	6	Without this parameter
7	Ten frequency subbands are selected from the available frequency subbands using the spatio-spectral method [12] for information embedding	7	Frequency subbands for information embedding $\Omega_1 = (0.02, 0.04)$, $\Omega_2 = (0.07, 0.09)$, $\Omega_3 = (0.12, 0.14)$, $\Omega_4 = (0.17, 0.19)$, $\Omega_5 = (0.22, 0.24)$, $\Omega_6 = (0.27, 0.29)$, $\Omega_7 = (0.32, 0.34)$, $\Omega_8 = (0.37, 0.39)$, $\Omega_9 = (0.42, 0.44)$, $\Omega_{10} = (0.47, 0.49)$, $\Omega_{11} = (0.52, 0.54)$, $\Omega_{12} = (0.57, 0.59)$, $\Omega_{13} = (0.62, 0.64)$, $\Omega_{14} = (0.67, 0.69)$, $\Omega_{15} = (0.72, 0.74)$, $\Omega_{16} = (0.77, 0.79)$, $\Omega_{17} = (0.82, 0.84)$, $\Omega_{18} = (0.87, 0.89)$, $\Omega_{19} = (0.92, 0.94)$, $\Omega_{20} = (0.97, 0.99)$

4. Discussion

(1) A method for designing an MIMO DFRC signal is proposed to enhance signal detection and communication in the presence of signal-dependent clutter. The proposed method optimizes the beampattern to create nullings in clutter directions, effectively suppressing clutter based on prior knowledge of clutter locations. Additionally, the beam direction is optimized to precisely detect the target location, obtained through a cognitive paradigm. This ensures good detection performance.

(2) A novel modulation technique called SPIA modulation, which combines spectral passband and stop-band selection with amplitude modulation, is proposed to achieve simultaneous detection and communication. Unlike the existing approaches [32–34], the SPIA modulation methodology is proposed instead of using a conventional communication sequence. Additionally, previous works on DFRC signal design have focused on index modulation [14–19], primarily utilizing intra-pulse modulation without altering the conventional radar baseband signal. However, in this paper, we focus on index modulation via inter-pulse modulation, where the baseband signal is modified as an alternative to intra-pulse modulation. However, it is worth mentioning that, in future work, we plan to explore the combination of intra- and inter-pulse modulation techniques to design a dual-function signal.

(3) The IBE-DSADMM is exploited to solve the original nonconvex NP-hard problem. The algorithm guarantees that the SINR value monotonically increases and converges to a finite value.

5. Conclusions

In this paper, an MIMO DFRC signal design method resorting to SPIA modulation was proposed to realize radar detection and communication in signal-dependent clutter. To effectively suppress signal-dependent clutter and improve radar detection performance, SINR was used as the criterion to construct an optimization problem with practical constraints. We also exploited an IBE-DSADMM algorithm to monotonically increase the SINR. The numerical results verified that the designed integrated signal ensures the detection performance in the signal-dependent clutter and simultaneously implements multi-user communication.

Author Contributions: Conceptualization, X.Y. (Xue Yao); Formal analysis, X.Y. (Xue Yao); Funding acquisition, X.N.; Investigation, B.P. and T.F.; Methodology, X.Y. (Xue Yao); Project administration, X.N.; Supervision, G.C.; Validation, X.Y. (Xianxiang Yu); Writing—original draft, B.P., T.F. and G.C.; Writing—review and editing, X.Y. (Xue Yao) and X.Y. (Xianxiang Yu). All authors have read and agreed to the published version of the manuscript.

Funding: The work was supported by the National Natural Science Foundation of China (62101097 and 62271126). Natural Science Foundation Project of Chongqing (2022NSCQ-MSX3815). National key research and development program (2021YFB3901400).

Acknowledgments: The authors appreciate the valuable comments and constructive suggestions from the anonymous referees and the editors who helped improve the manuscript.

Appendix A

Given $\mathbf{s} = \mathrm{vec}([\mathbf{s}_1, \cdots, \mathbf{s}_N]^T) \in \mathbb{C}^{NM}, \mathbf{A} \in {}^{NM}$, the quadratic function $g(\mathbf{s}) = \mathbf{s}^H \mathbf{A}\mathbf{s}$ can be recast as a function of the specific block $\mathbf{s}_n$, i.e.,

$$
\begin{aligned}
g(\mathbf{s}_n; \bar{\mathbf{s}}_{-n}) &= (\bar{\mathbf{s}}_{-n} + \Lambda_n \mathbf{s}_n)^H \mathbf{A}(\bar{\mathbf{s}}_{-n} + \Lambda_n \mathbf{s}_n) \\
&= \mathbf{s}_n^H \Lambda_n^H \mathbf{A} \Lambda_n \mathbf{s}_n + 2\Re\left\{ \bar{\mathbf{s}}_{-n}^H \mathbf{A} \Lambda_n \mathbf{s}_n \right\} \\
&\quad + \bar{\mathbf{s}}_{-n}^H \mathbf{A} \bar{\mathbf{s}}_{-n},
\end{aligned}
\tag{A1}
$$

where $\bar{\mathbf{s}}_{-n} = \mathbf{s} - \Lambda_n \mathbf{s}_n \in \mathbb{C}^{NM}$, and $\Lambda_n \in \mathbb{C}^{NM \times M}$ is defined as

$$
\Lambda_n(i, \bar{j}) = \begin{cases} 1, & \text{if } i = n + (\bar{j} - 1)N \\ 0, & \text{otherwise} \end{cases}
$$

with $i \in \{1, \cdots, NM\}, \bar{j} \in \{1, \cdots, M\}, \mathbf{s}_{-n} = \mathrm{vec}([\mathbf{s}_1, \ldots, \mathbf{s}_{n-1}, \mathbf{s}_{n+1}, \ldots, \mathbf{s}_N]^T) \in \mathbb{C}^{(N-1)M}$.

Let $\mathbf{s}^{(n_i)} = \mathrm{vect}([\mathbf{s}_1^{(i)}, \cdots, \mathbf{s}_{n-1}^{(i)}, \mathbf{s}_n, \mathbf{s}_{n+1}^{(i-1)}, \cdots, \mathbf{s}_{N_t}^{(i-1)}]^T) \in \mathbb{C}^{N_t M}$, the objective function $f(\mathbf{s}_1^{(i)}, \cdots, \mathbf{s}_{n-1}^{(i)}, \mathbf{s}_n, \mathbf{s}_{n+1}^{(i-1)}, \cdots, \mathbf{s}_{N_t}^{(i-1)})$ can be rewritten as

$$
f(\mathbf{s}_n; \bar{\mathbf{s}}_{-n}^{(i)}) = \frac{\mathbf{s}_n^H \mathbf{X}_n \mathbf{s}_n + \Re\{\mathbf{x}_n^H \mathbf{s}_n\} + x_n}{\mathbf{s}_n^H \mathbf{Y}_n \mathbf{s}_n + \Re\{\mathbf{y}_n^H \mathbf{s}_n\} + y_n},
\tag{A2}
$$

where $\mathbf{X}_n = \Lambda_n^H \mathbf{X}_0 \Lambda_n$, $\mathbf{x}_n = 2\Lambda_n^H \mathbf{X}_0^H \bar{\mathbf{s}}_{-n}^{(i)}$ and $x_n = \bar{\mathbf{s}}_{-n}^{(i)H} \mathbf{X}_0 \bar{\mathbf{s}}_{-n}^{(i)}$. $\mathbf{Y}_n = \Lambda_n^H \mathbf{Y}_0 \Lambda_n$, $\mathbf{y}_n = 2\Lambda_n^H \mathbf{Y}_0^H \bar{\mathbf{s}}_{-n}^{(i)}$ and $y_n = \bar{\mathbf{s}}_{-n}^{(i)H} \mathbf{Y}_0 \bar{\mathbf{s}}_{-n}^{(i)}$. $\mathbf{s}_{-n}^{(i)} = \mathrm{vect}([\mathbf{s}_1^{(i)}, \ldots, \mathbf{s}_{n-1}^{(i)}, \mathbf{s}_{n+1}^{(i-1)}, \ldots, \mathbf{s}_{N_t}^{(i-1)}]^T) \in \mathbb{C}^{(N_t - 1)M}$, $\bar{\mathbf{s}}_{-n}^{(i)} = \mathbf{s}^{(n_i)} - \Lambda_n \mathbf{s}_n \in \mathbb{C}^{N_t M}$.

Constraint ① in Problem $\mathcal{P}_1$ can be recast

$$
\mathbf{s}_n^H \mathbf{A}_{sn} \mathbf{s}_n + \Re\left\{ \mathbf{a}_{sn}^H \mathbf{s}_n \right\} + a_{sn} \le \varepsilon \left(\mathbf{s}_n^H \mathbf{A}_{mn} \mathbf{s}_n + \Re\left\{ \mathbf{a}_{mn}^H \mathbf{s}_n \right\} + a_{mn} \right)
\tag{A3}
$$

where $\mathbf{A}_{sn} = \Lambda_n^H \mathbf{A}_s \Lambda_n$, $\mathbf{a}_{sn} = 2\Lambda_n^H \mathbf{A}_s^H \bar{\mathbf{s}}_{-n}^{(i)}$ and $a_{sn} = \bar{\mathbf{s}}_{-n}^{(i)H} \mathbf{A}_s \bar{\mathbf{s}}_{-n}^{(i)}$, $\mathbf{A}_{mn} = \Lambda_n^H \mathbf{A}_m \Lambda_n$, $\mathbf{a}_{mn} = 2\Lambda_n^H \mathbf{A}_m^H \bar{\mathbf{s}}_{-n}^{(i)}$ and $a_{mn} = \bar{\mathbf{s}}_{-n}^{(i)H} \mathbf{A}_m \bar{\mathbf{s}}_{-n}^{(i)}$.

Constraint ② can be written to

$$
\mathbf{s}_n^H \mathbf{R}_{sn} \mathbf{s}_n + \Re\left\{ \mathbf{r}_{sn}^H \mathbf{s}_n \right\} + r_{sn} \le 0,
\tag{A4}
$$

where $\mathbf{R}_{sn} = \Lambda_n^H \mathbf{R}_s \Lambda_n$, $\mathbf{r}_{sn} = 2\Lambda_n^H \mathbf{R}_s^H \bar{\mathbf{s}}_{-n}^{(i)}$, $r_{sn} = \bar{\mathbf{s}}_{-n}^{(i)H} \mathbf{R}_s \bar{\mathbf{s}}_{-n}^{(i)} - \eta_s$.

Constraint ③ can be rewritten to

$$
\eta_{pc} \le \mathbf{s}_n^H \mathbf{R}_{cn} \mathbf{s}_n + \Re\left\{ \mathbf{r}_{cn}^H \mathbf{s}_n \right\} + r_{cn}, c = 1, \cdots, C,
\tag{A5}
$$

where $\mathbf{R}_{cn} = \boldsymbol{\Lambda}_n^H \mathbf{R}_c \boldsymbol{\Lambda}_n$, $\mathbf{r}_{cn} = 2\boldsymbol{\Lambda}_n^H \mathbf{R}_c^H \bar{\mathbf{s}}_{-n}^{(i)}$, $r_{cn} = \bar{\mathbf{s}}_{-n}^{(i)H} \mathbf{R}_c \bar{\mathbf{s}}_{-n}^{(i)}$.

Constraint ④ can be converted to

$$(P_L - \delta)\left(\mathbf{s}_n^H \mathbf{A}_{0n}\mathbf{s}_n + \Re\left\{\mathbf{a}_{0n}^H \mathbf{s}_n\right\} + a_{0n}\right) \le \mathbf{s}_n^H \mathbf{A}_{kn}\mathbf{s}_n + \Re\left\{\mathbf{a}_{kn}^H \mathbf{s}_n\right\} + a_{kn}, k = 1,2, \quad (A6)$$

$$\mathbf{s}_n^H \mathbf{A}_{kn}\mathbf{s}_n + \Re\left\{\mathbf{a}_{kn}^H \mathbf{s}_n\right\} + a_{kn} \le (P_L + \delta)\left(\mathbf{s}_n^H \mathbf{A}_{0n}\mathbf{s}_n + \Re\left\{\mathbf{a}_{0n}^H \mathbf{s}_n\right\} + a_{0n}\right), k = 1,2, \quad (A7)$$

where $\mathbf{A}_{kn} = \boldsymbol{\Lambda}_n^H \mathbf{A}_k \boldsymbol{\Lambda}_n$, $\mathbf{a}_{kn} = 2\boldsymbol{\Lambda}_n^H \mathbf{A}_k^H \bar{\mathbf{s}}_{-n}^{(i)}$, $a_{kn} = \bar{\mathbf{s}}_{-n}^{(i)H} \mathbf{A}_k \bar{\mathbf{s}}_{-n}^{(i)}$, $\mathbf{A}_{0n} = \boldsymbol{\Lambda}_n^H \mathbf{A}_0 \boldsymbol{\Lambda}_n$, $\mathbf{a}_{0n} = 2\boldsymbol{\Lambda}_n^H \mathbf{A}_0^H \bar{\mathbf{s}}_{-n}^{(i)}$, $a_{0n} = \bar{\mathbf{s}}_{-n}^{(i)H} \mathbf{A}_0 \bar{\mathbf{s}}_{-n}^{(i)}$.

Appendix B

Next, we are going to find an approximation solution to Problem (36) by solving its approximation version. Interestingly, we observe the objective function $\chi(w_{(t)}, \mathbf{s}_n)$ is the difference between two convex functions. To this end, the objective function can be approximated by a lower bound function. Specifically,

$$\chi(w_{(t)}, \mathbf{s}_n) \ge \mathbf{s}_{n(t-1)}^H \mathbf{X}_n \mathbf{s}_{n(t-1)} + 2\Re\left\{\mathbf{s}_{n(t-1)}^H \mathbf{X}_n (\mathbf{s}_n - \mathbf{s}_{n(t-1)})\right\} +$$

$$\Re\left\{\mathbf{x}_n^H \mathbf{s}_n\right\} + x_n - w_{(t)}(\mathbf{s}_n^H \mathbf{Y}_n \mathbf{s}_n + \Re\left\{\mathbf{y}_n^H \mathbf{s}_n\right\} + y_n) = \mathbf{s}_n^H \mathbf{D}_n \mathbf{s}_n + \Re\left\{\mathbf{d}_n^H \mathbf{s}_n\right\} + d_n, \quad (A8)$$

where $\mathbf{D}_n = -w_{(t)}\mathbf{Y}_n$, $\mathbf{d}_n = -w_{(t)}\mathbf{y}_n + 2\mathbf{X}_n \mathbf{s}_{n(t-1)} + \mathbf{x}_n$, $d_n = -w_{(t)}y_n - \mathbf{s}_{n(t-1)}^H \mathbf{X}_n \mathbf{s}_{n(t-1)} + x_n$.

Similarly, constraints ①, ④, ⑤ are also the difference between two convex quadratic functions. Specifically, if $\mathbf{A}_{sn} \succeq \varepsilon \mathbf{A}_{mn}$, constraint ① is convex. Otherwise, it is approximated to [50,51]

$$(\mathbf{s}_n^H \mathbf{A}_{sn}\mathbf{s}_n + \Re\left\{\mathbf{a}_{sn}^H \mathbf{s}_n\right\} + a_{sn}) -$$

$$\varepsilon\left(\mathbf{s}_{n(t-1)}^H \mathbf{A}_{mn}\mathbf{s}_{n(t-1)} + 2\Re\left\{\mathbf{s}_{n(t-1)}^H \mathbf{A}_{mn}(\mathbf{s}_n - \mathbf{s}_{n(t-1)})\right\} + \Re\left\{\mathbf{a}_{mn}^H \mathbf{s}_n\right\} + a_{mn}\right) \le 0. \quad (A9)$$

A similar operation can be implemented on constraints ④ and ⑤.

As to constraint ③, it can be approximated to

$$\eta_{pc} - \left(\mathbf{s}_{n(t-1)}^H \mathbf{R}_{cn}\mathbf{s}_{n(t-1)} + 2\Re\left\{\mathbf{s}_{n(t-1)}^H \mathbf{R}_{cn}(\mathbf{s}_n - \mathbf{s}_{n(t-1)})\right\} + \Re\left\{\mathbf{r}_{cn}^H \mathbf{s}_n\right\} + r_{cn}\right) \le 0. \quad (A10)$$

And constraint ⑦ is treated in the same way.

In this respect, we can approximately tackle the maximization of $\chi(w_{(t)}, \mathbf{s}_n)$ through solving the following convex problem

$$\mathcal{P}_{\mathbf{s}_{n(t)}} \begin{cases} \max\limits_{\mathbf{s}_n} & \mathbf{s}_n^H \mathbf{D}_n \mathbf{s}_n + \Re\left\{\mathbf{d}_n^H \mathbf{s}_n\right\} + d_n \\ \text{s.t.} & \mathbf{s}_n^H \mathbf{B}_n \mathbf{s}_n + \Re\left\{\mathbf{b}_n^H \mathbf{s}_n\right\} + b_n \le 0, \\ & \mathbf{s}_n^H \mathbf{R}_{sn} \mathbf{s}_n + \Re\left\{\mathbf{r}_{sn}^H \mathbf{s}_n\right\} + r_{sn} \le 0, \\ & \mathbf{s}_n^H \bar{\mathbf{A}}_{kn} \mathbf{s}_n + \Re\left\{\bar{\mathbf{a}}_{kn}^H \mathbf{s}_n\right\} + \bar{a}_{kn} \le 0, k = 1,2, \\ & \mathbf{s}_n^H \tilde{\mathbf{A}}_{kn} \mathbf{s}_n + \Re\left\{\tilde{\mathbf{a}}_{kn}^H \mathbf{s}_n\right\} + \tilde{a}_{kn} \le 0, k = 1,2, \\ & \Re\left\{\bar{\mathbf{r}}_{cn}^H \mathbf{s}_n\right\} + \bar{r}_{cn} \le 0, c = 1,\ldots,C, \\ & \mathbf{s}_n(m)^H \mathbf{s}_n(m) - 1/N_t - \kappa \le 0, m = 1,\cdots,M \\ & \Re\left\{\bar{p}_1 \mathbf{s}_n(m)\right\} + \bar{p}_2 \le 0, m = 1,\cdots,M, \end{cases} \quad (A11)$$

where $\bar{\mathbf{r}}_{cn} = -2\mathbf{R}_{cn}\mathbf{s}_{n(t-1)} - \mathbf{r}_{cn}$, $\bar{r}_{cn} = \mathbf{s}_{n(t-1)}^H \mathbf{R}_{cn}\mathbf{s}_{n(t-1)} + \eta_{pc} - r_{cn}$, $\bar{p}_1 = -2\mathbf{s}_{n(t-1)}(m)$, $\bar{p}_2 = 1/N_t - \kappa + \mathbf{s}_{n(t-1)}(m)^H \mathbf{s}_{n(t-1)}(m)$. If constraints ④ and ⑤ are convex, $\mathbf{B}_n = \mathbf{A}_{sn} - \varepsilon \mathbf{A}_{mn}$, $\bar{\mathbf{A}}_{kn} = (P_L - \delta)\mathbf{A}_{0n} - \mathbf{A}_{kn}$, $\tilde{\mathbf{A}}_{kn} = \mathbf{A}_{kn} - (P_L + \delta)\mathbf{A}_{0n}$, $\mathbf{b}_n = \mathbf{a}_{sn} - \varepsilon \mathbf{a}_{mn}$, $\bar{\mathbf{a}}_{kn} = (P_L - \delta)\mathbf{a}_{0n} - \mathbf{a}_{kn}$, $\tilde{\mathbf{a}}_{kn} = \mathbf{a}_{kn} - (P_L + \delta)\mathbf{a}_{0n}$, $b_n = a_{sn} - \varepsilon a_{mn}$, $\bar{a}_{kn} = (P_L - \delta)a_{0n} - a_{kn}$, $\tilde{a}_{kn} = a_{kn} -$

$(P_L + \delta)a_{0n}$. Otherwise, if constraints ④ and ⑤ are non-convex, $\mathbf{B}_n = \mathbf{A}_{sn}$, $\bar{\mathbf{A}}_{kn} = (P_L - \delta)\mathbf{A}_{0n}$, $\tilde{\mathbf{A}}_{kn} = \mathbf{A}_{kn}$, $b_n = a_{sn} - \varepsilon a_{mn} - 2\varepsilon \mathbf{A}_{mn}\mathbf{s}_{n(t-1)}$, $\tilde{a}_{kn} = (P_L - \delta)a_{0n} - a_{kn} - 2\mathbf{A}_{kn}\mathbf{s}_{n(t-1)}$, $\bar{\tilde{a}}_{kn} = a_{kn} - (P_L + \delta)(2\mathbf{A}_{0n}\mathbf{s}_{n(t-1)} + a_{0n})$, $b_n = a_{sn} - \varepsilon a_{mn} + \varepsilon \mathbf{s}_{n(t-1)}^H \mathbf{A}_{mn}\mathbf{s}_{n(t-1)}$, $\bar{a}_{kn} = (P_L - \delta)a_{0n} - a_{kn} + \mathbf{s}_{n(t-1)}^H \mathbf{A}_{kn}\mathbf{s}_{n(t-1)}$, $\tilde{a}_{kn} = a_{kn} - (P_L + \delta)(a_{0n} - \mathbf{s}_{n(t-1)}^H \mathbf{A}_{0n}\mathbf{s}_{n(t-1)})$.

Appendix C

Assume that the l-th iteration results of $\mathbf{h}, \mathbf{z}, \{\mathbf{h}_{\bar{c}}\}, \{\mathbf{v}_c\}, \{\boldsymbol{\mu}_{\bar{c}}\}, \{\boldsymbol{\mu}_c\}, \boldsymbol{\mu}_z$ are respectively $\mathbf{h}^{(l)}, \mathbf{z}^{(l)}, \{\mathbf{h}_{\bar{c}}^{(l)}\}, \{\mathbf{v}_c^{(l)}\}, \{\boldsymbol{\mu}_{\bar{c}}^{(l)}\}, \{\boldsymbol{\mu}_c^{(l)}\}, \boldsymbol{\mu}_z^{(l)}$. The ADMM [52,53] procedure can be reported in Algorithm A1.

Algorithm A1: ADMM for solving $\mathcal{P}_{\mathbf{s}_{n(t)}}$

Input: $\mathbf{h}^{(0)}, \mathbf{z}^{(0)}, \{\mathbf{h}_{\bar{c}}^{(0)}\}, \{\mathbf{v}_c^{(0)}\}, \{\boldsymbol{\mu}_{\bar{c}}^{(0)}\}, \{\boldsymbol{\mu}_c^{(0)}\}, \boldsymbol{\mu}_z^{(0)}$;

Output: An optimized solution $\mathbf{s}_{n(t)}^*$ to $\mathcal{P}_{\mathbf{s}_{n(t)}}$;

1: $l = 0$;

2: $l := l + 1$;

3: Update $\mathbf{h}^{(l)}, \mathbf{z}^{(l)}, \{\mathbf{h}_{\bar{c}}^{(l)}\}, \{\mathbf{v}_c^{(l)}\}, \{\boldsymbol{\mu}_{\bar{c}}^{(l)}\}, \{\boldsymbol{\mu}_c^{(l)}\}, \boldsymbol{\mu}_z^{(l)}$ by solving the following problems.

$$\mathbf{h}_1^{(l)} := \arg\min_{\mathbf{h}_1} \|\mathbf{h}_1 - \mathbf{h}^{(l-1)} + \boldsymbol{\mu}_1^{(l-1)}/\varrho\|^2$$

$$\text{s.t. } \mathbf{h}_1^H \mathbf{R}_{sn}\mathbf{h}_1 + \Re\left\{\mathbf{r}_{sn}^H \mathbf{h}_1\right\} + r_{sn} \le 0,$$

$$\mathbf{h}_2^{(l)} := \arg\min_{\mathbf{h}_2} \|\mathbf{h}_2 - \mathbf{h}^{(l-1)} + \boldsymbol{\mu}_2^{(l-1)}/\varrho\|^2$$

$$\text{s.t. } \mathbf{h}_2^H \mathbf{B}_n\mathbf{h}_2 + \Re\left\{\mathbf{b}_n^H \mathbf{h}_2\right\} + b_n \le 0,$$

$$\{\mathbf{h}_k^{(l)}\} := \arg\min_{\{\mathbf{h}_k\}} \sum_{k=3}^{4} \|\mathbf{h}_k - \mathbf{h}^{(l-1)} + \boldsymbol{\mu}_k^{(l-1)}/\varrho\|^2$$

$$\text{s.t. } \mathbf{h}_k^H \bar{\mathbf{A}}_{n,k_0}\mathbf{h}_k + \Re\left\{\bar{\mathbf{a}}_{n,k_0}^H \mathbf{h}_k\right\} + \bar{a}_{n,k_0} \le 0, \quad k = 3, 4, k_0 = k - 2,$$

$$\{\mathbf{h}_{\bar{k}}^{(l)}\} := \arg\min_{\{\mathbf{h}_{\bar{k}}\}} \sum_{\bar{k}=5}^{6} \|\mathbf{h}_{\bar{k}} - \mathbf{h}^{(l-1)} + \boldsymbol{\mu}_{\bar{k}}^{(l-1)}/\varrho\|^2$$

$$\text{s.t. } \mathbf{h}_{\bar{k}}^H \tilde{\mathbf{A}}_{n,k_1}\mathbf{h}_{\bar{k}} + \Re\left\{\tilde{\mathbf{a}}_{n,k_1}^H \mathbf{h}_{\bar{k}}\right\} + \tilde{a}_{n,k_1} \le 0, \quad \bar{k} = 5, 6, k_1 = \bar{k} - 4,$$

$$\{\mathbf{v}_c^{(l)}\} := \arg\min_{\{\mathbf{v}_c\}} \sum_{c=1}^{C} \|\mathbf{v}_c - \mathbf{h}^{(l-1)} + \boldsymbol{\mu}_c^{(l-1)}/\varrho\|^2$$

$$\text{s.t. } \Re\left\{\bar{\mathbf{r}}_{cn}^H \mathbf{v}_c\right\} + \bar{r}_{cn} \le 0, \quad c = 1, \cdots, C,$$

$$\mathbf{z}^{(l)} := \arg\min_{\mathbf{z}} \|\mathbf{z} - \mathbf{h}^{(l-1)} + \boldsymbol{\mu}_z^{(l-1)}/\varrho\|^2$$

$$\text{s.t. } \mathbf{z}(m)^H \mathbf{z}(m) - 1/N_t - \kappa \le 0, m = 1, \cdots, M$$

$$\Re\{\bar{p}_1 \mathbf{z}(m)\} + \bar{p}_2 \le 0, m = 1, \cdots, M$$

$$\mathbf{h}^{(l)} := \arg\min_{\mathbf{h}} L_\varrho(\mathbf{h}, \mathbf{z}^{(l)}, \{\mathbf{h}_{\bar{c}}^{(l)}\}, \{\mathbf{v}_c^{(l)}\}, \{\boldsymbol{\mu}_{\bar{c}}^{(l-1)}\}, \{\boldsymbol{\mu}_c^{(l-1)}\}, \{\boldsymbol{\mu}_z^{(l-1)}\})$$

4: Update $\{\boldsymbol{\mu}_{\bar{c}}^{(l)}\}, \{\boldsymbol{\mu}_c^{(l)}\} \boldsymbol{\mu}_z^{(l)}$ by: $\boldsymbol{\mu}_{\bar{c}}^{(l)} = \boldsymbol{\mu}_{\bar{c}}^{(l-1)} + \varrho(\mathbf{h}_{\bar{c}}^{(l)} - \mathbf{h}^{(l)})$, $\boldsymbol{\mu}_c^{(l)} = \boldsymbol{\mu}_c^{(l-1)} + \varrho(\mathbf{v}_c^{(l)} - \mathbf{h}^{(l)})$, $\boldsymbol{\mu}_z^{(l)} = \boldsymbol{\mu}_z^{(l-1)} + \varrho(\mathbf{h}_z^{(l)} - \mathbf{h}^{(l)})$;

5: If a pre-set exit condition is met, output $\mathbf{s}_{n(t)}^* = \mathbf{h}^{(l)}$. Otherwise, go to Step 2.

Obviously, we can parallelly update $\mathbf{z}, \mathbf{h}_{\bar{c}}, \bar{c} = 1, \cdots, 6, \mathbf{v}_c, c = 1, \cdots, C$. In particular, the optimization problems concerning $\mathbf{h}_{\bar{c}}, \bar{c} = 1, \cdots, 6$ can split into 6 subproblems falling into the convex QCQP with only one constraint (QCQP-1) whose closed-form solution has been derived using KKT conditions in [54]. As to the optimization problem with respect to $\mathbf{z}, \mathbf{v}_c, c = 1, \cdots, C$ can be solved via the KKT technique in [46]. Finally, the problem for $\mathbf{h}$ is equivalent to

$$\min_{\mathbf{h}} \mathbf{h}^H \mathbf{D} \mathbf{h} + \Re\{\mathbf{d}^H \mathbf{h}\}, \tag{A12}$$

where $\mathbf{D} = -\mathbf{D}_n + 8\varrho/2\mathbf{I}_M$ and $\mathbf{d} = -\mathbf{d}_n - \sum_{\bar{c}=1}^{6} \varrho(\mathbf{h}_{\bar{c}}^{(l)} + \boldsymbol{\mu}_{\bar{c}}^{(l-1)}/\varrho) - \sum_{c=1}^{C} \varrho(\mathbf{v}_c^{(l)} + \boldsymbol{\mu}_c^{(l-1)}/\varrho) - \varrho(\mathbf{z}^{(l)} + \boldsymbol{\mu}_z^{(l-1)}/\varrho)$. By letting its first derivative be zero, the closed-form solution to Problem (38) is $-\mathbf{D}^{-1}\mathbf{d}/2$.

References

1. Saddik, G.N.; Singh, R.S.; Brown, E.R. Ultra-wideband multifunctional communications/radar system. *IEEE Trans. Microw. Theory Tech.* **2007**, *55*, 1431–1437. [CrossRef]
2. Liu, Z. Waveform Research on the Integration of Radar and Communication. Ph.D. Thesis, Beijing Institute of Technology, Beijing, China, 12 June 2015. (In Chinese)
3. Cui, G.; Yang, J.; Lu, S.; Yu, X.; Kong, L. Dual-use unimodular sequence design via frequency nulling modulation. *IEEE Access* **2018**, *6*, 62470–62481. [CrossRef]
4. Yang, J.; Cui G.; Yu, X.; Kong, L. Dual-use signal design for radar and communication via ambiguity function sidelobe control. *IEEE Trans. Veh. Technol.* **2020**, *69*, 9781–9794. [CrossRef]
5. Yang, J.; Tan, Y.; Yu, X.; Cui, G.; Zhang, D. Waveform Design for Watermark Framework Based DFRC System With Application on Joint SAR Imaging and Communication. *IEEE Trans. Geosci. Remote Sens.* **2022**, *61*, 1–14. [CrossRef]
6. Eedara, I.P.; Hassamien, A.; Amin, M.G. Performance analysis of dual-function multiple-input multiple-output radar-communications using frequency hopping waveforms and phase shift keying signalling. *IET Radar Sonar Navig.* **2021**, *15*, 402–418. [CrossRef]
7. Eedara, I.P.; Amin, M.G.; Fabrizio, G.A. Target detection in frequency hopping MIMO dual-function radar-communication systems. In Proceedings of the 2021 IEEE International Conference on Acoustics, Speech and Signal Processing, Toronto, ON, Canada, 6 June 2021; pp. 8458–8462.
8. Eedara, I.P.; Amin, M.G.; Hoorfar, A.; Chalise, B.K. Dual-function frequency-hopping MIMO radar system with CSK signaling. *IEEE Trans. Aerosp. Electron. Syst.* **2022**, *58*, 1501–1513. [CrossRef]
9. Hassanien, A.; Amin, M.G.; Zhang, Y.D.; Ahmad, F. Dual-function radar-communications: Information embedding using sidelobe control and waveform diversity. *IEEE Trans. Signal Process.* **2016**, *64*, 2168–2181. [CrossRef]
10. Hassanien, A.; Amin, M.G.; Zhang, Y.D.; Ahmad, F. Signaling strategies for dual-function radar communications: An overview. *IEEE Aerosp. Electron. Syst. Mag.* **2016**, *31*, 36–45. [CrossRef]
11. Hassanien, A.; Amin, M.G.; Zhang, Y.D.; Ahmad, F. Phase-modulation based dual-function radar-communications. *IET Radar Sonar Navig.* **2016**, *10*, 1411–1421. [CrossRef]
12. Yu, X.; Yao, X.; Yang, J.; Zhang, L.; Kong, L.; Cui, G. Integrated waveform design for MIMO radar and communication via spatio-spectral modulation. *IEEE Trans. Signal Process.* **2022**, *70*, 2293–2305. [CrossRef]
13. Tang, B.; Stoica, P. MIMO multifunction RF systems: Detection performance and waveform design. *IEEE Trans. Signal Process.* **2022**, *70*, 4381–4394. [CrossRef]
14. Hassanien, A.; Aboutanios, E.; Amin, M.G.; Fabrizio, G.A. A dual-function MIMO radar-communication system via waveform permutation. *Digit. Signal Process.* **2018**, *83*, 118–128. [CrossRef]
15. Wang, X.; Hassanien, A.; Amin, M.G. Dual-function MIMO radar communications system design via sparse array optimization. *IEEE Trans. Aerosp. Electron. Syst.* **2019**, *55*, 1213–1226. [CrossRef]
16. Baxter, W.; Aboutanios, E.; Hassanien, A. Joint radar and communications for frequency-hopped MIMO systems. *IEEE Trans. Signal Process.* **2022**, *70*, 729–742. [CrossRef]
17. Huang, T.; Shlezinger, N.; Xu, X.; Liu, Y.; Eldar, Y.C. MAJoRCom: A dual-function radar communication system using index modulation. *IEEE Trans. Signal Process.* **2020**, *68*, 3423–3438. [CrossRef]
18. Ma, D.; Shlezinger, N.; Huang, T.; Liu, Y.; Eldar, Y.C. FRaC: FMCW-based joint radar-communications system via index modulation. *IEEE J. Sel. Top. Signal Process.* **2021**, *15*, 1348–1364. [CrossRef]
19. Xu, J.; Wang, X.; Aboutanios, E.; Cui, G. Hybrid index modulation for dual-functional radar communications systems. *IEEE Trans. Veh. Technol.* **2022**, *72*, 3186–3200. [CrossRef]
20. Wen, M.; Zheng, B.; Kim, K.; Di, R.M.; Tsiftsis, T.A.; Chen, K.C.; Al-Dhahir, N. A Survey on Spatial Modulation in Emerging Wireless Systems: Research Progresses and Applications. *IEEE J. Sel. Areas Commun.* **2019**, *37*, 1949–1972. [CrossRef]

21. Li, J.; Dang, S.; Wen, M.; Li, Q.; Chen, Y.; Huang, Y.; Shang, W. Index Modulation Multiple Access for 6G Communications: Principles, Applications, and Challenges. *IEEE Netw.* **2023**, *37*, 52–60. [CrossRef]
22. Wen, M.; Basar, E.; Li, Q.; Zheng, B.; Zhang, M. Multiple-Mode Orthogonal Frequency Division Multiplexing with Index Modulation. *IEEE Trans. Commun.* **2017**, *65*, 3892–3906. [CrossRef]
23. Duly, A.J.; Love, D.J.; Krogmeier, J.V. Time-division beamforming for MIMO radar waveform design. *IEEE Trans. Aerosp. Electron. Syst.* **2013**, *49*, 1210–1223. [CrossRef]
24. Wang, Y.; Li, W.; Sun, Q.; Huang, G. Robust joint design of transmit waveform and receive filter for MIMO radar space-time adaptive processing with signal-dependent interferences. *IET Radar Sonar Navig.* **2017**, *11*, 1321–1332. [CrossRef]
25. Yu, X.; Cui, G.; Kong, L.; Li, J.; Gui, G. Constrained waveform design for colocated MIMO radar with uncertain steering matrices. *IEEE Trans. Aerosp. Electron. Syst.* **2019**, *55*, 356–370. [CrossRef]
26. Karbasi, S.M.; Aubry, A.; Carotenuto, V.; Naghsh, M.M.; Bastani, M.H. Knowledge-based design of space-time transmit code and receive filter for a multiple-input-multiple-output radar in signal-dependent interference. *IET Radar Sonar Navig.* **2015**, *9*, 1124–1135. [CrossRef]
27. Cui, G.; Yu, X.; Carotenuto, V.; Kong, L. Space-time transmit code and receive filter design for colocated MIMO radar. *IEEE Trans. Signal Process.* **2017**, *65*, 1116–1129. [CrossRef]
28. Yang, J.; Aubry, A.; De Maio, A.; Yu, X.; Cui, G. Multi-spectrally constrained transceiver design against signal-dependent interference. *IEEE Trans. Signal Process.* **2022**, *70*, 1320–1332. [CrossRef]
29. Tang, B.; Tang, J. Joint design of transmit waveforms and receive filters for MIMO radar space-time adaptive processing. *IEEE Trans. Signal Process.* **2016**, *64*, 4707–4722. [CrossRef]
30. Tang, B.; Tuck, J.; Stoica, P. Polyphase waveform design for MIMO radar space time adaptive processing. *IEEE Trans. Signal Process.* **2020**, *68*, 2143–2154. [CrossRef]
31. Tang, B.; Stoica, P. Information-theoretic waveform design for MIMO radar detection in range-spread clutter. *Signal Process.* **2021**, *182*, 107961. [CrossRef]
32. Wu, W.; Tang, B.; Tang, J. Waveform design for dual-function radar-communication systems in clutter. *J. Radars* **2022**, *11*, 570–580. (In Chinese) [CrossRef]
33. Tsinos, C.G.; Arora, A.; Chatzinotas, S.; Ottersten, B. Joint transmit waveform and receive filter design for dual-function radar-communication systems. *IEEE J. Sel. Top. Signal Process.* **2021**, *15*, 1378–1392.
34. Liu, R.; Li, M.; Liu, Q.; Swindlehurst, A.L. Joint waveform and filter designs for STAP-SLP-based MIMO-DFRC systems. *IEEE J. Sel. Areas Commun.* **2022**, *40*, 1918–1931. [CrossRef]
35. Khalaj-Amirhosseini, M. Synthesis of linear and planar arrays with sidelobes of individually arbitrary levels. *Int. J. RF Microw. Comput. Eng.* **2019**, *29*. [CrossRef]
36. Kang, M.; Baek, J. Efficient and Accurate Synthesis for Array Pattern Shaping. *Sensors* **2022**, *22*, 5537. [CrossRef]
37. Dicandia, F.A.; Genovesi, S. Wide-Scan and Energy-Saving Phased Arrays by Exploiting Penrose Tiling Subarrays. *IEEE Trans. Antennas Propag.* **2022**, *70*, 7524–7537. [CrossRef]
38. Raei, E.; Alaee-Kerahroodi, M.; Mysore Rand, B.S. Spatial- and Range- ISLR Trade-Off in MIMO Radar Via Waveform Correlation Optimization. *IEEE Trans. Signal Process.* **2021**, *69*, 3283–3298. [CrossRef]
39. Aubry, A.; De Maio, A.; Farina, A.; Wicks, M. Knowledge-aided (potentially cognitive) transmit signal and receive filter design in signal-dependent clutter. *IEEE Trans. Aerosp. Electron. Syst.* **2013**, *49*, 93–117. [CrossRef]
40. Yu, X.; Cui, G.; Yang, J.; Li, J.; Kong, L. Quadratic optimization for unimodular sequence design via an ADPM framework. *IEEE Trans. Signal Process.* **2020**, *68*, 3619–3634. [CrossRef]
41. He, H.; Li, J.; Stoica, P. *Waveform Design for Active Sensing Systems: A Computational Approach*; Cambridge University Press: New York, NY, USA, 2012. [CrossRef]
42. Aubry, A.; De Maio, A.; Piezzo, M.; Farina, A. Radar waveform design in a spectrally crowded environment via nonconvex quadratic optimization. *IEEE Trans. Aerosp. Electron. Syst.* **2014**, *50*, 1138–1152.
43. Yu, X.; Alhujaili, K.; Cui, G.; Monga, V. MIMO radar waveform design in the presence of multiple targets and practical constraints. *IEEE Trans. Signal Process.* **2020**, *68*, 1974–1989. [CrossRef]
44. Seber, G.A.F. *A Matrix Handbook for Statisticians*; John Wiley & Sons: Hoboken, NJ, USA, 2007. [CrossRef]
45. Zhao, L.; Song, J.; Babu, P.; Palomar, D.P. A unified framework for low autocorrelation sequence design via majorization-minimization. *IEEE Trans. Signal Process.* **2017**, *65*, 438–453.
46. Yu, X.; Cui, G.; Yang, J.; Kong, L.; Li, J. Wideband MIMO radar waveform design. *IEEE Trans. Signal Process.* **2019**, *67*, 3487–3501. [CrossRef]
47. Bertsekas, D.P.; Tsitsiklis, J.N. *Parallel and Distributed Computation: Numerical Methods*; Athena Scientific: New York, NY, USA, 1997. [CrossRef]
48. Mehanna, O.; Huang, K.; Gopalakrishnan, B.; Konar, A.; Sidiropoulos, N.D. Feasible point pursuit and successive approximation of non-convex QCQPs. *IEEE Signal Process. Lett.* **2015**, *22*, 804–808.
49. Yu, X.; Yao, X.; Qiu, H.; Cui, G. Integrated MIMO signal design via spatio-spectral modulation. In Proceedings of the 2022 30th European Signal Processing Conference (EUSIPCO), Belgrade, Serbia, 29 August 2022; pp. 1911–1915. [CrossRef]
50. Razaviyayn, M.; Hong, M.; Luo, Z.-Q. A unified convergence analysis of block successive minimization methods for nonsmooth optimization. *SIAM J. Optimiz.* **2013**, *23*, 1126–1153.

51. Aubry, A.; De Maio, A.; Zappone, A.; Razaviyayn, M.; Luo, Z. A new sequential optimization procedure and its applications to resource allocation for wireless systems. *IEEE Trans. Signal Process.* **2018**, *66*, 6518–6533. [CrossRef]
52. Yu, X.; Qiu, H.; Yang, J.; Wei, W.; Cui, G.; Kong, L. Multispectrally constrained MIMO radar beampattern design via sequential convex approximation. *IEEE Trans. Aerosp. Electron. Syst.* **2022**, *58*, 2935–2949. [CrossRef]
53. Chen, J.; Zhang, Y.; Guo, S.; Cui, G.; Wu, P.; Jia, C.; Kong, L. Joint estimation of NLOS building layout and targets via sparsity-driven approach. *IEEE Trans. Geosci. Remote Sens.* **2022**, *60*, 1–13. [CrossRef]
54. Huang, K.; Sidiropoulos, N.D. Consensus-ADMM for general quadratically constrained quadratic programming. *IEEE Trans. Signal Process.* **2016**, *64*, 5297–5310. [CrossRef]

 *remote sensing*

Technical Note

Joint Design of Transmitting Waveform and Receiving Filter via Novel Riemannian Idea for DFRC System

Yinan Zhao [1], Zhongqing Zhao [2], Fangqiu Tong [2], Ping Sun [2], Xiang Feng [2,*] and Zhanfeng Zhao [2]

[1] School of Communication Engineering, Hangzhou Dianzi University, Hangzhou 310018, China; zhaoyinan@hdu.edu.cn

[2] School of Information Science and Engineering, Harbin Institute of Technology, Weihai 264209, China

* Correspondence: fengxiang@hit.edu.cn

Abstract: Recently, the problem of target detection in noisy environments for the Dual-Functional Radar Communication (DFRC) integration system has been a hot topic. In this paper, to suppress the noise and further enhance the target detection performance, a novel manifold Riemannian Improved Armijo Search Conjugate Gradient algorithm (RIASCG) framework has been proposed which jointly optimizes the integrated transmitting waveform and receiving filter. Therein, the reference waveform is first designed to achieve excellent pattern matching of radar beamforming. Furthermore, to ensure the quality of system information transmission, the energy of multi-user interference (MUI) of communication signals is incorporated as the constraint. Additionally, the typical similarity constraint is introduced to ensure the transmitting waveform with a good ambiguity function. Finally, simulation results demonstrate that the designed waveform not only enhances the system's target detection performance in noisy environments but also achieves a relatively good multi-user communication ability when compared with other prevalent waveforms.

Keywords: radar communication integration system; waveform design; manifold optimization; target detection

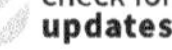 check for
updates

Citation: Zhao, Y.; Zhao, Z.; Tong, F.; Sun, P.; Feng, X.; Zhao, Z. Joint Design of Transmitting Waveform and Receiving Filter via Novel Riemannian Idea for DFRC System. *Remote Sens.* **2023**, *15*, 3548. https://doi.org/10.3390/rs15143548

Academic Editors: Guolong Cui, Bin Liao, Yong Yang and Xianxiang Yu

Received: 22 April 2023
Revised: 3 July 2023
Accepted: 11 July 2023
Published: 14 July 2023

1. Introduction

In the past decades, with the rapid development of commercial wireless communication and remote sensing image processing [1–3], the demand for available frequency bands has also increased dramatically where most of the frequency resources are allocated to radar requirements and also limit the communication throughput. Therein, the integration of sensing and communication to improve spectrum utilization has become a research hotspot [4–6].

At present, the radar communication integration system can be divided into two categories according to whether the spectrum is shared: radar communication coexistence (RCC) [6] and dual functional radar communication (DFRC) [7]. For RCC, it means that radar and communication work independently on the same platform through some resource diversity, including frequency division multiplexing, power multiplexing and time division multiplexing [8]. Authors in [9] presented the power allocation and subcarrier selection scheme to minimize the transmission power while ensuring the presence of mutual information between radar and communication. Authors in [10] discussed the integration of radar communication by orthogonalizing radar and communication signals. Unlike RCC, DFRC integrates radar and communication functions through a single integrated waveform and allocates its power to a specific spatial area to detect targets while transmitting the user communication signals. Early research focused on the integrated waveform design where the digital information was embedded into the radar waveform by modifying the traditional radar waveform and controlling its sidelobe in the direction of the objective user [11,12]. Typically, when the communication user is in the mainlobe of the radar waveform, the communication rate would drop greatly, and the communication function would fail. Moreover, the communication symbols can also modulate the radar waveform to achieve coexistence for dual functions [13–15]. Authors in [16,17] tried to use the existing communication waveforms to achieve radar's tasks. From the communication view, DFRC signals need high-quality communication performance, such as a high communication rate, which can be improved by minimizing multi-user interference (MUI).

Considering the scenarios involving multiple communication users and multiple radar sensing targets, the DFRC waveform has faced more challenges [18]. For communication, the signal-to-interference-plus-noise ratio (SINR) of each user should be considered. For radar, the SINR should also be considered. For DFRC systems, the performance of communication and sensing is coupled together, which means that any improvement of communication may deteriorate the radar performance, and vice versa. Therefore, in fact, when DFRC systems work in multi-user and multi-target scenarios, they inevitably face multiple performance tradeoffs between multi-users, multi-targets and also communication and perception. To solve this problem, authors in [19] designed the transmitting waveform by minimizing the joint least squares of weighted squared error and total MUI, where the weighting factor is used to balance two systems. Moreover, in this integration, most communication signals are modulated by multiple carriers, which will inevitably lead to a high peak average power ratio (PAPR) and incur distortion at the RF end. To improve the power efficiency of transmitters, PAPR constraints or constant modulus (CM) are widely used. Authors in [20] tried to jointly optimize the MUI and radar SINR by alternating the minimization and gradient projection frameworks. Authors in [21,22] designed the radar transmitting waveform under a given PAPR and similarity constraints. Note that radar waveforms as well as DFRC waveforms with constant modulus are also in need [23,24]. Authors in [25] designed a CM-integrated waveform by synthesizing different signals in the direction of communication and radar and also proposed an iterative optimization amplitude weighting method. No matter the PAPR/CM constraint or similarity constraint, these prior works would have to tackle the non-convex optimization problem with a heavy computation burden. How to design DFRC waveforms within the non-convex framework has been a hot topic. Furthermore, authors in [26] designed the CM DFRC waveform to minimize MUI and maximize the similarity between integrated signals and reference radar waveforms. Authors in [27] considered the joint design of the receiving filter and transmitting waveform with the maximum signal-to-noise ratio (SNR) and proposed a novel algorithm based on manifold ideas which give us lots of inspiration.

In this paper, to improve the detection performance of the integration system in noisy environments, the joint design of the system's transmitting waveform and receiving filter has been proposed to enhance the output SNR, which is based on the Riemannian Improved Armijo Search Conjugate Gradient algorithm (RIASCG) framework. This framework could transform the non-convex optimization problem into a novel convex one within Riemannian manifold space. Firstly, the MIMO radar waveform, with the constant modulus and similarity (CM&S) constraint as well as good directivity, was designed. Furthermore, the waveform with the minimum MUI was also considered. By using the manifold principle, the CM&S constraint was transformed into an unconstrained Riemannian space. Particularly in the Riemannian space, the final solution can be obtained through the iterative closed-form. Finally, we compare their performance with several existing ones.

The organization of this paper is as follows. The system model and problem formulation are presented in Sections 2 and 3. The DFRC waveform design by the novel algorithm is proposed in Section 4. Section 5 presents the numerical results. Finally, conclusions are drawn in Section 6.

Notation: Lower-case letters x and upper-case letters $\mathbf{X}$ denote vectors and matrices, respectively. The symbols $(\cdot)^{\mathrm{T}}$, $(\cdot)^{\mathrm{H}}$ and $(\cdot)^{*}$ stand for the transpose, the conjugate transpose and the conjugate operators, respectively. The set of $N \times N$ complex matrices and the set of n-dimensional complex numbers vectors are denoted by $\mathbb{C}^{N \times N}$ and $\mathbb{C}^{n}$, respectively. The l_2 norm is denoted by the symbol $||\cdot||_2$, and the Frobenius norm is represented by the symbol $||\cdot||_F$. $\mathbf{I}_N$ stands for the identity matrix of size $N \times N$. Finally, the notations $\mathrm{Re}\{x\}$

and Im$\{x\}$ are denoted as the real and imaginary part of x, $\mathbb{E}(\cdot)$ represents the expectation operator, the symbol $\otimes$ denotes the Kronecker product and the symbol $\odot$ represents the Hadamard product.

2. System Model

Cognitive radar adjusts its waveforms via artificial intelligence or machine learning as shown in Figure 1, which is regarded as a closed-loop feedback cycle. This adaptive system makes it more intelligent and offers higher robustness in waveform optimization compared with the traditional one. This paper focuses on the design of transmitting waveforms and receiving filters in DFRC systems where the integrated waveform is suitable for both target detection and information transmission. Namely, to achieve this, the optimization problem should satisfy the transmitted beampattern favorable for target detection while generating minimal MUI to multiple users in the downlink. The DFRC system needs to perform the following two tasks simultaneously: (i) target detection and (ii) communication with single-antenna users in the downlink.

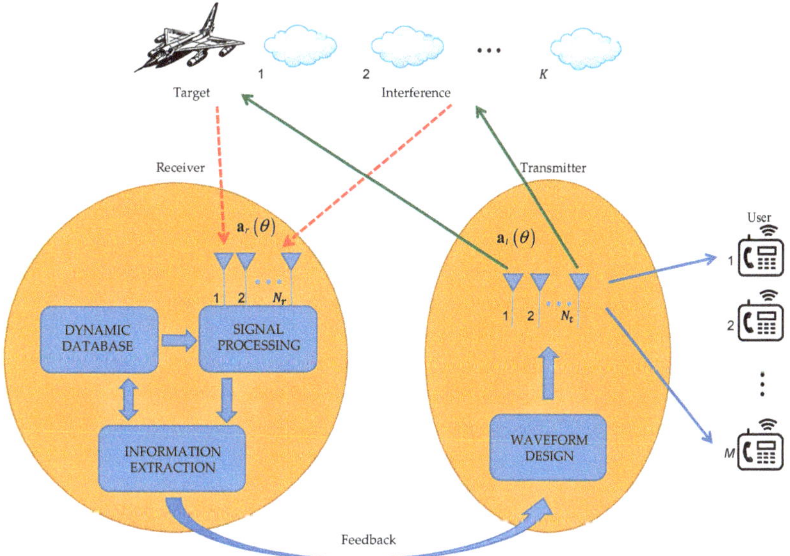

Figure 1. MIMO DFRC system based on cognitive radar idea.

2.1. Communication Model

Assume that the channel between the bifunctional base-station and the communication user is flat Rayleigh fading and the channel characteristics remain unchanged for a certain period of time. The integrated MIMO radar-antenna system has N_t transmitting antennas and N_r receiving antennas, and the frame length of the dual-functional waveform is assumed to be N. Suppose that there are M communication users subjected to the interference from K irrelevant signals when detecting the target. Specifically, the signals received by M communication users can be represented as

$$\mathbf{Y} = \mathbf{HX} + \mathbf{W} \tag{1}$$

where $\mathbf{H} = [\mathbf{h}_1, \mathbf{h}_2, \ldots, \mathbf{h}_M]^{\mathrm{T}} \in \mathbb{C}^{M \times N_t}$ is the channel matrix. The transmission signal matrix is $\mathbf{X} = [x_1, x_2, \ldots, x_N] \in \mathbb{C}^{N_t \times N}$, where $x_i \in \mathbb{C}^{N_t \times 1}$ denotes the i-th transmitted symbol vector, and $\mathbf{W} = [\mathbf{w}_1, \mathbf{w}_2, \ldots, \mathbf{w}_N] \in \mathbb{C}^{M \times N}$ is the white Gaussian noise matrix of the receiver.

Furthermore, as the integrated system would transmit a signal matrix $\mathbf{S} \in \mathbb{C}^{M \times N}$ to M users, (1) can be rewritten as

$$\mathbf{Y} = \mathbf{S} + \underbrace{(\mathbf{HX} - \mathbf{S})}_{\text{MUI}} + \mathbf{W} \tag{2}$$

where the second term in (2) represents the multi-user interference [19], and its energy can be expressed as $\varphi(\mathbf{X}) = \|\mathbf{HX} - \mathbf{S}\|_F^2$ with separable property $\|\mathbf{HX} - \mathbf{S}\|_F^2 = \sum_{i=1}^{N} \|\mathbf{H}x_i - s_i\|^2$. According to [20], the achievable sum-rate of the users can be defined as

$$\vartheta \triangleq \sum_{m=1}^{M} \log_2(1 + \gamma_m) \tag{3}$$

where γ_m represents the SINR$_t$ of per-frame received by the m-th user, i.e.,

$$\gamma_m = \frac{\mathbb{E}\left(|s_{m,i}|^2\right)}{\mathbb{E}\left(\left|\mathbf{h}_m^T x_i - s_{m,i}\right|^2\right) + N_0} \tag{4}$$

where $s_{m,i}$ represents the i-th code-unit for the m-th user, N_0 is the power of the received noise, and the energy of the (m, i) term of MUI can be expressed as $\mathbb{E}(\left|\mathbf{h}_m^T x_i - s_{m,i}\right|)$. By minimizing the energy of MUI, the achievable rate of the system can be maximized, which is equivalent to minimizing $\varphi(\mathbf{X})$.

2.2. Detection Model

This section primarily aims to synthesize waveforms to achieve beampattern matching while maximizing SINR$_r$. The SINR$_r$ of the MIMO system is determined by the transmitted waveform and its covariance matrix $\mathbf{R}$. By optimizing the transmitting waveform vector, the quality of the radar-output signal can also be enhanced, which in turn improves the SNR and anti-interference ability of radar and also improves the sensitivity and accuracy of target detection. Assuming that a point target exists in the direction θ_0, and K independent interference sources are located at $\theta_k(\theta_k \neq \theta_0, k = 1, 2, \cdots, K)$, then the received signals at the i-th frame $(i = 1, 2, \ldots, N)$ is formulated by

$$\mathbf{y}_i = \alpha_0 \mathbf{a}_r(\theta_0)\mathbf{a}_t^T(\theta_0)x_i + \sum_{k=1}^{K} \alpha_k \mathbf{a}_r(\theta_k)\mathbf{a}_t^T(\theta_k)x_i + \mathbf{v}_i \tag{5}$$

where $\alpha_0, \alpha_1, \ldots, \alpha_K$ are the amplitudes of target and interference sources, and $\mathbf{v}_i$ denotes the receiver noise. The $\mathbf{a}_r(\theta) \in \mathbb{C}^{N_r}$ and $\mathbf{a}_t(\theta) \in \mathbb{C}^{N_t}$ are the propagation vector and steering vector for the direction θ where the transmit and receive arrays are assumed to be linear uniform ones with $\mathbf{a}_t(\theta) = \frac{1}{\sqrt{N_t}}\left[e^{-j\pi 0 \sin\theta}, \ldots, e^{-j\pi(N_t-1)\sin\theta}\right]^T$ and $\mathbf{a}_r(\theta) = \frac{1}{\sqrt{N_r}}\left[e^{-j\pi 0 \sin\theta}, \ldots, e^{-j\pi(N_r-1)\sin\theta}\right]^T$. To simplify the expression, the N vectors corresponding to $\mathbf{y}_i$ in Equation (5) can be represented as

$$\tilde{\mathbf{y}} = \alpha_0 \mathbf{A}_0 \tilde{x} + \sum_{k=1}^{K} \alpha_k \mathbf{A}_k \tilde{x} + \tilde{\mathbf{v}} \tag{6}$$

where $\mathbf{A}_k = \mathbf{I}_N \otimes \left(\mathbf{a}_r(\theta_k)\mathbf{a}_t^T(\theta_k)\right)$, $\tilde{\mathbf{y}} = [\mathbf{y}_1^T, \ldots, \mathbf{y}_N^T]^T$, $\tilde{x} = [x_1^T, \ldots, x_N^T]^T$, $\tilde{\mathbf{v}} = [\mathbf{v}_1^T, \ldots, \mathbf{v}_N^T]^T$.

To improve the detection performance, it is required to process the received signal. Regarding the receiving filter w, the resulting output can be expressed as

$$r = w^{\mathrm{H}}\widetilde{y} = \underbrace{\alpha_0 w^{\mathrm{H}}\mathbf{A}_0\widetilde{x}}_{\text{Target}} + \underbrace{w^{\mathrm{H}}\sum_{k=1}^{K}\alpha_k\mathbf{A}_k\widetilde{x}}_{\text{Interference}} + \underbrace{w^{\mathrm{H}}\widetilde{v}}_{\text{Noise}} \tag{7}$$

In Equation (7), the first term represents the desired signal, the second term represents the interference signal, and the third term represents the noise. Thus, the SINR$_{\mathrm{r}}$ of the filter output can be expressed as

$$\mathrm{SINR_r} = \frac{\sigma_0^2\left|w^{\mathrm{H}}\mathbf{A}_0\widetilde{x}\right|^2}{w^{\mathrm{H}}\left[\sum\limits_{k=1}^{K}\sigma_k^2\mathbf{A}_k\widetilde{x}\widetilde{x}^{\mathrm{H}}\mathbf{A}_k^{\mathrm{H}}\right]w + \sigma_v^2 w^{\mathrm{H}}w} \tag{8}$$

where $\mathbb{E}\lfloor|\alpha_k|_2\rfloor = \sigma_k^2$ represents the complex amplitude of α_k. Generally, in a Gaussian noise environment, the larger the SINR$_{\mathrm{r}}$, the better the detection performance would be. Equation (8) can be transformed into a convex problem for a fixed $\widetilde{x}$.

$$\min_{w} w^{\mathrm{H}}\left[\sum_{k=1}^{K}\ddot{\sigma}_k\mathbf{A}_k\widetilde{x}\widetilde{x}^{\mathrm{H}}\mathbf{A}_k^{\mathrm{H}} + \mathbf{I}\right]w$$
$$s.t.\,w^{\mathrm{H}}\mathbf{A}_0\widetilde{x} = 1 \tag{9}$$

where $\ddot{\sigma}_k = \sigma_k^2/\sigma_0^2$. Furthermore, to accurately obtain some information of target or interference in the environment, waveforms in all directions should be equipped with some low sidelobes so as to reduce mutual interference. The beampattern power located at θ can be expressed as

$$P(\theta) = \mathbf{a}_t^{\mathrm{H}}(\theta)\mathbf{R}\mathbf{a}_t(\theta) \tag{10}$$

Considering that each array has the same emission energy, i.e., unitary power, the covariance matrix $\mathbf{R}$ can be designed as

$$R_{\mathrm{bb}} = \frac{1}{N_t}, \mathrm{b} = 1, \cdots, N_t \tag{11}$$

where R_{bb} represents the (b,b)-th element of covariance matrix $\mathbf{R}$. $\phi(\theta)$ denotes an expected transmit beampattern where $\{\theta_k\}_{k=1}^{K}$ is a fine grid covering the points of interest. Assuming that there are $\widetilde{K}$ expected target locations, the objective is to detect the target at locations $\{\widetilde{\theta}_k\}_{k=1}^{\widetilde{K}}$, which can be confirmed by calculating the Capon spatial spectrum or generalized likelihood ratio test (GLRT) [28]. Here, the dominant peak position has been calculated by the GLRT pseudo-spectrum so as to form the desired beampattern $\{\breve{\theta}_k\}_{k=1}^{\breve{K}}$, and $\breve{K}$ is the resulting estimate of $\widetilde{K}$, i.e.,

$$\phi(\theta) = \begin{cases} 1, & \theta \in \left[\widetilde{\theta}_k - \frac{\Delta}{2}, \widetilde{\theta}_k + \frac{\Delta}{2}\right], k \in \left\{1, \cdots, \breve{K}\right\} \\ 0, & \text{others} \end{cases} \tag{12}$$

where Δ is the beamwidth selected by each target.

Based on the aforementioned discussions, it is necessary to design a matrix $\mathbf{R}$ that minimizes the least squares error between the transmitted beampattern $P(\theta)$ and the expected beampattern $\phi(\theta)$, while also minimizing the cross-correlation terms from different backscattered signals. Consequently, the corresponding problem could be expressed as

$$
\begin{aligned}
\min_{\mathbf{R},\alpha} J(\mathbf{R},\alpha) \;=\;& \frac{1}{K}\sum_{k=1}^{K}\omega_k\left|\mathbf{a}_t^{\mathrm{H}}(\theta_k)\mathbf{R}\mathbf{a}_t(\theta_k)-\alpha\phi(\theta_k)\right|^2 \\
&+\frac{2\omega_c}{\widetilde{K}(\widetilde{K}-1)}\sum_{p=1}^{\widetilde{K}-1}\sum_{q=p+1}^{\widetilde{K}}\left|\mathbf{a}_t^{\mathrm{H}}\left(\widetilde{\theta}_p\right)\mathbf{R}\mathbf{a}_t\left(\widetilde{\theta}_q\right)\right|^2
\end{aligned}
\tag{13}
$$

where ω_k represents the weight factor of the k-th source, ω_c represents the weight factor of the cross-correlation term, and α is a scaling factor that needs to be optimized.

Moreover, some appropriate constraints are further imposed on the covariance matrix $\mathbf{R}$, which also considers the need of the low cross-correlation beampattern. The designed $\mathbf{R}$ must be positive semidefinite, and all diagonal elements of $\mathbf{R}$ must be equal to the uniform antenna power while satisfying the uniform basic-power constraint. The covariance optimization problem can be formulated as

$$
\begin{aligned}
&\min_{\mathbf{R},\alpha} J(\mathbf{R},\alpha) \\
&s.t.\,\mathbf{R}\succcurlyeq 0 \\
&R_{\mathrm{bb}}=\tfrac{1}{N_t},\mathrm{b}=1,\cdots,N_t \\
&\alpha>0
\end{aligned}
\tag{14}
$$

To solve the optimization problem of variables $(\mathbf{R},\alpha)$, the convex optimization toolbox CVX can be employed [29]. Note that when designing MIMO radar systems with low cross-correlation beampatterns, it is crucial to make a balance between detection and communication performance. It is necessary to define the optimization objectives and constraints based on the specific application requirements. This ensures that the design effectively meets the desired performance criteria.

3. Optimization Modeling of Radar and Communication Integrated System

In this section, the optimization objective is to maximize SINR_r while considering the signal-dependent interference. The design of the covariance matrix and transmitting waveform of MIMO radar will be addressed simultaneously. In practical radar applications, to improve the detection performance while maintaining the ability of multi-user communication, good ambiguity function and range resolution characteristics, it is necessary to introduce similarity constraints, such as

$$
\frac{1}{\sqrt{NN_t}}\|\widetilde{x}-\widetilde{x}_0\|_\infty \leq \xi
\tag{15}
$$

where $\widetilde{x}_0$ is the reference waveform and ξ is the similarity coefficient. To enhance detection performance and facilitate multi-user communication, the permissible range of ξ is usually set as $0\leq\xi\leq 2c_m$, where $c_m=1/\sqrt{NN_t}$. It should be noted that the communication performance may be severely degraded under the above constraints. Once the output result is obtained as the integrated referenced waveform $\mathbf{X}_0$, then it is equivalent to add the communication information into the integrated signal. Therefore, the optimization problem for the design of directional beamforming can be formulated as

$$
\begin{aligned}
&\min_{\mathbf{X}_0}\|\mathbf{HX}_0-\mathbf{S}\|_F^2 \\
&s.t.\,\tfrac{1}{N}\mathbf{X}_0\mathbf{X}_0^{\mathrm{H}}=\mathbf{R}
\end{aligned}
\tag{16}
$$

According to [19], the problem described in (16) can be characterized as an Orthogonal Procrustes problem (OPP) with a closed-form solution $\mathbf{X}_0=\sqrt{N}\mathbf{FUI}_{N_t\times N}\overset{\smile}{\mathbf{V}}{}^{\mathrm{H}}$, where $\mathbf{R}=\mathbf{FF}^{\mathrm{H}}$ represents the Cholesky decomposition or other valid square-root decomposition, while $\mathbf{U\Sigma V}^{\smile H} = \mathbf{F}^H\mathbf{H}^H\mathbf{S}$ represents the singular value decomposition (SVD). Consequently, the benchmark radar waveform $\mathbf{X}_0$ is uncorrelated with the power of the desired constellation $\mathbf{S}$.

Next, a compromise constraint, i.e., $\rho\|\mathbf{HX}-\mathbf{S}\|_F^2 + (1-\rho)\|\mathbf{X}-\mathbf{X}_0\|_F^2 \leq \mathrm{Y}$, is proposed, where Y defines the maximum permissible level for the communication performance metric and radar waveform similarity error. By taking the limited transmission energy into account, the optimization problem can be formulated as

$$
\min_{\tilde{x}} w^H\left[\sum_{k=1}^{K} \ddot{\sigma}_k \mathbf{A}_k\widetilde{\mathbf{x}}\widetilde{\mathbf{x}}^H\mathbf{A}_k^H + \mathbf{I}\right]w
$$

$$
s.t.\,\rho\|\mathbf{HX}-\mathbf{S}\|_F^2 + (1-\rho)\|\mathbf{X}-\mathbf{X}_0\|_F^2 \leq \mathrm{Y}
$$

$$
\frac{1}{\sqrt{NN_t}}\|\widetilde{\mathbf{x}}-\widetilde{\mathbf{x}}_0\|_\infty \leq \zeta
$$

$$
\frac{1}{N}\|\mathbf{X}\|_F^2 = P_T
$$

(17)

where $0 \leq \rho \leq 1$ denote the weight factor, and P_T represents the total power of all N_t antennas per symbol. The constraint conditions can be transformed into the following composite form [19], i.e.,

$$
\begin{aligned}
&\rho\|\mathbf{HX}-\mathbf{S}\|_F^2 + (1-\rho)\|\mathbf{X}-\mathbf{X}_0\|_F^2 \\
&= \left\|[\sqrt{\rho}\mathbf{H}^T, \sqrt{1-\rho}\mathbf{I}_{N_t}]^T\mathbf{X} - [\sqrt{\rho}\mathbf{S}^T, \sqrt{1-\rho}\mathbf{X}_0^T]^T\right\|_F^2
\end{aligned}
$$

(18)

Denoting $\mathbf{C} = [\sqrt{\rho}\mathbf{H}^T, \sqrt{1-\rho}\mathbf{I}_{N_t}]^T$ and $\mathbf{D} = [\sqrt{\rho}\mathbf{S}^T, \sqrt{1-\rho}\mathbf{X}_0^T]^T$, Equation (18) can be rewritten and extended to

$$
\begin{aligned}
\|\mathbf{CX}-\mathbf{D}\|_F^2 &= \mathrm{tr}\left((\mathbf{CX}-\mathbf{D})^H(\mathbf{CX}-\mathbf{D})\right) \\
&= \mathrm{tr}\left(\mathbf{X}^H\mathbf{C}^H\mathbf{CX}\right) - \mathrm{tr}\left(\mathbf{X}^H\mathbf{C}^H\mathbf{D}\right) - \mathrm{tr}(\mathbf{D}^H\mathbf{CX}) + \mathrm{tr}(\mathbf{D}^H\mathbf{D})
\end{aligned}
$$

(19)

Further defining $\mathbf{Q} = \mathbf{C}^H\mathbf{C}$ and $\mathbf{G} = \mathbf{C}^H\mathbf{D}$, Equation (19) can be rewritten as

$$
\mathrm{tr}\left(\mathbf{X}^H\mathbf{QX}\right) - 2\mathrm{Re}\left(\mathrm{tr}\left(\mathbf{X}^H\mathbf{G}\right)\right)
$$

(20)

where $\mathbf{Q}$ is a Hermitian matrix, Equation (20) can be written in the form of a Lagrange multiplier with respect to the total power as follows

$$
L(\mathbf{X},\lambda) = \mathrm{tr}\left(\mathbf{X}^H\mathbf{QX}\right) - 2\mathrm{Re}\left(\mathrm{tr}\left(\mathbf{X}^H\mathbf{G}\right)\right) + \lambda\left(\|\mathbf{X}\|_F^2 - NP_T\right)
$$

(21)

where λ is the dual variable of the equality constraint. Defining $\dot{\mathbf{X}}$ and $\dot{\lambda}$ as the optimal point and the dual optimal point with zero duality gap, according to the trust-region subproblem (TRS) optimality conditions [30], the following conclusions show

$$
\begin{aligned}
\dot{\mathbf{X}} &= \left(\mathbf{Q}+\dot{\lambda}\mathbf{I}_{N_t}\right)^\dagger\mathbf{G} \\
\left\|\left(\mathbf{Q}+\dot{\lambda}\mathbf{I}_{N_t}\right)^\dagger\mathbf{G}\right\|_F^2 &= \left\|\mathbf{V}\left(\mathbf{\Lambda}+\dot{\lambda}\mathbf{I}_{N_t}\right)^{-1}\mathbf{V}^H\mathbf{G}\right\|_F^2 \\
&= NP_T \\
\dot{\lambda} &\geq -\lambda_{min}
\end{aligned}
$$

(22)

where the notation $(\cdot)^{\dagger}$ refers to the Moore–Penrose pseudoinverse of the matrix. Furthermore, the matrix $\mathbf{Q}$ can be decomposed into $\mathbf{Q} = \mathbf{V}\boldsymbol{\Lambda}\mathbf{V}^{H}$ where λ_{min} denotes the minimum eigenvalue of $\mathbf{Q}$. It can be further proved that formula (22) has a unique solution, i.e.,

$$
\begin{aligned}
P(\lambda) &= \left\| \mathbf{V}(\boldsymbol{\Lambda} + \lambda \mathbf{I}_{N_t})^{-1} \mathbf{V}^{H} \mathbf{G} \right\|_{F}^{2} \\
&= \sum_{n=1}^{N_t} \sum_{j=1}^{N} \frac{\left(\left[\mathbf{V}^{H}\mathbf{G} \right]_{n,j} \right)^{2}}{(\lambda + \lambda_n)^{2}}
\end{aligned}
\tag{23}
$$

From deduction of (23), it is noted that when $\dot{\lambda} \geq -\lambda_{min}$, the function $P(\lambda)$ is strictly decreasing and convex. Therefore, the golden-section search method can be employed to determine the optimal solution for $\dot{\lambda}$.

4. Waveform Optimization Algorithm

In this section, a novel RIASCG algorithm is proposed to optimize the objective function. The Riemann gradient of function $h(\widetilde{x})$ is defined as $\mathrm{grad}h(\widetilde{x})$, which can be obtained by projecting the gradient on the Euclidean space. Here, $\mathrm{Grad}h(\widetilde{x})$ represents the Euclidean gradient of $h(\widetilde{x})$, and the contraction operator $\mathrm{Retr}(\cdot)$ maps the vector on the tangent space $\mathcal{T}_{\widetilde{x}}\mathcal{M}_s$ at the vicinity of manifold $\widetilde{x} \in \mathcal{M}_s$. The next iteration point $\widetilde{x}^{(l+1)}$ is considered when the objective value satisfies the descent condition. In each descent process, a more accurate step size $d^{(l)}$ needs to be selected to ensure faster convergence. To achieve this, an improved Riemannian manifold conjugate gradient algorithm based on the Armijo back-tracking line-search idea is proposed, offering several advantages over the first-order conjugate gradient algorithm [27]:

(1) Faster convergence speed: The second-order conjugate gradient algorithm, utilizing second-order derivative information, could more accurately determine the search direction and step size compared with the first-order conjugate gradient one, resulting in better results in the same number of iterations.
(2) More effective optimization for high-dimensional data: The first-order conjugate gradient algorithm may have a slow convergence speed when optimizing high-dimensional data, while the second-order conjugate gradient algorithm can better overcome this problem.
(3) Stronger numerical stability: The second-order conjugate gradient algorithm can better avoid numerical instability, which is particularly prominent in optimizing high-dimensional data.
(4) Fewer iterations: Due to faster convergence, the second-order conjugate gradient algorithm typically requires fewer iterations to achieve the same optimization effect, which is particularly important for optimizing large-scale data.

Suppose that the above constraints can be denoted as $\mathcal{M}_s$. For any $\mathbf{z} \in \mathbb{C}^{NN_t}$, its projection operator of the sequence $\widetilde{x}^{(l)} \in \mathcal{M}_s$ can be expressed as

$$
\mathrm{Proj}_{\widetilde{x}^{(l)}}^{\mathcal{M}_s}(\mathbf{z}) = \mathbf{z} - \mathrm{Re}\left\{ \mathbf{z}^{*} \odot \widetilde{x}^{(l)} \right\} \odot \widetilde{x}^{(l)}
\tag{24}
$$

Once the input $\mathbf{z} \in \mathbb{C}^{NN_t}$ is given, it is possible to apply a universal compression function to effectively address common constraints, i.e.,

$$
\mathrm{Retr}(\mathbf{z}) = \arg \min_{\widetilde{x} \in \mathcal{M}_s} \| \widetilde{x} - \mathbf{z} \|_{2}
\tag{25}
$$

The closed-form solution of w in Equation (17) can be derived, i.e.,

$$w = \frac{\left[\sum_{k=1}^{K} \ddot{\sigma}_k \mathbf{A}_k \widetilde{x} \widetilde{x}^{\mathrm{H}} \mathbf{A}_k^{\mathrm{H}} + \mathbf{I}\right]^{-1} \mathbf{A}_0 \widetilde{x}}{\widetilde{x}^{\mathrm{H}} \mathbf{A}_0^{\mathrm{H}} \left[\sum_{k=1}^{K} \ddot{\sigma}_k \mathbf{A}_k \widetilde{x} \widetilde{x}^{\mathrm{H}} \mathbf{A}_k^{\mathrm{H}} + \mathbf{I}\right]^{-1} \mathbf{A}_0 \widetilde{x}} \tag{26}$$

As a result, a subproblem of the optimization problem has been obtained, i.e.,

$$\min_{\widetilde{x}} \; -\widetilde{x}^{\mathrm{H}} \mathbf{A}_0^{\mathrm{H}} \left[\sum_{k=1}^{K} \ddot{\sigma}_k \mathbf{A}_k \widetilde{x} \widetilde{x}^{\mathrm{H}} \mathbf{A}_k^{\mathrm{H}} + \mathbf{I}\right]^{-1} \mathbf{A}_0 \widetilde{x} \tag{27}$$
$$s.t. \, \widetilde{x} \epsilon \mathcal{M}_s$$

The subproblem can be further expressed as

$$h(\widetilde{x}) = -\widetilde{x}^{\mathrm{H}} \left(\mathbf{A}_0^{\mathrm{H}} \left[\sum_{k=1}^{K} \ddot{\sigma}_k \mathbf{A}_k \widetilde{x} \widetilde{x}^{\mathrm{H}} \mathbf{A}_k^{\mathrm{H}} + \mathbf{I}\right]^{-1} \mathbf{A}_0 \right) \widetilde{x} \tag{28}$$

The Euclidean gradient of the smooth extension $\dot{h}(\widetilde{x})$ can be denoted as $\mathrm{Grad}\dot{h}(\widetilde{x})$, i.e.,

$$\begin{aligned}
\mathrm{Grad}\dot{h}(\widetilde{x}) \;=\; & -2\left(\mathbf{A}_0^{\mathrm{H}} \left(\sum_{k=1}^{K} \ddot{\sigma}_k \mathbf{A}_k \widetilde{x} \widetilde{x}^{\mathrm{H}} \mathbf{A}_k^{\mathrm{H}} + \mathbf{I}\right)^{-1} \mathbf{A}_0 \widetilde{x} \right) - \left(\widetilde{x}^{\mathrm{H}} \tfrac{\partial}{\partial \widetilde{x}} \left(\mathbf{A}_0^{\mathrm{H}} \left(\sum_{k=1}^{K} \ddot{\sigma}_k \mathbf{A}_k \widetilde{x} \widetilde{x}^{\mathrm{H}} \mathbf{A}_k^{\mathrm{H}} + \mathbf{I}\right)^{-1} \mathbf{A}_0 \right) \right) \widetilde{x} \\
=\; & -2\left(\mathbf{A}_0^{\mathrm{H}} \left(\sum_{k=1}^{K} \ddot{\sigma}_k \mathbf{A}_k \widetilde{x} \widetilde{x}^{\mathrm{H}} \mathbf{A}_k^{\mathrm{H}} + \mathbf{I}\right)^{-1} \mathbf{A}_0 \widetilde{x} \right) - \left(1_{NN_t} \otimes \widetilde{x}^{\mathrm{H}} \mathbf{A}_0^{\mathrm{H}} \left(\sum_{k=1}^{K} \ddot{\sigma}_k \mathbf{A}_k \widetilde{x} \widetilde{x}^{\mathrm{H}} \mathbf{A}_k^{\mathrm{H}} + \mathbf{I}\right)^{-1} \right) \\
& \times \sum_{k=1}^{K} \left(\ddot{\sigma}_k \mathbf{I}_{NNt} \otimes \mathbf{A}_k \left[\tfrac{\partial \widetilde{x} \widetilde{x}^{\mathrm{H}}}{\partial \widetilde{x}_1} \cdots \tfrac{\partial \widetilde{x} \widetilde{x}^{\mathrm{H}}}{\partial \widetilde{x}_{NN_t}}\right]^{\mathrm{T}} \mathbf{A}_k^{\mathrm{H}} \right) \times \left(\sum_{k=1}^{K} \ddot{\sigma}_k \mathbf{A}_k \widetilde{x} \widetilde{x}^{\mathrm{H}} \mathbf{A}_k^{\mathrm{H}} + \mathbf{I}\right)^{-1} \mathbf{A}_0 \widetilde{x}
\end{aligned} \tag{29}$$

The steps of manifold RIASCG for DFRC waveform design can be summarized in Algorithm 1.

Algorithm 1: The Manifold RIASCG for DFRC Waveform Design.

Input: $l = 0, \widetilde{x}^{(0)}, \beta, d, \omega, \eta, \mathbf{X}_0, \mathbf{H}, \mathbf{S}, P_T, \sigma_1 \in (0,1), \sigma_2 \in (0,1)$, weight factor $0 \le \rho \le 1$.
Output: $\widetilde{x}^{(l)}, w$.
While $\left| h\left(\widetilde{x}^{(i+1)}\right) - h\left(\widetilde{x}^{(i)}\right) \right| \ge \mathcal{J}$ **do**

    1. Compute $\mathbf{C} = [\sqrt{\rho}\mathbf{H}^{\mathrm{T}}, \sqrt{1-\rho}\mathbf{I}_{N_i}]^{\mathrm{T}}$, $\mathbf{D} = [\sqrt{\rho}\mathbf{S}^{\mathrm{T}}, \sqrt{1-\rho}\mathbf{X}_0^{\mathrm{T}}]^{\mathrm{T}}$, $\mathbf{Q} = \mathbf{C}^{\mathrm{H}}\mathbf{C}$ and $\mathbf{G} = \mathbf{C}^{\mathrm{H}}\mathbf{D}$.

    2. Compute the eigenvalue decomposition $\mathbf{Q} = \mathbf{V}\mathbf{\Lambda}\mathbf{V}^{\mathrm{H}}$ of $\mathbf{Q}$, set the searching interval as $[-\lambda_{min}, b]$, where $b \ge 0$ is a searching upper-bound.

    3. Find the optimal solution $\dot{\lambda}$ to (22) using golden-section search.

    4. Compute $\mathbf{X} = \left(\mathbf{Q} + \dot{\lambda}\mathbf{I}_N\right)^{\dagger} \mathbf{G}$.

    5. Once the global minimizer $\mathbf{X}$ is obtained, given its separability property, it can be employed as the reference waveform for the similarity constraint, denoted as $\widetilde{x}_0$.

    6. Compute $\mathrm{Grad}\dot{h}\left(\widetilde{x}^{(i)}\right)$ according to (29).

    7. Compute $\mathrm{Proj}_{\widetilde{x}^{(i)}}\left(\mathrm{Grad}h\left(\widetilde{x}^{(i)}\right)\right)$ according to (24).

    8. Compute the improved Armijo back-tracking line-search parameter $d^{(i)} = \omega\eta^\beta$, β is the smallest non-negative integer satisfying

$$h\left(\widetilde{x}^{(l)}\right) - h\left(\widetilde{x}^{(l)} - \omega\eta^\beta \mathrm{Grad}h\left(\widetilde{x}^{(l)}\right)\right) \ge \sigma_1 d^{(l)} \left\| \mathrm{Proj}_{\widetilde{x}^{(l)}}\left(\mathrm{Grad}h\left(\widetilde{x}^{(l)}\right)\right) \right\|_2^2$$
$$+ \sigma_2 \left(d^{(l)}\right)^2 \left\| \mathrm{Proj}_{\widetilde{x}^{(l)}}\left(\mathrm{Grad}h\left(\widetilde{x}^{(l)}\right)\right) \right\|_2^2$$

    9. Perform the projection step $\dot{\widetilde{x}}^{(l+1)} = \widetilde{x}^{(l)} - d^{(l)} \mathrm{Proj}_{\widetilde{x}^{(l)}}\left(\mathrm{Grad}h\left(\widetilde{x}^{(l)}\right)\right)$.

    10. Obtain $\widetilde{x}^{(l+1)} = \mathrm{Retr}\left(\widetilde{x}^{(l)}\right)$ according to (25).

    11. $l = l + 1$.
End while
Compute w according to (26).

---

5. Numerical Result

To evaluate the performance of the designed waveforms, we first formulated the simulation scenarios. A uniform linear array configuration is assumed for both the transmitting and receiving arrays. Units in these arrays have an element spacing of half a wavelength. The number of transmitting antennas is $N_t = 10$, the number of receiving antennas is $N_r = 10$, and the target echo power is 10 dB. Additionally, the interference power is set to 20 dB, the noise power is 0 dB and the code-length of the waveform is $N = 8$, while ϖ and η are drawn from $[0, 0.5]$ and $[0, 1]$, respectively. For convenience, $P_T = 1$ and assume that each entry of the channel matrix **H** is modeled as flat fading one and also is independently and identically distributed with a standard complex Gaussian distribution $h_{i,j} \sim \mathcal{CN}(0, 1)$. The constellation selected for the communication users is the unit-power QPSK alphabet, and the threshold value of $\left| h\left(\widetilde{x}^{(i+1)} \right) - h\left(\widetilde{x}^{(i)} \right) \right|$ is set as 10^{-4}.

By employing Equation (14), the objective is to construct the covariance matrix **R** for problem (16), and then its performance would be compared with the omni-directional waveform. Assume that there are four users, and the direction of arrival (DoA) information for $\widetilde{K} = 3$ targets with unit complex amplitude is approximately $\{-50°, 0°, 50°\}$, which can be obtained by the Capon or GLRT method. Three symmetrical beampatterns of interest are denoted as $\widetilde{\theta}_1 = -50°, \widetilde{\theta}_2 = 0°, \widetilde{\theta}_3 = 50°$, with a beampattern width of $\Delta = 20°$. Next, the performance of the omni-directional and directional beampattern would be discussed, and a radar-communication compromise waveform would be designed to achieve some flexible trade-off between radar and communication for practical needs. Considering the trade-off design for radar communication, the Pareto weight factor $\rho = 0.2$ is introduced, and 'Omni' and 'Directional' are denoted as the omnidirectional and directional beam-patterns. Further, the waveforms with strict equality constraints are denoted as 'Strict', while the trade-off designs are denoted as 'Tradeoff'. In Figure 2, it is evident that the proposed method exhibits significantly lower sidelobe levels compared to the method of [19].

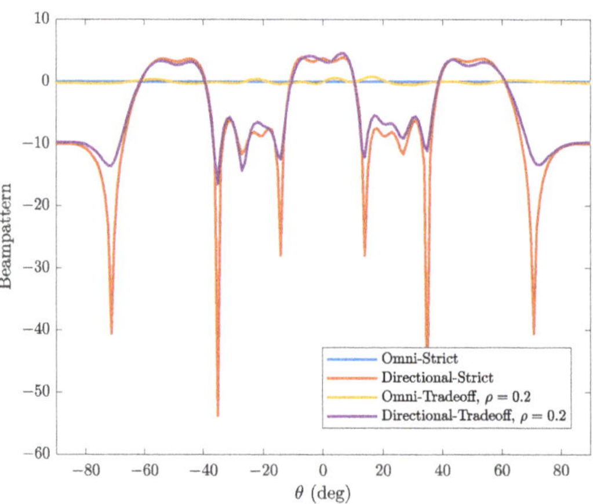

Figure 2. Comparison of radar beampattern corresponding to different cases.

Moreover, to further assess the robustness of different algorithms, the proposed RIASCG would be compared with RCG [31], MM [32], MM-SQUAREM [33] and RCG-Armijo algorithms. To enhance the comparability, two additional constraints are also incorporated: the constant modulus constraint [34] and the e-uncertainty constant modulus constraint [35]. As shown in Figure 3, in comparison to the first-order RCG-Armijo algorithm, the second-order conjugate RIASCG algorithm used in this paper demonstrates faster convergence. The first-order conjugate gradient algorithm needs some precise line searching, which could increase computational costs and lead to some direction inconsistency in certain cases. Particularly in ill-conditioned problems, this algorithm might encounter direction loss within iterations and fail to converge to the minimum value. In such scenarios, the second-order conjugate gradient algorithm appears to be more advantageous than the first-order one, as it utilizes more information to determine the search direction and reduces the direction inconsistency. The RIASCG algorithm has demonstrated fewer iterations and a faster convergence rate compared with other algorithms.

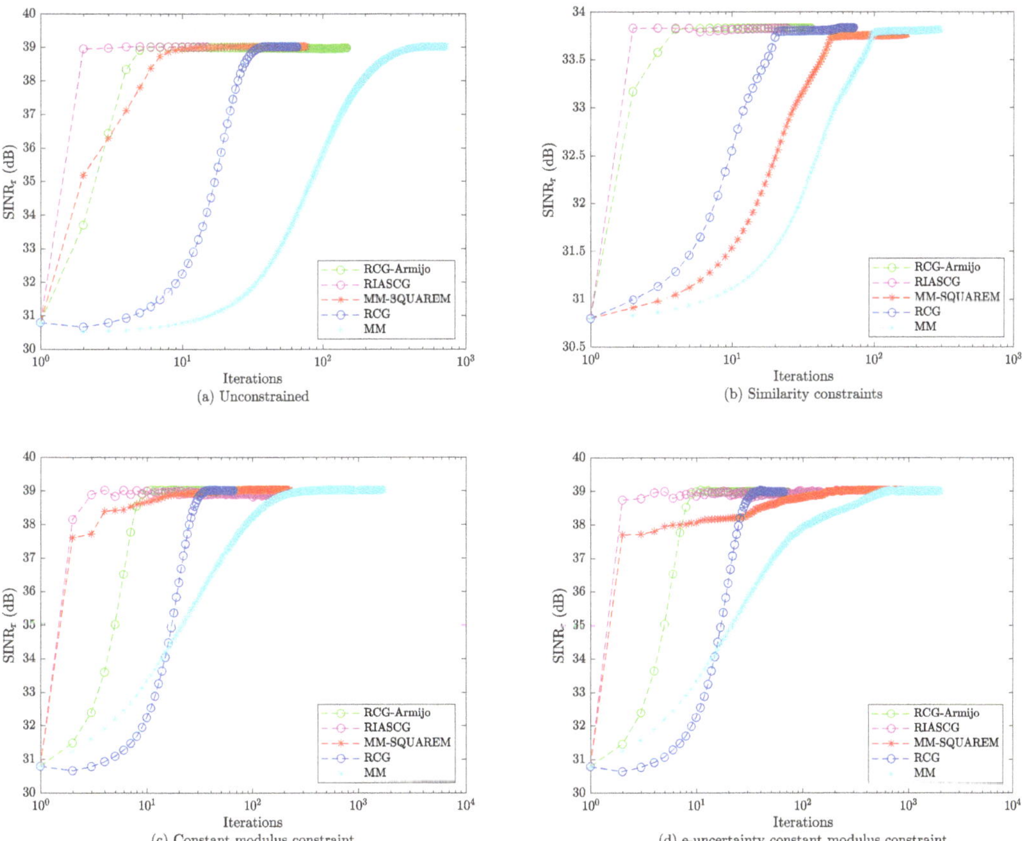

Figure 3. Comparison of convergence rates under different constraints. (**a**) Unconstrained; (**b**) Similarity constraints; (**c**) Constant modulus constraint; (**d**) e-uncertainty constant modulus constraint.

To evaluate the scalability and robustness of algorithms, the computation runtime comparisons were also conducted for different N in Table 1 considering the CM&S constraint. Obviously, the practicality and scalability of RIASCG has effectively demonstrated that the comparative runtime performance has outperformed other prior works, such as RCG, RCG-Armijo, MM and MM-SQUAREM, especially for larger N. These comparisons could underscore the effectiveness and applicability of RIASCG.

Next, the effect of the similarity constraint has been discussed. As shown in Figure 4, when the similarity coefficient between two users is lower, it indicates a lower similarity between their signals, which can result in a greater degree of interference between them. When multiple users transmit signals simultaneously, the interference will affect the quality of the received signal and also result in a lower SINR_r. In Figure 5, waveforms optimized by RIASCG algorithm as ζ decreasing have shown different shapes of ambiguity function. The similarity between the designed waveform and the reference waveform would be gradually increased, which would result in a better-formed ambiguity function. However, this also leads to a reduction in the degrees of freedom for waveform design.

Table 1. Runtime comparisons of different methods under CM&S constraint.

Algorithm	*N*=4	*N*=8	*N*=16	*N*=32
RIASCG	0.6875 s	2.7031 s	14.1719 s	100.0785 s
RCG	4.3751 s	9.9843 s	188.2811 s	709.3284 s
RCG-Armijo	0.8281 s	3.6718 s	40.1406 s	232.0167 s
MM	6.7968 s	14.4375 s	328.3755 s	2562.1734 s
MM-SQUAREM	1.5937 s	5.5156 s	113.8751 s	630.3911 s

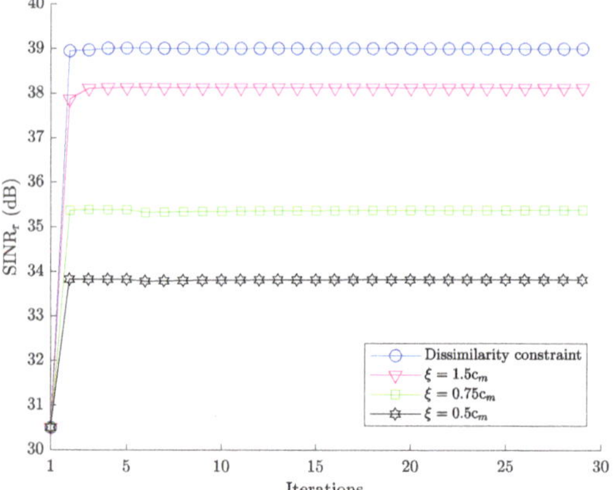

Figure 4. $\mathrm{SINR_r}$ vs. iteration number for different ξ.

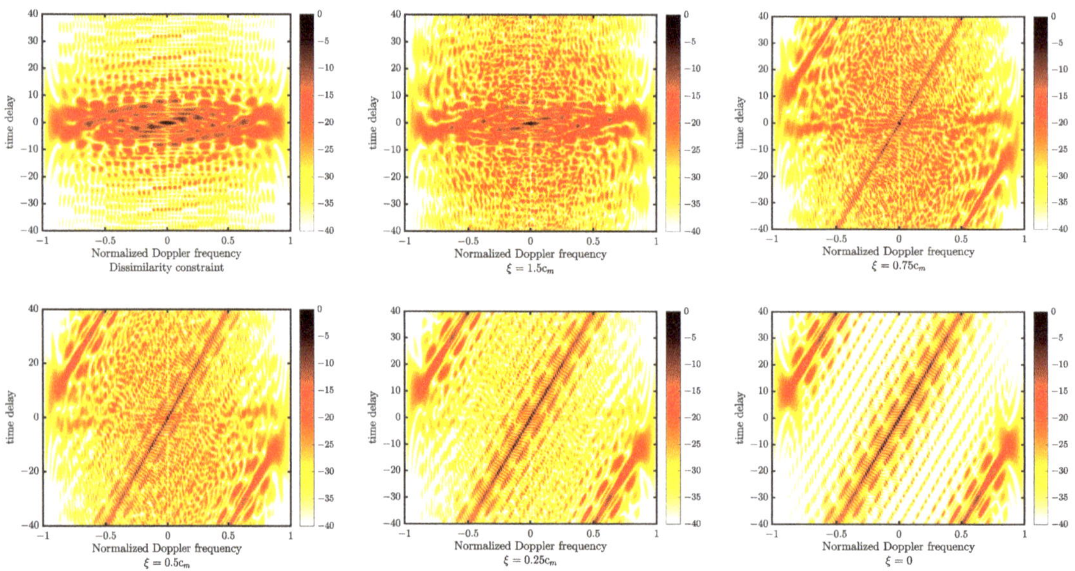

Figure 5. Ambiguity functions of the presented waveforms as ξ decreasing.

Furthermore, given a specified radar beampattern, the design of dual-function waveforms is further analyzed. As demonstrated in Figure 6, we define the transmit SNR as SNR $= P_T/N_0$, and then examine the relationship between the transmit SNR of the communication signal and the average achievable sum-rate. Here, ϵ represents the stopping criterion for the golden-section search iteration. As the transmit SNR increases, the effect of the signal synthesis error on the average achievable sum-rate increases. Therefore, the sum-rate of the synthesized signal is slightly lower than that of a single communication signal.

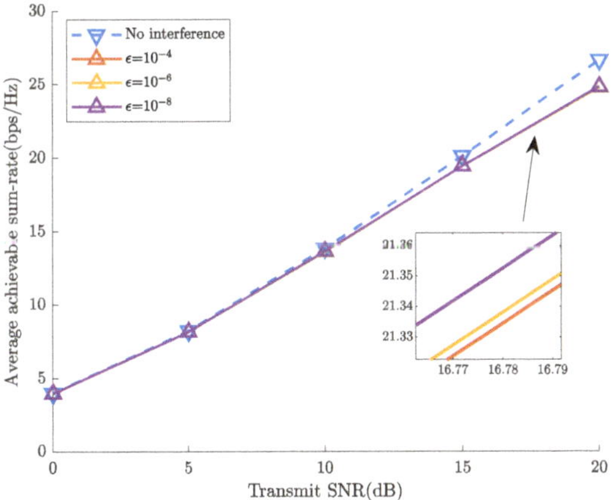

Figure 6. SNR comparison of different stop conditions for golden-section search iterations.

Moreover, in Figure 7, the relationship between the average achievable rate per-user vs. the detection probability is illustrated when the receive SNR has -6 dB and the false-alarm probability has $P_{FA} = 10^{-7}$. With an increase in the number of users, the detection probability would be decreased for a fixed average achievable rate per-user, which suggests that the increasing degrees of freedom could further minimize MUI energy. In Figure 8, the relationship between the average achievable rate per-user vs. the number of iterations has been demonstrated. Once the number of communication users increases, the feasible solution space decreases, which leads to a decline in the average achievable rate.

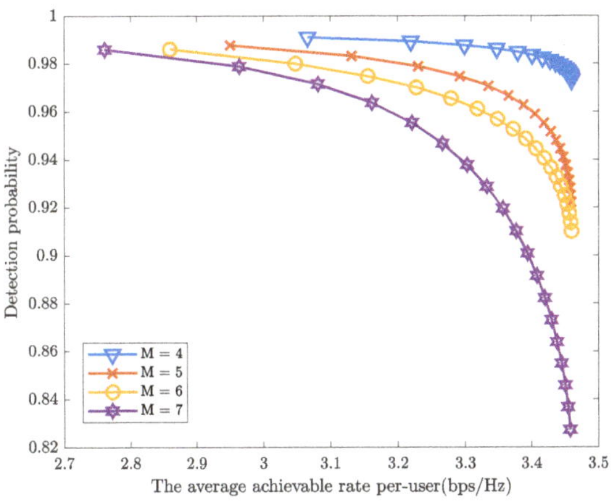

Figure 7. The average achievable rate per-user vs. radar detection probability.

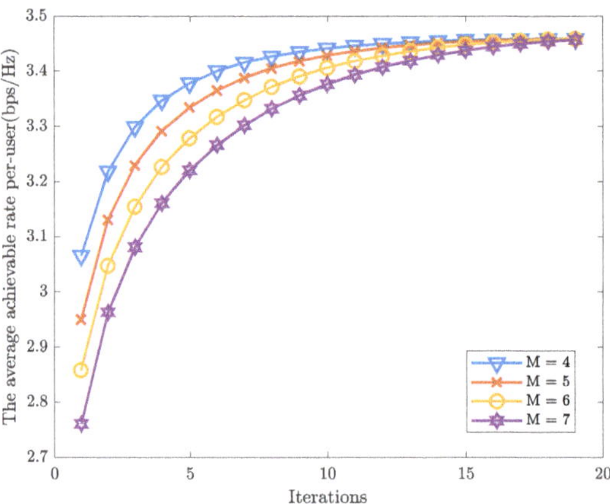

Figure 8. The average achievable rate per-user vs. number of iterations.

6. Conclusions

In this paper, we aimed to design an integrated waveform for DFRC using an RI-ASCG framework, which offers a flexible trade-off between radar and communication performance. To accomplish this, the manifold principle was leveraged to transform the constrained CM&S problem into unconstrained Riemann spaces, ensuring radar beampattern constraints and their trade-offs. Simulations have demonstrated the convergence performance and superiority of RIASCG when compared with other existing algorithms. Moreover, by adjusting the similarity coefficient, the designed waveform exhibited desirable properties in terms of the ambiguity function. As this paper demonstrated, we mainly focused on presenting a novel optimizing idea to tackle the joint optimization of integrated transmitting waveforms and receiving filters, which ignores the relative motion or Doppler shift. Next, in our future research, we will investigate the influence of relative motion and the Doppler shift and try to assess the robustness and adaptablity of the proposed techniques in realistic scenarios. Additionally, we will also investigate the impact of varying channel conditions and different fading models, such as frequency-selective fading or multipath fading.

Author Contributions: Conceptualization, Y.Z. and X.F.; data curation, Z.Z. (Zhongqing Zhao) and F.T.; formal analysis, X.F. and P.S.; funding acquisition, X.F. and Y.Z.; investigation, Y.Z., X.F. and Z.Z. (Zhanfeng Zhao); methodology, X.F., Z.Z. (Zhanfeng Zhao) and P.S.; project administration, Y.Z., X.F. and Z.Z. (Zhongqing Zhao); resources, Z.Z. (Zhanfeng Zhao) and P.S.; supervision, X.F.; validation, Y.Z. and F.T.; writing—original draft, X.F. and Z.Z. (Zhongqing Zhao); writing—review and editing, X.F. and F.T. All authors have read and agreed to the published version of the manuscript.

Funding: This research was supported in part by the major scientific and technological innovation projects of Shandong Province of China (Grant No. 2021ZLGX05 and 2022ZLGX04).

Data Availability Statement: Not applicable.

Conflicts of Interest: The authors declare no conflict of interest.

References

1. Liu, R.; Li, M.; Liu, Y.; Wu, Q.; Liu, Q. Joint Transmit Waveform and Passive Beamforming Design for RIS-Aided DFRC Systems. *IEEE J. Sel. Top. Signal Process.* **2022**, *16*, 995–1010. [CrossRef]
2. Qian, X.; Zeng, Y.; Wang, W.; Zhang, Q. Co-Saliency Detection Guided by Group Weakly Supervised Learning. *IEEE Trans. Multimed.* **2023**, *25*, 1810–1818. [CrossRef]
3. Lin, S.; Zhang, M.; Cheng, X.; Wang, L.; Xu, M.; Wang, H. Hyperspectral Anomaly Detection via Dual Dictionaries Construction Guided by Two-Stage Complementary Decision. *Remote Sens.* **2022**, *14*, 1784. [CrossRef]

4. Yuan, W.; Wei, Z.; Li, S.; Yuan, J.; Ng, D.W.K. Integrated Sensing and Communication-Assisted Orthogonal Time Frequency Space Transmission for Vehicular Networks. *IEEE J. Sel. Top. Signal Process.* **2021**, *15*, 1515–1528. [CrossRef]
5. Liu, R.; Li, M.; Liu, Q.; Swindlehurst, A.L. Joint Waveform and Filter Designs for STAP-SLP-Based MIMO-DFRC Systems. *IEEE J. Sel. Areas Commun.* **2022**, *40*, 1918–1931. [CrossRef]
6. Labib, M.; Marojevic, V.; Martone, A.F.; Reed, J.H.; Zaghloui, A.I. Coexistence between Communications and Radar Systems: A Survey. *URSI Radio Sci. Bull.* **2017**, *2017*, 74–82. [CrossRef]
7. Ma, D.; Shlezinger, N.; Huang, T.; Liu, Y.; Eldar, Y.C. Joint Radar-Communication Strategies for Autonomous Vehicles: Combining Two Key Automotive Technologies. *IEEE Signal Process. Mag.* **2020**, *37*, 85–97. [CrossRef]
8. Chen, Y.; Gu, X. Time Allocation for Integrated Bi-Static Radar and Communication Systems. *IEEE Commun. Lett.* **2021**, *25*, 1033–1036. [CrossRef]
9. Tian, T.; Li, G.; Zhou, T. Power Distribution for an OFDM-Based Dual-Function Radar-Communication Sensor. *IEEE Sens. Lett.* **2020**, *4*, 1–4. [CrossRef]
10. Yu, X.; Yao, X.; Yang, J.; Zhang, L.; Kong, L.; Cui, G. Integrated Waveform Design for MIMO Radar and Communication via Spatio-Spectral Modulation. *IEEE Trans. Signal Process.* **2022**, *70*, 2293–2305. [CrossRef]
11. Hassanien, A.; Amin, M.G.; Zhang, Y.D.; Ahmad, F. Dual-Function Radar-Communications: Information Embedding Using Sidelobe Control and Waveform Diversity. *IEEE Trans. Signal Process.* **2016**, *64*, 2168–2181. [CrossRef]
12. Cheng, Z.; Wu, L.; Wang, B.; Shankar, M.R.B.; Ottersten, B. Double-Phase-Shifter Based Hybrid Beamforming for MmWave DFRC in the Presence of Extended Target and Clutters. *IEEE Trans. Wirel. Commun.* **2023**, *22*, 3671–3686. [CrossRef]
13. Ahmed, A.; Zhang, Y.D.; Gu, Y. Dual-Function Radar-Communications Using QAM-Based Sidelobe Modulation. *Digit. Signal Process.* **2018**, *82*, 166–174. [CrossRef]
14. Huang, T.; Xu, X.; Liu, Y.; Shlezinger, N.; Eldar, Y.C. A Dual-Function Radar Communication System Using Index Modulation. In Proceedings of the 2019 IEEE 20th International Workshop on Signal Processing Advances in Wireless Communications (SPAWC), Cannes, France, 2–5 July 2019; pp. 1–5.
15. Yang, J.; Cui, G.; Yu, X.; Kong, L. Dual-Use Signal Design for Radar and Communication via Ambiguity Function Sidelobe Control. *IEEE Trans. Veh. Technol.* **2020**, *69*, 9781–9794. [CrossRef]
16. Sen, S. PAPR-Constrained Pareto-Optimal Waveform Design for OFDM-STAP Radar. *IEEE Trans. Geosci. Remote Sens.* **2014**, *52*, 3658–3669. [CrossRef]
17. Zhou, S.; Liang, X.; Yu, Y.; Liu, H. Joint Radar-Communications Co-Use Waveform Design Using Optimized Phase Perturbation. *IEEE Trans. Aerosp. Electron. Syst.* **2019**, *55*, 1227–1240. [CrossRef]
18. Hassanien, A.; Amin, M.G.; Aboutanios, E.; Himed, B. Dual-Function Radar Communication Systems: A Solution to the Spectrum Congestion Problem. *IEEE Signal Process. Mag.* **2019**, *36*, 115–126. [CrossRef]
19. Liu, F.; Zhou, L.; Masouros, C.; Li, A.; Luo, W.; Petropulu, A. Toward Dual Functional Radar-Communication Systems: Optimal Waveform Design. *IEEE Trans. Signal Process.* **2018**, *66*, 4264–4279. [CrossRef]
20. Tsinos, C.G.; Arora, A.; Chatzinotas, S.; Ottersten, B. Joint Transmit Waveform and Receive Filter Design for Dual-Function Radar-Communication Systems. *IEEE J. Sel. Top. Signal Process.* **2021**, *15*, 1378–1392. [CrossRef]
21. De Maio, A.; Huang, Y.; Piezzo, M.; Zhang, S.; Farina, A. Design of Optimized Radar Codes With a Peak to Average Power Ratio Constraint. *IEEE Trans. Signal Process.* **2011**, *59*, 2683–2697. [CrossRef]
22. Cheng, Z.; He, Z.; Liao, B.; Fang, M. MIMO Radar Waveform Design With PAPR and Similarity Constraints. *IEEE Trans. Signal Process.* **2018**, *66*, 968–981. [CrossRef]
23. Cui, G.; Li, H.; Rangaswamy, M. MIMO Radar Waveform Design With Constant Modulus and Similarity Constraints. *IEEE Trans. Signal Process.* **2014**, *62*, 343–353. [CrossRef]
24. BouDaher, E.; Hassanien, A.; Aboutanios, E.; Amin, M.G. Towards a Dual-Function MIMO Radar-Communication System. In Proceedings of the 2016 IEEE Radar Conference (RadarConf), Philadelphia, PA, USA, 2–6 May 2016; pp. 1–6.
25. Jiang, M.; Liao, G.; Yang, Z.; Liu, Y.; Chen, Y. Integrated Radar and Communication Waveform Design Based on a Shared Array. *Signal Process.* **2021**, *182*, 107956. [CrossRef]
26. Shi, S.; Wang, Z.; He, Z.; Cheng, Z. Constrained Waveform Design for Dual-Functional MIMO Radar-Communication System. *Signal Process.* **2020**, *171*, 107530. [CrossRef]
27. Liu, F.; Masouros, C.; Li, A.; Sun, H.; Hanzo, L. MU-MIMO Communications With MIMO Radar: From Co-Existence to Joint Transmission. *IEEE Trans. Wirel. Commun.* **2018**, *17*, 2755–2770. [CrossRef]
28. Bose, A.; Khobahi, S.; Soltanalian, M. Efficient Waveform Covariance Matrix Design and Antenna Selection for MIMO Radar. *Signal Process.* **2021**, *183*, 107985. [CrossRef]
29. Grant, M.; SP, B. CVX: MATLAB Software for Disciplined Convex Programming. 2014. Available online: http://cvxr.com/cvx (accessed on 19 March 2023).
30. Rendl, F.; Wolkowicz, H. A Semidefinite Framework for Trust Region Subproblems with Applications to Large Scale Minimization. *Math. Program.* **1997**, *77*, 273–299. [CrossRef]
31. Alhujaili, K.; Monga, V.; Rangaswamy, M. Transmit MIMO Radar Beampattern Design via Optimization on the Complex Circle Manifold. *IEEE Trans. Signal Process.* **2019**, *67*, 3561–3575. [CrossRef]
32. Hunter, D.R.; Lange, K. A Tutorial on MM Algorithms. *Am. Stat.* **2004**, *58*, 30–37. [CrossRef]

33. Song, J.; Babu, P.; Palomar, D. Sequence Design to Minimize the Weighted Integrated and Peak Sidelobe Levels. *IEEE Trans. Signal Process.* **2015**, *64*, 2051–2064. [CrossRef]
34. He, H.; Li, J.; Stoica, P. *Waveform Design for Active Sensing Systems: A Computational Approach*; Cambridge University Press: Cambridge, UK; New York, NY, USA, 2012.
35. Zhao, L.; Song, J.; Babu, P.; Palomar, D.P. A Unified Framework for Low Autocorrelation Sequence Design via Majorization–Minimization. *IEEE Trans. Signal Process.* **2017**, *65*, 438–453. [CrossRef]

remote sensing

Article

Joint Design of Complementary Sequence and Receiving Filter with High Doppler Tolerance for Simultaneously Polarimetric Radar

Yun Chen [1,2], Yunhua Zhang [1,2,*], Dong Li [1,2] and Jiefang Yang [1,2]

[1] CAS Key Laboratory of Microwave Remote Sensing, National Space Science Center, Chinese Academy of Sciences, Beijing 100190, China; chenyun191@mails.ucas.ac.cn (Y.C.); lidong@mirslab.cn (D.L.); yangjiefang@mirslab.cn (J.Y.)

[2] School of Electronic, Electrical, and Communication Engineering, University of Chinese Academy of Sciences, Beijing 100049, China

* Correspondence: zhangyunhua@mirslab.cn

Abstract: Simultaneously polarimetric radar (SPR) realizes the rapid measurement of a target's polarimetric scattering matrix by transmitting orthogonal radar waveforms of good ambiguity function (AF) properties and receiving their echoes via two orthogonal polarimetric channels at the same time, e.g., horizontal (H) and vertical (V) channels (antennas) sharing the same phase center. The orthogonality of the transmitted waveforms can be realized using low-correlated phase-coded sequences in the H and V channels. However, the Doppler tolerances of the waveforms composed by such coded sequences are usually quite low, and it is hard to meet the requirement of accurate measurement regarding moving targets. In this paper, a joint design approach for unimodular orthogonal complementary sequences along with the optimal receiving filter is proposed based on the majorization–minimization (MM) method via alternate iteration for obtaining simultaneously polarimetric waveforms (SPWs) of good orthogonality and of the desired AF. During design, the objective function used for minimizing the sum of the complementary integration sidelobe level (CISL) and the complementary integration isolation level (CIIL) is constructed under the mismatch constraint of signal-to-noise ratio (SNR) loss. Different SPW examples are given to show the superior performance of our design in comparison with other designs. Finally, practical experiments implemented with different SPWs are conducted to show our advantages more realistically.

Keywords: simultaneously polarimetric radar (SPR); orthogonal waveforms; complementary sequences; Doppler tolerance; majorization–minimization (MM); mismatch filter

Citation: Chen, Y.; Zhang, Y.; Li, D.; Yang, J. Joint Design of Complementary Sequence and Receiving Filter with High Doppler Tolerance for Simultaneously Polarimetric Radar. *Remote Sens.* **2023**, *15*, 3877. https://doi.org/10.3390/rs15153877

Academic Editors: Guolong Cui, Yong Yang, Xianxiang Yu and Bin Liao

Received: 16 June 2023
Revised: 29 July 2023
Accepted: 31 July 2023
Published: 4 August 2023

1. Introduction

Developments in radar technology have promoted the application of the polarimetric scattering information of targets, which can be characterized by the polarimetric scattering matrix (PSM) that connects the Jones vector of the incident wave with that of the scattering wave. The performance of radar on imaging, detection, and classification can be greatly improved by making full use of PSM in many fields, such as terrain observation, battlefield investigation, and disaster monitoring [1–4].

To accurately measure the PSM of moving targets, two typical fully polarimetric measurement schemes have been extensively investigated, i.e., the alternate polarimetric scheme and the simultaneously polarimetric scheme [5–7]. For the first scheme, the polarization state of the transmitted waveform is alternately switched between the vertical (V) and horizontal (H) polarizations, while both polarization states are received simultaneously. For high-speed targets, the two columns of the measured PSM can be decorrelated in the time domain and therefore the accuracy of the measurement results may be influenced. By contrast, for the second scheme, a pair of different waveforms of H and V polarization states are transmitted within one pulse simultaneously. As a result, the decorrelation impact can be avoided, but a stringent orthogonality requirement is put forward for the transmitted waveforms [6–9].

In addition to the orthogonality, low sidelobes of pulse compression in co-polarization channels are often required for weak target detection [10], i.e., good pulse compression characteristics of the waveforms are also desired for simultaneously polarimetric radar (SPR), as well as good orthogonality. Meanwhile, in the consideration of moving targets, the radar waveform should have good Doppler tolerance, i.e., it should be Doppler resilient [7,11]. These waveform properties can usually be evaluated by utilizing the ambiguity function (AF) [12,13]. Therefore, the AFs of ideal SPR waveforms must be of a two-dimensional "thumbtack" shape in co-polarization and an all-zero "plane" in cross-polarization channels, respectively [10,14].

Considering the nonlinear effects in practical hardware, it is preferable to make the waveform be of unimodular property, i.e., constant modulus [15–17]. In the early stage, a pair of linear frequency modulation (LFM) waveforms with opposite slopes are used to realize the simultaneously polarimetric measurement [5,18]. However, the peak sidelobe level (PSL) is just −13.26 dB after matched filtering. According to the principle of stationary phase, the orthogonality between the positive and negative slope LFM waveform is limited by the time–bandwidth product [19]. In recent years, phase-coded waveforms (PCWs) with good orthogonality are attractive for multichannel radar, such as SPR and MIMO (multiple-input–multiple-output) radar [7,20–22]. In [15,23], Stoica et al. proposed a series of cyclic algorithms for minimizing the integral sidelobe level (ISL) of unimodular waveforms, including cyclic algorithm-pruned (CAP), CA new (CAN), weighted CAN (We CAN), and CA direct (CAD). It is worth noting that Palomar et al. developed the majorization–minimization (MM) method to solve the nonconvex problem in unimodular sequence design by fast Fourier transform (FFT) [24,25]. This method is computationally attractive in the optimization of phase-coded sequences. Although many efforts have been put into the design of orthogonal PCWs with both low PSL and ISL. Nonetheless, it is impossible to obtain ideal correlation properties (ICPs), i.e., impulse-like autocorrelation and all-zero cross-correlation, for all the time delays using single pulse phase code, let alone the situation of various Doppler shifts [26–28].

The aforementioned difficulties motivate researchers to use the complete complementary sequence (CCS) in waveform design by taking advantage of the complementarity between CCS, which is a generalization of the well-known Golay code having been widely applied in the MIMO and CDMA communication systems because of its ICPs [29,30]. However, the ICPs will be seriously degraded even if a slight Doppler shift exists. This is the main obstacle to the wide application of CCS waveforms in radar [31]. In contrast to the waveform design suppressing the sidelobes of correlation functions, the waveform design minimizing the sidelobes of AFs can achieve the desired Doppler tolerance [12,32]. In [33], a construction method was proposed based on the Generalized Prouhet–Thue–Morse (GPTM) sequence by reordering the expanded version of an existing complementary sequence. The complementary properties can be kept over a modest Doppler frequency range (DFR). Nevertheless, the construction method has a strict limitation on the pulse number and code length of the complementary sequence. In [34], a set of almost complementary sequences was designed using the MM method. In their work, the pulse number and code length of CCS are no longer restricted, but there still is room to improve the Doppler tolerance. Recently, the limited-memory Broyden–Fletcher–Goldfarb–Shanno (L-BFGS) algorithm was used to optimize the AF of CCS [35–37]; however, the efficiency and the obtained optimal result still can be further improved.

The above studies mainly aim at the waveform design based on the matched filter scheme. It has been demonstrated that the pulse compression sidelobes can be suppressed effectively via mismatched filtering with a caveat of inducing an appropriate signal-to-noise ratio (SNR) loss [38–40]. In recent work, particularly [41,42], the joint design algorithm of the waveform and the receiving filter under the SNR loss constraint based on the MM method was reported with the purpose of minimizing the pulse compression sidelobes. However, for the SPR waveform, not only the low sidelobe of co-polarization channels but also the good orthogonality between cross-polarization channels and the Doppler tolerance should be considered simultaneously. Therefore, the applicability of these algorithms in the waveform design of SPR is limited.

In this paper, we focus on the joint design of orthogonal CCSs and receiving filters with desired Doppler tolerance for SPR waveforms. In addition, the main contributions can be summarized as follows:

(1) Based on the AF, the joint design of unimodular orthogonal CCS (UOCCS) and receiving filter is proposed for SPR waveforms. Specifically, the complementary integrated sidelobe level (CISL) of Auto-AFs, the complementary integrated isolation level (CIIL) of Cross-AFs, and the mismatch constraint with controllable SNR loss are all considered simultaneously in the objective function formulated for optimization. By setting the predefined SNR loss, a trade-off between the suppression of CISL/CIIL and actual SNR loss can be achieved. In other words, the work in [41,42] was extended, i.e., the proposed scheme not only considers the low sidelobe of the pulse compression of CCS but also takes into account the orthogonality for all time delays within appropriate DFR.

(2) The joint design problem is decomposed into subproblems of waveform design and receiving filter design via theoretical derivation, which is solved via an alternatively iterative approach. Concretely, the two subproblems are transformed into nonconvex quadratic terms containing the Hermitian matrix. The MM method is then applied to transforming these two nonconvex quadratic terms into linear programming problems with closed solutions. By utilizing the characteristics of Toeplitz matrix-vector multiplication, the main computation step can be completed via FFT. For further improvement, the convergence speed of the algorithm, an acceleration scheme of the squared iterative method (SQUAREM) is introduced. Compared with the representative and latest MM-CCS method [34] and the L-BFGS algorithm [35], better performance is achieved by the proposed algorithm, benefiting from both the joint design and the application of the MM framework.

The remainder of the paper is organized as follows. In Section 2, the problem of joint design CCSs and filters is formulated for the optimization of both the Auto-AFs and the Cross-AFs. In Section 3, the algorithm is developed with an acceleration scheme incorporated. In Section 4, several design examples are presented to show superior performance, and in Section 5, the designed waveforms using our method and that using the MM-CCS and the L-BFGS algorithm are implemented using an experimental hardware system, and extensive tests are conducted to demonstrate the actual performance. Finally, conclusions are drawn in Section 6.

Table 1 lists all used operators with explanations, and the bold lowercase and uppercase letters are used to denote the column vector of a matrix and a matrix throughout the paper.

Table 1. Mathematical Notations.

Notation	Meaning
$(\cdot)^T$	the transposition of a vector/matrix
$(\cdot)^*$	the conjugate of a complex number/vector/matrix
$(\cdot)^\dagger$	the conjugate transpose of a vector/matrix
$\lvert \cdot \rvert$	the modulus of a complex number
$\lVert \cdot \rVert$	the l_2-norm of a vector
$Re(\cdot)$	the real part of a complex number
$Tr(\cdot)$	the trace of a matrix
$vec(\cdot)$	the stacking vectorization of a matrix
$Diag(x)$	a diagonalized matrix of x
$\odot$	Hadamard product
$\mathbf{0}_N$	an all-zero column vector with dimension N
$\mathbf{1}_N$	an all-one column vector with dimension N
$\mathbf{0}_{L \times L}$	an all-zero matrix with dimension $L \times L$
$\mathbf{I}_L$	the identity matrix with dimension $L \times L$

2. Problem Statement

As we know, the unimodular constraint can avoid the nonlinear effect [43]. In the following, we first introduce two metrics, i.e., CISL and CIIL, for measuring the performance of the AFs of the unimodular waveforms. Then, the design problem is formulated as the objective function used for minimizing these two metrics under the constraint of controllable SNR loss.

Let us consider an SPR that transmits two orthogonal CCSs with H and V polarization, as shown in Figure 1a, and the signal-processing procedure at the radar receiver is shown in Figure 1b. In the transmitted signals of length N, each coherent processing interval (CPI) contains K pulses. In this case, the waveforms are diverse not only in different polarization states but also in different pulse-repetition intervals (PRIs).

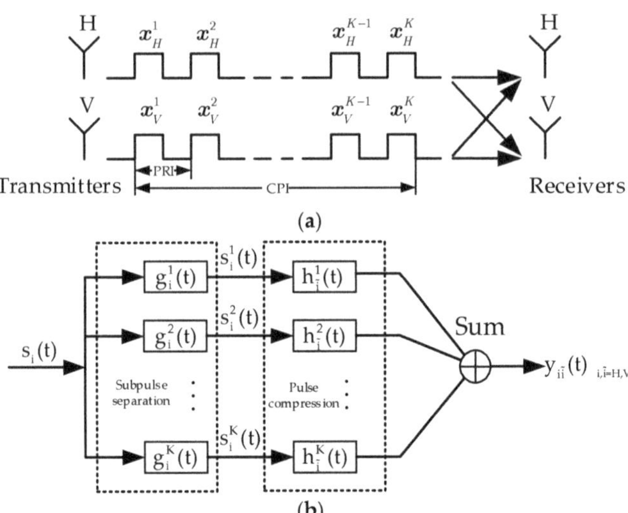

Figure 1. (**a**) Transmitting and receiving radar signal by simultaneously polarimetric radar. (**b**) Signal-processing procedure [28].

Let us denote a pair of UOCCSs containing K pulses with length N in each pulse as [33]:

$$\mathbf{X}_H = \begin{bmatrix} x_H^1 & x_H^2 & \cdots & x_H^K \end{bmatrix}_{N \times K} \tag{1}$$

and

$$X_V = \begin{bmatrix} x_V^1 & x_V^2 & \cdots & x_V^K \end{bmatrix}_{N \times K} \tag{2}$$

where

$$x_i^k = \begin{bmatrix} x_i^k(1) & x_i^k(2) & \cdots & x_i^k(N) \end{bmatrix}^T, \ x_i^k(n) = e^{j\varnothing_i^k(n)}, \ k = 1,\dots,K, \ i = H, V. \tag{3}$$

Let $\mathbf{H}_H = \begin{bmatrix} h_H^1 & h_H^2 & \cdots & h_H^K \end{bmatrix}_{N \times K}$, $\mathbf{H}_V = \begin{bmatrix} h_V^1 & h_V^2 & \cdots & h_V^K \end{bmatrix}_{N \times K}$ denote the receiving filters, so, the discrete AF of sequence $\mathbf{X}_i$ after filtered by $\mathbf{H}_{\underset{i}{\sim}}$ can be expressed as [33]:

$$A_{\mathbf{X}_i,\mathbf{H}_{\underset{i}{\sim}}}(n,f) = \sum_{k=1}^{K} h_{\underset{i}{\sim}}^{k\,\dagger} \mathbf{J}_n x_i^k e^{j2\pi kf} \tag{4}$$

where

$$\mathbf{J}_n[p,q] = \begin{cases} 1, & q - p = n \\ 0, & q - p \neq n \end{cases} \ p,q = 1,\dots,N, \ n = -N+1,\dots,N-1, \tag{5}$$

and $f = f_d T_r$ is the normalized Doppler frequency, where f_d and T_r represent the nominal Doppler shift and PRI, respectively. When $i = \tilde{i}$, (4) is the Auto-AF of $\mathbf{X}_i$; otherwise, it is the Cross-AF. The key to designing high-Doppler-tolerance waveforms for SPR is to synthesize a pair of orthogonal CCS $\mathbf{X}_H$ and $\mathbf{X}_V$ with the desired AF properties within a certain DFR. In other words, our goal is to obtain the waveforms whose AFs are of low sidelobe and good orthogonality over a certain DFR.

The metrics CISL and CIIL should be considered when designing UOCCS for SPR waveforms, which are used to assess the performance of the Auto- and Cross-AF, respectively. The CISL and CIIL for a pair of orthogonal CCSs are defined by

$$CISL(\mathbf{X}_i, \mathbf{H}_i) = \sum_{n=1-N, n \neq 0}^{N-1} \int_{f_1}^{f_2} \left| \sum_{k=1}^{K} h_i^{k\dagger} \mathbf{J}_n x_i^k e^{j2\pi k f} \right|^2 \mathrm{d}f \tag{6}$$

and

$$CIIL\left(\mathbf{X}_i, \mathbf{H}_{\tilde{i}}\right) = \sum_{n=1-N}^{N-1} \int_{f_1}^{f_2} \left| \sum_{k=1}^{K} h_{\tilde{i}}^{k\dagger} \mathbf{J}_n x_i^k e^{j2\pi k f} \right|^2 \mathrm{d}f, i \neq \tilde{i} \tag{7}$$

where $[f_1, f_2]$ denotes the interested DFR.

Compared with the matched filtering, the receiving filter in the mismatched scheme is no longer the conjugated and reversed transmitted waveform. In this case, the SNR loss caused by mismatch filtering is inevitable, and the SNR loss of two co-polarization channels of the SPR can be written as [44]

$$\mathrm{SNRL}_i = 10\log_{10} \frac{\sum_{k=1}^{K} \left\| x_i^k \right\|^2 \sum_{k=1}^{K} \left\| h_i^k \right\|^2}{\left| \sum_{k=1}^{K} h_i^{k\dagger} x_i^k \right|^2} \tag{8}$$

As shown in (8), the SNR loss is related to the peak values of pulse compression, the energy of the sequences, and the receiving filters. According to (3), the sequences satisfy unimodular constraint, so $\sum_{k=1}^{K} \left\| x_i^k \right\|^2 = KN$. For the purpose of normalization, we assume the receiving filters have a constraint of $\sum_{k=1}^{K} \left\| h_i^k \right\|^2 = KN$, the SNRL_i can be constrained by a cost function

$$g(\mathbf{X}_i, \mathbf{H}_i) = \left| \sum_{k=1}^{K} h_i^{k\dagger} x_i^k - a_{max} \right|^2 \tag{9}$$

where a_{max} is the predefined peak value of pulse compression. To guarantee an expected SNR loss μ (in dB), the predefined a_{max} can be formulated as $a_{max} = KN10^{-\mu/20}$.

Thus, based on the Pareto weighting [45], the joint design problem under the mismatch constraint can be formulated as

$$\min_{\mathbf{X}, \mathbf{H} \in \mathbb{C}^{N \times K}} \Gamma(\mathbf{X}, \mathbf{H}) = \varepsilon[CISL(\mathbf{X}_H, \mathbf{H}_H) + CIIL(\mathbf{X}_H, \mathbf{H}_V) + CIIL(\mathbf{X}_V, \mathbf{H}_H) + CISL(\mathbf{X}_V, \mathbf{H}_V)]$$

$$+ (1 - \varepsilon)[g(\mathbf{X}_H, \mathbf{H}_H) + g(\mathbf{X}_V, \mathbf{H}_V)]$$

$$s.\,t. \sum_{k=1}^{K} \left\| h_i^k \right\|^2 = KN, \ \left| x_i^k(n) \right| = 1, \ n = 1, \ldots, N, \ k = 1, \ldots, K, i = H, V \tag{10}$$

where ε is the Pareto weight used to balance the metrics and the cost functions.

3. Joint Design of UOCCS and Receiving Filters via MM Method

3.1. Reformulation of the Problem

In this section, we shall reformulate the design problem in (10) for ease of solution. Let us define auxiliary sequences $\mathbf{u}$ and $\mathbf{v}$ of length $2L$ ($L = K(2N - 1)$) as follows:

$$\mathbf{u} = \left[\mathbf{x}_H^1{}^{\mathrm{T}} \ \mathbf{0}_{N-1}^{\mathrm{T}} \ \cdots \ \mathbf{x}_H^K{}^{\mathrm{T}} \ \mathbf{0}_{N-1}^{\mathrm{T}} \ \mathbf{x}_V^1{}^{\mathrm{T}} \ \mathbf{0}_{N-1}^{\mathrm{T}} \cdots \mathbf{x}_V^K{}^{\mathrm{T}} \ \mathbf{0}_{N-1}^{\mathrm{T}} \right]^{\mathrm{T}} \tag{11}$$

and

$$\mathbf{v} = \left[\mathbf{h}_H^1{}^{\mathrm{T}} \ \mathbf{0}_{N-1}^{\mathrm{T}} \ \cdots \ \mathbf{h}_H^K{}^{\mathrm{T}} \ \mathbf{0}_{N-1}^{\mathrm{T}} \ \mathbf{h}_V^1{}^{\mathrm{T}} \ \mathbf{0}_{N-1}^{\mathrm{T}} \cdots \mathbf{h}_V^K{}^{\mathrm{T}} \ \mathbf{0}_{N-1}^{\mathrm{T}} \right]^{\mathrm{T}} \tag{12}$$

where $\mathbf{u}$ and $\mathbf{v}$ denote the transmitted CCSs and receiving filters, respectively. Then we have

$$\mathbf{u}_m = \mathbf{S}_m \mathbf{u}, \ \mathbf{v}_{\tilde{m}} = \mathbf{S}_{\tilde{m}} \mathbf{v}, \ m, \tilde{m} = 1, 2 \tag{13}$$

where 1 and 2 represent H and V polarization, respectively, and $\mathbf{S}_m$ is an $L \times 2L$ block selection matrix defined as

$$\mathbf{S}_1 = [\mathbf{I}_L \ \mathbf{0}_L], \ \mathbf{S}_2 = [\mathbf{0}_L \ \mathbf{I}_L] \tag{14}$$

Equation (6) can be rewritten as

$$CISL(\mathbf{u}_m, \mathbf{v}_m) = \sum_{n=1-L}^{L-1} \int_{f_1}^{f_2} w_1(n, f) \left| \mathbf{v}_m{}^\dagger \mathbf{T}_n \mathrm{Diag}(\mathbf{a}(f)) \mathbf{u}_m \right|^2 \mathrm{d}f \tag{15}$$

where $\mathbf{T}_n$, $n = 1 - L, \ldots, L - 1$ denotes $L \times L$ Toeplitz shift matrix similar to Equation (5),

$$\mathbf{a}(f) = \left[e^{j2\pi f} \mathbf{1}_N^{\mathrm{T}} \ \mathbf{0}_{N-1}^{\mathrm{T}} \cdots \ e^{j2\pi K f} \mathbf{1}_N^{\mathrm{T}} \ \mathbf{0}_{N-1}^{\mathrm{T}} \right]^{\mathrm{T}} \tag{16}$$

and

$$w_1(n, f) = \begin{cases} 1, \ 1 \leq |n| \leq N - 1, \ f_1 \leq f \leq f_2 \\ 0, \ else. \end{cases} \tag{17}$$

Similarly, (7) can be rewritten as

$$CIIL\left(\mathbf{u}_m, \mathbf{v}_{\tilde{m}}\right) = \sum_{n=1-L}^{L-1} \int_{f_1}^{f_2} w_2(n, f) \left| \mathbf{v}_{\tilde{m}}{}^\dagger \mathbf{T}_n \mathrm{Diag}(\mathbf{a}(f)) \mathbf{u}_m \right|^2 \mathrm{d}f, m \neq \tilde{m} \tag{18}$$

where

$$w_2(n, f) = \begin{cases} 1, \ |n| \leq N - 1, \ f_1 \leq f \leq f_2 \\ 0, \ else. \end{cases} \tag{19}$$

The objective function in (10) can then be rewritten as

$$\Gamma(\mathbf{u}, \mathbf{v}) = \varepsilon \sum_{m=1}^{2} \sum_{n=1-L}^{L-1} \int_{f_1}^{f_2} w_1(n, f) \left| \mathbf{v}_m{}^\dagger \mathbf{T}_n \mathrm{Diag}(\mathbf{a}(f)) \mathbf{u}_m \right|^2 df +$$
$$\varepsilon \sum_{m=1}^{2} \sum_{\substack{\tilde{m}=1 \\ \tilde{m} \neq m}}^{2} \sum_{n=1-L}^{L-1} \int_{f_1}^{f_2} w_2(n, f) \left| \mathbf{v}_{\tilde{m}}{}^\dagger \mathbf{T}_n \mathrm{Diag}(\mathbf{a}(f)) \mathbf{u}_m \right|^2 df + \tag{20}$$
$$(1 - \varepsilon) \left[\left| \mathbf{v}_1^\dagger \mathbf{u}_1 - a_{max} \right|^2 + \left| \mathbf{v}_2^\dagger \mathbf{u}_2 - a_{max} \right|^2 \right]$$

In (20), the first term contains the Auto-AF integrated sidelobes of two sequences for H and V polarizations. The second term denotes the integration of the Cross-AFs. The third term accounts for the differences in the peak values by mismatch filtering and the predefined peak values. For the convenience of discussion, we discretize the Doppler interval $[f_1, f_2]$ into Q bins evenly with the grid size $\Delta f = (f_2 - f_1)/(Q - 1)$, and ignore the constant terms. Equation (20) can then be reformulated as

$$\Gamma(\mathbf{u}, \mathbf{v}) = \sum_{m=1}^{2} \sum_{\tilde{m}=1}^{2} \sum_{n=1-L}^{L-1} \sum_{q=1}^{Q} w_{m,\tilde{m}}(n, f_q) \left| \mathbf{v}_{\tilde{m}}{}^\dagger \mathbf{T}_n \mathrm{Diag}(\mathbf{a}(f_q)) \mathbf{u}_m \right|^2 -$$
$$2a_{max}\lambda \left(Re\left(\mathbf{v}_2^\dagger \mathbf{u}_2\right) + Re\left(\mathbf{v}_1^\dagger \mathbf{u}_1\right) \right) \tag{21}$$

where $f_q = f_1 + (q-1)\Delta f, 1 \leq q \leq Q$ and $\lambda = (1-\varepsilon)/\varepsilon$, and the weighting factor changes to

$$w_{m,\tilde{m}}(n,f) = \begin{cases} w_1(n,f) + \lambda, & m = \tilde{m}, \ n = 0, \ and \ f = 0 \\ w_2(n,f), & m \neq \tilde{m} \\ w_1(n,f), & else \end{cases} \tag{22}$$

and

$$\mathbf{v}_{\tilde{m}}^{\dagger} \mathbf{T}_n \mathrm{Diag}\left(\mathbf{a}(f_q)\right) \mathbf{u}_m = \mathrm{Tr}\left(\mathbf{T}_n \mathbf{P}_{m\tilde{m}q}\right) \tag{23}$$

where $\mathbf{P}_{m\tilde{m}q} = \mathbf{u}_{mq} \mathbf{v}_{\tilde{m}}^{\dagger}$ and $\mathbf{u}_{mq} = \mathrm{Diag}\left(\mathbf{a}(f_q)\right)\mathbf{u}_m$.

Since $\mathrm{Tr}\left(\mathbf{T}_n \mathbf{P}_{m\tilde{m}q}\right) = \mathrm{vec}\left(\mathbf{P}_{m\tilde{m}q}^{\dagger}\right)^{\dagger} \mathrm{vec}(\mathbf{T}_n)$, so the problem (10) can be finally formulated as

$$\min_{\mathbf{u},\mathbf{v}\in\mathbb{C}^{2L\times 1}} \Gamma(\mathbf{u},\mathbf{v}) - \sum_{m=1}^{2}\sum_{\tilde{m}=1}^{2}\sum_{q=1}^{Q} \mathrm{vec}\left(\mathbf{P}_{m\tilde{m}q}^{\dagger}\right)^{\dagger} \mathbf{Q}_{m\tilde{m}q} \mathrm{vec}\left(\mathbf{P}_{m\tilde{m}q}^{\dagger}\right) - 2a_{max}\lambda\left(\mathrm{Re}\left(\mathbf{v}_2^{\dagger}\mathbf{u}_2\right) + \mathrm{Re}\left(\mathbf{v}_1^{\dagger}\mathbf{u}_1\right)\right) \tag{24}$$

$$s.\ t.\ \mathbf{v}_m^{\dagger}\mathbf{v}_m = KN, \ |x_k(n)| = 1, \ n = 1,\ldots,N, \ k = 1,\ldots,2K, \ m,\tilde{m} = 1,2$$

where

$$\mathbf{Q}_{m\tilde{m}q} = \sum_{n=1-L}^{L-1} w_{m,\tilde{m}}(n,f_q)\mathrm{vec}(\mathbf{T}_n)\mathrm{vec}(\mathbf{T}_n)^{\dagger} \tag{25}$$

3.2. Joint Optimization via the MM Method

Due to the unimodular constraint, the optimization problem in (24) is nonconvex, so it cannot be straightforwardly used to solve for the transmitted sequence $\mathbf{u}$ and the receiving filter $\mathbf{v}$ at the same time. Thus, the alternately iterative scheme is introduced to transfer this complex problem (24) into two sub-optimization problems. In addition, for each subproblem, one variable is optimized while keeping the other variable fixed, which can be formulated as

$$\mathbf{v}^{(l)} = \arg\min_{\mathbf{v}} \Gamma\left(\mathbf{u}^{(l-1)}, \mathbf{v}\right) \tag{26a}$$

$$\mathbf{u}^{(l)} = \arg\min_{\mathbf{u}} \Gamma\left(\mathbf{u}, \mathbf{v}^{(l)}\right) \tag{26b}$$

where $\mathbf{v}^{(l)}$ and $\mathbf{u}^{(l)}$ are the solutions to the two sub-optimization problems of (26a) and (26b) at the lth iteration, respectively.

Lemma 1 [24]. *Let both $\mathbf{L}$ and $\mathbf{M}$ denote an $n \times n$ Hermitian matrix such that $\mathbf{M} \succeq \mathbf{L}$. Then for any point $\mathbf{x}_0 \in \mathbb{C}^n$, the quadratic function $\mathbf{x}^{\dagger}\mathbf{L}\mathbf{x}$ can be majorized by $\mathbf{x}^{\dagger}\mathbf{M}\mathbf{x} + 2\mathrm{Re}\left(\mathbf{x}^{\dagger}(\mathbf{L} - \mathbf{M})\mathbf{x}_0\right) + \mathbf{x}_0^{\dagger}(\mathbf{M} - \mathbf{L})\mathbf{x}_0$ at $\mathbf{x}_0$.*

The proof of Lemma 1 was given in [24], and it is omitted here. One can see from (24) that the objective function is composed of a quadratic term of $\mathrm{vec}\left(\mathbf{P}_{m\tilde{m}q}^{\dagger}\right)$, and a Hermitian matrix $\mathbf{Q}_{m\tilde{m}q}$. Thus, by applying Lemma 1 and selecting $\mathbf{M} = \gamma_{max}\left(\mathbf{Q}_{m\tilde{m}q}\right)\mathbf{I}_{L^2}$, we have

$$\mathrm{vec}\left(\mathbf{P}_{m\tilde{m}q}^{\dagger}\right)^{\dagger} \mathbf{Q}_{m\tilde{m}q} \mathrm{vec}\left(\mathbf{P}_{m\tilde{m}q}^{\dagger}\right) \leq \mathrm{vec}\left(\mathbf{P}_{m\tilde{m}q}^{\dagger}\right)^{\dagger} \gamma_{max}\left(\mathbf{Q}_{m\tilde{m}q}\right)\mathrm{vec}\left(\mathbf{P}_{m\tilde{m}q}^{\dagger}\right) +$$
$$2\mathrm{Re}\left(\mathrm{vec}\left(\mathbf{P}_{m\tilde{m}q}^{\dagger}\right)^{\dagger} \left(\mathbf{Q}_{m\tilde{m}q} - \gamma_{max}\left(\mathbf{Q}_{m\tilde{m}q}\right)\mathbf{I}_{L^2}\right) \mathrm{vec}\left(\mathbf{P}_{m\tilde{m}q}^{(l)\dagger}\right)\right) +$$
$$\mathrm{vec}\left(\mathbf{P}_{m\tilde{m}q}^{(l)\dagger}\right)^{\dagger} \left(\gamma_{max}\left(\mathbf{Q}_{m\tilde{m}q}\right)\mathbf{I}_{L^2} - \mathbf{Q}_{m\tilde{m}q}\right) \mathrm{vec}\left(\mathbf{P}_{m\tilde{m}q}^{(l)\dagger}\right) \tag{27}$$

where $\gamma_{max}\left(Q_{m\tilde{m}q}\right)$ is the maximum eigenvalue of $Q_{m\tilde{m}q}$, which has been deduced in detail in [24], and can be written as

$$\gamma_{max}\left(Q_{m\tilde{m}q}\right) = \max_{n\in\{-N,\dots,N\}} \left\{ w_{m,\tilde{m}}(n, f_q)(L - |n|) \right\} \tag{28}$$

Since the elements of $\mathbf{u}$ are of unimodular or zero and $\mathbf{v}^\dagger \mathbf{v} = KN$, the first term of the right side of (27) is just a constant and it is true for the last term. After ignoring the constants, the optimization of (24) can be written as

$$\min_{\mathbf{u},\mathbf{v}\in\mathbb{C}^{2L\times 1}} \Gamma(\mathbf{u}, \mathbf{v}) = \sum_{m=1}^{2} \sum_{\tilde{m}=1}^{2} \sum_{q=1}^{Q} 2\mathrm{Re}\left(\mathrm{vec}\left(\mathbf{P}^\dagger_{m\tilde{m}q}\right)^\dagger \left(Q_{m\tilde{m}q} - \gamma_{max}\left(Q_{m\tilde{m}q}\right)\mathbf{I}_{L^2}\right) \mathrm{vec}\left(\mathbf{P}^{(l)\dagger}_{m\tilde{m}q}\right)\right) - $$
$$2a_{max}\lambda\left(\mathrm{Re}\left(\mathbf{v}_2^\dagger \mathbf{u}_2\right) + \mathrm{Re}\left(\mathbf{v}_1^\dagger \mathbf{u}_1\right)\right) \tag{29}$$
$$s.\,t.\ \mathbf{v}_m{}^\dagger \mathbf{v}_m = KN,\ |x_k(n)| = 1,\ n = 1,\dots, N,\ k = 1,\dots, 2K,\ m, \tilde{m} = 1, 2$$

For subproblem (26a), it optimizes the receiving filter $\mathbf{v}$ under the condition of fixing CCS $\mathbf{u}$. By substituting $Q_{m\tilde{m}q}$ of (25) and $\mathbf{P}^{(l)}_{m\tilde{m}q} = \mathbf{u}_{mq}\mathbf{v}_{\tilde{m}}^{(l)\dagger}$ back into (29), we have (detailed derivation is provided in Appendix A)

$$\mathrm{Re}\left(\mathrm{vec}\left(\mathbf{P}^\dagger_{m\tilde{m}q}\right)^\dagger Q_{m\tilde{m}q} \mathrm{vec}\left(\mathbf{P}^{(l)\dagger}_{m\tilde{m}q}\right)\right) = \mathrm{Re}\left(\mathbf{v}_{\tilde{m}}^\dagger \mathbf{R}_{m\tilde{m}q}^{(l)} \mathbf{u}_{mq}\right) \tag{30}$$

where

$$\mathbf{R}_{m\tilde{m}q}^{(l)} = \sum_{n=1-L}^{L-1} w_{m,\tilde{m}}(n, f_q)\left(A_{\mathbf{u}_m, \mathbf{v}_{\tilde{m}}^{(l)}}(n, f_q)\right)^* \mathbf{T}_n \tag{31}$$

and we have

$$\mathrm{Re}\left(\mathrm{vec}\left(\mathbf{P}^\dagger_{m\tilde{m}q}\right)^\dagger \gamma_{max}\left(Q_{m\tilde{m}q}\right) \mathrm{vec}\left(\mathbf{P}^{(l)\dagger}_{m\tilde{m}q}\right)\right) = \mathrm{Re}\left(\gamma_{max}\left(Q_{m\tilde{m}q}\right) \mathbf{v}_{\tilde{m}}^\dagger \mathbf{v}_{\tilde{m}}^{(l)} \mathbf{u}_{mq}^\dagger \mathbf{u}_{mq}\right) \tag{32}$$

The optimization in (29) is a linear program problem that is quadratically constrained, which can be readily solved using the Lagrange multiplier [46]. Therefore, the problem (26a) can be simplified as

$$\min_{\mathbf{v}\in\mathbb{C}^{2L\times 1}} \sum_{\tilde{m}=1}^{2} \left(2\mathrm{Re}\left(\mathbf{v}_{\tilde{m}}^\dagger \mathbf{y}_{\tilde{m}}\right)\right) \tag{33}$$
$$s.\,t.\ \mathbf{v}_{\tilde{m}}^\dagger \mathbf{v}_{\tilde{m}} = KN,\ |x_k(n)| = 1,\ n = 1,\dots, N,\ k = 1,\dots, 2K,$$

where

$$\mathbf{y}_{\tilde{m}} = \sum_{m=1}^{2} \sum_{q=1}^{Q} \left(\mathbf{R}_{m\tilde{m}q}^{(l)} \mathbf{u}_{mq} - \gamma_{max}\left(Q_{m\tilde{m}q}\right)(NK)\mathbf{v}_{\tilde{m}}^{(l)}\right) - a_{max}\lambda\mathbf{u}_{\tilde{m}} \tag{34}$$

According to the constraint of $\sum_{k=1}^{K}\left\|h_i^k\right\|^2 = KN$ on receiving filters, the update of the optimal solution to receiving filter is given by

$$\mathbf{v}_{\tilde{m}}^{(l+1)} = -\sqrt{\frac{KN}{\left\|\mathbf{y}_{\tilde{m}}\right\|^2}}\,\mathbf{y}_{\tilde{m}} \tag{35}$$

Then, for subproblem (26b), it optimizes the CCS $\mathbf{u}$ under the condition of fixing receiving filter $\mathbf{v}$. By substituting $Q_{m\tilde{m}q}$ of (25) and $\mathbf{P}^{(l)}_{m\tilde{m}q} = \mathbf{u}_{mq}^{(l)}\mathbf{v}_{\tilde{m}}^\dagger$ back into (29) again, we obtain

$$\mathrm{Re}\left(\mathrm{vec}\left(\mathbf{P}^{\dagger}_{m\tilde{m}q}\right)^{\dagger}\boldsymbol{Q}_{m\tilde{m}q}\,\mathrm{vec}\left(\mathbf{P}^{(l)\dagger}_{\tilde{m}mq}\right)\right)=\mathrm{Re}\left(\mathbf{u}^{\dagger}_{m}\mathrm{Diag}\left(\mathbf{a}(f_q)\right)^{\dagger}\mathbf{R}_{m^{(l)}\tilde{m}q}\mathbf{v}_{\tilde{m}}\right) \tag{36}$$

where

$$\mathbf{R}_{m^{(l)}\tilde{m}q}=\sum_{n=1-L}^{L-1}w_{m,\tilde{m}}(n,f_q)\left(A_{\mathbf{u}^{(l)}_{m},\mathbf{v}_{\tilde{m}}}(n,f_q)\right)\mathbf{T}_n \tag{37}$$

and obtain

$$\mathrm{Re}\left(\mathrm{vec}\left(\mathbf{P}^{\dagger}_{m\tilde{m}q}\right)^{\dagger}\gamma_{max}\left(\boldsymbol{Q}_{m\tilde{m}q}\right)\mathrm{vec}\left(\mathbf{P}^{(l)\dagger}_{\tilde{m}mq}\right)\right)=\mathrm{Re}\left(\mathbf{u}^{\dagger}_{m}\mathrm{Diag}\left(\mathbf{a}(f_q)\right)^{\dagger}\gamma_{max}\left(\boldsymbol{Q}_{m\tilde{m}q}\right)NK\mathbf{u}^{(l)}_{mq}\right) \tag{38}$$

Similarly, problem (26b) can be simplified as

$$\min_{\mathbf{u}\in\mathbb{C}^{2L\times1}}\sum_{m=1}^{2}\left(2\mathrm{Re}\left(\mathbf{u}^{\dagger}_{m}\mathbf{z}_{m}\right)\right) \\ s.\,t.\ \mathbf{v}^{\dagger}_{m}\mathbf{v}_{m}=KN,\ |x_k(n)|=1,\ n=1,\ldots,N,\ k=1,\ldots,2K \tag{39}$$

where

$$\mathbf{z}_{m}=\sum_{\tilde{m}=1}^{2}\sum_{q=1}^{Q}\mathrm{Diag}\left(\mathbf{a}(f_q)\right)^{\dagger}\left(\mathbf{R}_{m^{(l)}\tilde{m}q}\mathbf{v}_{\tilde{m}}-\gamma_{max}\left(\boldsymbol{Q}_{m\tilde{m}q}\right)NK\mathbf{u}^{(l)}_{mq}\right)-a_{max}\lambda\mathbf{v}_{m} \tag{40}$$

According to the unimodular constraint, the closed-form solution to CCS can be expressed as

$$\mathbf{u}_{m}^{(l+1)}=-e^{jarg(\mathbf{z}_m)}\odot\mathbf{c} \tag{41}$$

and

$$\mathbf{c}=\begin{bmatrix}\mathbf{1}^{\mathrm{T}}_{\mathrm{N}}&\mathbf{0}^{\mathrm{T}}_{\mathrm{N-1}}&\cdots&\mathbf{1}^{\mathrm{T}}_{\mathrm{N}}&\mathbf{0}^{\mathrm{T}}_{\mathrm{N-1}}\end{bmatrix}^{\mathrm{T}} \tag{42}$$

However, it should be noted that the matrices $\mathbf{R}_{m\tilde{m}^{(l)}q}$ and $\mathbf{R}_{m^{(l)}\tilde{m}q}$ are Hermitian Toeplitz, so the matrix-vector multiplication terms of $\mathbf{R}_{m\tilde{m}^{(l)}q}\mathbf{u}_{mq}$ and $\mathbf{R}_{m^{(l)}\tilde{m}q}\mathbf{v}_{\tilde{m}}$ in (34) and (40) can be computed more efficiently via FFT. In the following, we introduce a simple conclusion regarding the Toeplitz matrices $\mathbf{T}$.

Lemma 2 [34]. *Let $\mathbf{T}$ denote an $n\times n$ Toeplitz matrix defined as follows:*

$$\mathbf{T}=\begin{bmatrix}t_0&t_1&\cdots&t_{N-1}\\t_{-1}&t_0&\ddots&\vdots\\\vdots&\ddots&\ddots&t_1\\t_{1-N}&\cdots&t_{-1}&t_0\end{bmatrix}$$

and $\mathbf{F}$ is another $2N\times2N$ FFT matrix with $F_{m,n}=e^{-j\frac{2mn\pi}{2N}},\ 0\le m,n\le 2N$. Then $\mathbf{T}$ can be decomposed as $\mathbf{T}=\frac{1}{2N}\mathbf{F}^{\dagger}_{:,1:N}\mathrm{Diag}(\mathbf{Fc})\mathbf{F}_{:,1:N}$, where $\mathbf{c}=[t_0,t_{-1},\ldots,t_{1-N},0,t_{N-1},\ldots,t_1]^{\mathrm{T}}$.

The detailed proof of Lemma 2 can be viewed in Appendix B of Reference [34]. By defining the first N columns of the $2N\times2N$ FFT matrix as an $2N\times N$ matrix $\mathbf{H}$, according to Lemma 2, we have

$$\mathbf{R}_{m\tilde{m}^{(l)}q}=\frac{1}{2\mathrm{N}}\mathbf{H}^{\dagger}\mathrm{Diag}\left(\mathbf{Fc}_{m\tilde{m}^{(l)}q}\right)\mathbf{H} \tag{43}$$

where

$$\mathbf{c}_{\underset{m\widetilde{m}}{\sim}(l)\,q} = \left[w_{m,\widetilde{m}}\left(0, f_q\right) A_{\mathbf{u}_m, \mathbf{v}_{\widetilde{m}}^{\sim}(l)}\left(0, f_q\right)^*, \; w_{m,\widetilde{m}}\left(-1, f_q\right) A_{\mathbf{u}_m, \mathbf{v}_{\widetilde{m}}^{\sim}(l)}\left(-1, f_q\right)^*, \ldots, \right.$$
$$w_{m,\widetilde{m}}\left(1 - L, f_q\right) A_{\mathbf{u}_m, \mathbf{v}_{\widetilde{m}}^{\sim}(l)}\left(1 - L, f_q\right)^*, 0, w_{m,\widetilde{m}}\left(L - 1, f_q\right) A_{\mathbf{u}_m, \mathbf{v}_{\widetilde{m}}^{\sim}(l)}\left(L - 1, f_q\right)^*, \ldots,$$
$$\left. w_{m,\widetilde{m}}\left(1, f_q\right) A_{\mathbf{u}_m, \mathbf{v}_{\widetilde{m}}^{\sim}(l)}, \left(1, f_q\right)^* \right]^T \tag{44}$$

Thus, the matrix-vector multiplication $\mathbf{R}_{\underset{m\widetilde{m}}{\sim}(l)\,q} \mathbf{u}_{mq}$ in (34) can be efficiently calculated by

$$\mathbf{R}_{\underset{m\widetilde{m}}{\sim}(l)\,q} \mathbf{u}_{mq} = \frac{1}{2N} \mathbf{H}^\dagger \mathrm{Diag}\left(\mathbf{Fc}_{\underset{m\widetilde{m}}{\sim}(l)\,q} \right) \mathbf{H} \mathbf{u}_{mq} \tag{45}$$

Similarly, $\mathbf{R}_{m(l)\,\widetilde{mq}} \mathbf{v}_{\widetilde{m}}$ in (40) can be calculated in the same fashion. Now, we are ready to summarize the algorithm for the joint design of the orthogonal CCSs and the receiving filters with a desired AF shape as Algorithm 1. Its computational complexity is of order $\mathcal{O}(4QKNlogKN)$ per iteration.

Algorithm 1. Joint Design CCSs and Receiving Filters with Expected AFs Shape Based on the MM Method via Alternately Iteration

---

Initialize: $l = 0$, pulse number K, sequence and filter length N, a predefine SNR loss μ, the DFR as $[f_1, f_2]$.
1: Compute the predefined a_{max}
2: Initialize $\mathbf{u}^{(0)}, \mathbf{v}^{(0)}$ of length $2L$ as (11) and (12)
3: compute $\gamma_{max}\left(Q_{m\widetilde{m}q}\right)$ by (28)
4: **repeat**
5: Compute $\mathbf{y}_{\widetilde{m}}$ by (34) with the designated $\mathbf{u}^{(l)}$
6: Update the receiving filters $\mathbf{v}_1^{(l+1)}$ and $\mathbf{v}_2^{(l+1)}$ by (35)
7: Compute $\mathbf{z}_m$ by (40) with the designated $\mathbf{v}^{(l+1)}$
8: Update the CCSs $\mathbf{u}_1^{(l+1)}$ and $\mathbf{u}_2^{(l+1)}$ by (41)
9: $l = l + 1$;
10: **untilconvergence**
Output: CCS $\mathbf{u}$ and receiving filter $\mathbf{v}$ with expected AFs.

3.3. Acceleration Scheme Using SQUAREM

The monotonicity of the optimization problem can be guaranteed by the MM method, but the convergence speed of which depends on the property of the majorization function. Generally speaking, the speed of the proposed algorithms is quite slow. Here, the SQUAREM is adopted to accelerate the optimization, which is based on the idea of the Cauchy–Barzilai–Borwein (CBB) [47], and it was originally applied to accelerating the Expectation-Maximization (EM) method [48]. For SQUAREM, it only requires the update rule of the optimization algorithm to be the same as the EM algorithm. Since the MM method is the generalization of the EM method, the update rules are all based on fixed-point iteration, so the SQUAREM scheme can be conveniently used to accelerate the convergence of the proposed algorithm after some minor modifications.

Based on the proposed Algorithm 1, let us denote the nonlinear fixed-point iteration map $\mathcal{F}_{MM}(\cdot)$ for minimizing $\Gamma(\mathbf{u}, \mathbf{v})$ as

$$\mathbf{x}^{(l+1)} = \mathcal{F}_{MM}\left(\mathbf{x}^{(l)}\right) \tag{46}$$

where $\mathbf{x}$ represents optimization variable $\mathbf{u}$ or $\mathbf{v}$ during using alternately iterative.

The iteration map $\mathcal{F}_{MM}(\cdot)$ of the proposed algorithm is given by (35) and (41). Then the SQUAREM scheme can be implemented as Algorithm 2. For the general SQUAREM, it should be pointed out that it may break the nonlinear constraints. Therefore, a projection transformation $\mathcal{P}_x(\cdot)$ is needed to project wayward points into the feasible domain in Algorithm 2. Considering two kinds of different constraints, i.e., the unimodular constraint on the sequences and the constraint of $\sum_{k=1}^{K}\left\|h_i^k\right\|^2 = KN$ on the receiving filter, the projection can be simply represented as $\mathcal{P}_x(\cdot) = e^{j\arg(\cdot)}$ for the first constraint, and as $\mathcal{P}_x(\cdot) = \sqrt{\frac{KN}{\|(\cdot)\|^2}}(\cdot)$ for the second constraint. Another problem is that the general SQUAREM may break the monotonicity of the initial MM method. Thus, a strategy has been taken in Algorithm 2 based on backtracking: repeatedly halves the distance between α and -1, i.e., $\alpha \leftarrow (\alpha - 1)/2$ until the monotonicity is maintained, and in practice the monotonicity of the algorithm can be maintained by taking only a few backtracking steps.

Algorithm 2. The Acceleration Scheme for Algorithm 1 using SQUAREM.

1: **Initialize:** $l = 0, x^{(0)}$
2: **repeat**
3: $\mathbf{x}_1 = \mathcal{F}_{MM}\left(\mathbf{x}^{(l)}\right)$
4: $\mathbf{x}_2 = \mathcal{F}_{MM}(\mathbf{x}_1)$
5: $y = \mathbf{x}_1 - \mathbf{x}^{(l)}$
6: $z = \mathbf{x}_2 - \mathbf{x}_1 - y$
7: Compute the step-length $\alpha = -\|y\|/\|z\|$
8: $\mathbf{x} = \mathcal{P}_x\left(\mathbf{x}^{(l)} - 2\alpha y + \alpha^2 \mathbf{z}\right)$
9: **while** $f(\mathbf{x}) > f\left(\mathbf{x}^{(l)}\right)$ **do**
10: $\alpha \leftarrow (\alpha - 1)/2$
11: $\mathbf{x} = \mathcal{P}_x\left(\mathbf{x}^{(l)} - 2\alpha y + \alpha^2 \mathbf{z}\right)$
12: **endwhile**
13: $\mathbf{x}^{(l+1)} = \mathbf{x}$
14: $l \leftarrow l + 1$
15: **until convergence.**

---

4. Simulations and Performance Analysis

In this section, simulations are carried out to show the effectiveness of the proposed algorithm and at the same time to investigate the effects of key parameters on the performance. The sequences and filters are initialized with randomly generated phase-coded sequences $\left\{e^{j\varnothing(n)}\right\}$ uniformly distributed within $[0, 2\pi]$. Meanwhile, since we try to optimize nonconvex problems, although the local minima of the objectives are approximately global [49], the difference in initial sequences has a slight effect on the performance of the algorithm. To guarantee the optimized result with the best performance, 100 Monte Carlo trials are carried out for all cases and the optimal solutions are obtained. For clarity, the accelerated algorithm for optimization problem (10), i.e., Algorithm 2, is denoted as MM CSRF (MM-Complementary Sequence and Receiving Filter). All simulations are conducted using a PC equipped with a 3.0-GHz Intel Core i7-9700 CPU and 16-GB RAM along with MATLAB R2020a.

4.1. Performance of the Proposed Method

In this subsection, we define the metric complementary integrated AF level (CIAL) as the sum of $\mathrm{CISL}(\mathbf{X}, \mathbf{H})$ and $CIIL(\mathbf{X}, \mathbf{H})$, and measure the performance of our algorithm and that of the L-BFGS algorithm [35] in minimizing the CIAL. For all cases, the transmitted CCS $\mathbf{u}^{(0)}$ and receiving filter $\mathbf{v}^{(0)}$ are initialized by independent random variables. The metric normalized CIAL (NCIAL) (in dB) at the ith iteration is defined as:

$$\mathrm{NCIAL}(i) = 10\log_{10}\left(\frac{\mathrm{CISL}\left(\mathbf{u}^{(i)}, \mathbf{v}^{(i)}\right) + CIIL\left(\mathbf{u}^{(i)}, \mathbf{v}^{(i)}\right)}{\mathrm{CISL}\left(\mathbf{u}^{(0)}, \mathbf{v}^{(0)}\right) + CIIL\left(\mathbf{u}^{(0)}, \mathbf{v}^{(0)}\right)}\right) \tag{47}$$

The bandwidth of the phase-codes waveform is approximately equal to $1/t_P$, where t_P is the subcode time duration [50]. If the code length is N, the pulse duration is then Nt_P. Therefore, the time–bandwidth product of a single pulse of the designed waveforms is equal to N. Hereafter, we assume that the code length of a single pulse in the designed waveforms and the filter are both N.

Example 1: In this example, we show the convergence performances of the MM-CSRF algorithm and that of the L-BFGS algorithm. Suppose $N = 64$, $K = 16$, and the normalized DFR is $f \in [-0.2, 0.2]$, which are the same as those in [35]. For the MM-CSRF algorithm, the weight factor is $\lambda = 10$, and the predefined SNR loss is $\mu = 0.5$ dB. For both algorithms, the iteration stops when the NCIAL in (47) is less than -35 dB or after 10,000 s of processing.

The evolutions of the NCIAL along with the running time are presented in Figure 2, from which we can observe that our algorithm exhibits significant superiority in the running time compared to the L-BFGS algorithm. More specifically, it takes only 27.62 s to drive the NCIAL close to -35 dB for our algorithm, while for the L-BFGS algorithm, it is still greater than -30 dB after 1700 s of runtime, and finally drives NCIAL to -35 dB after 4000 s. This is because the main steps are the computation of matrix-vector multiplication in the proposed algorithm, which can be completed via FFT using the characteristic of the Toeplitz matrix.

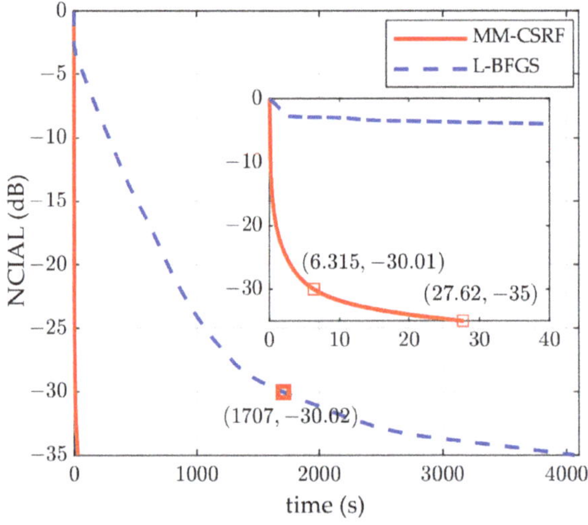

Figure 2. Evolutions of the NCIAL along with the running time.

Example 2: In this example, the CCSs and receiving filters for zero-Doppler shift are designed by the MM-CSRF algorithm with N = 64 and without the SNR loss, i.e., $\mu = 0$ dB, and the weight factor is $\lambda = 10$. The processing stops after 10,000 iterations. A group of zero-Doppler cuts of the AF of CCS sets with K = 2, 4, 5, 6 is given in Figure 3.

As we can see from Figure 3 that the zero-Doppler AF performance of the sequence is remarkably improved as K increases. This is because the larger K, the greater the degrees of freedom for the CSS design. For guaranteeing the complementarity property between multiple pulses, the CPI should not be too long, i.e., the pulse number K should be as small as possible to avoid the scattering fluctuation from the target. In this example particularly, when K = 5, the sidelobe levels have already been smaller than -130 dB, the sequences can be viewed as completely complementary in practice.

Example 3: In this example, the influence of the weighting factor λ on the NCIAL of the designed CCS is investigated with the predefined SNR loss $\mu = 0.5$ dB. Same as before, N = 64, K = 5, and the normalized DFR $f \in [-0.2, 0.2]$ are used. The waveform is optimized via 10,000 iterations. Here, $\lambda = [0.1\ 0.2\ 0.5\ 1\ 2\ 5\ 10\ 20\ 50\ 100\ 200\ 500\ 1000\ 2000\ 5000]$.

The influence of the weighting factor λ on the statistical mean of actual SNR loss and NCIAL is shown in Figure 4a, as can be seen, the actual SNR loss decreases as the weight λ increases and gradually approaches the predefined SNRL. Specifically, when $\lambda = 10$, the SNR loss is -0.53 dB, and when $\lambda = 5000$, it is -0.5003 dB. To precisely control the SNR loss, the weight factor λ should be chosen to be large enough. However, an excessively large weight factor limits the suppression performance of CIAL on the contrary. Therefore, the choice of weight factor λ should balance the control accuracy of SNR loss and CIAL suppression. Fortunately, for this example, the weight factor λ can be chosen between 1 and 1000.

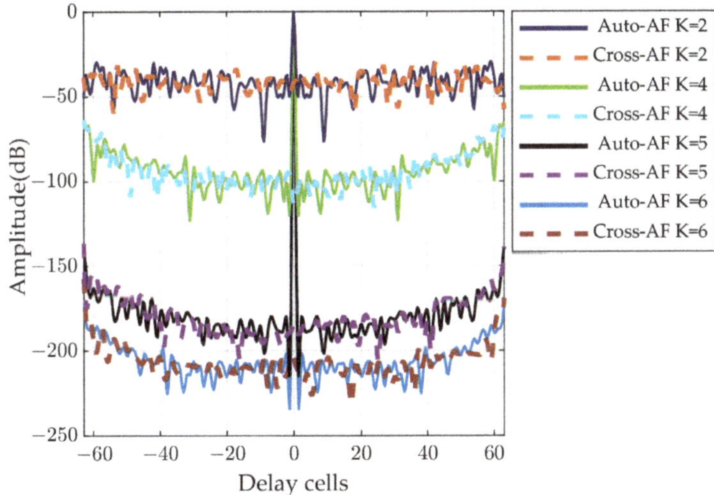

Figure 3. Zero-Doppler cuts of the Auto- and Cross-AF of CCSs with N = 64 and K = 2, 4, 5, 6.

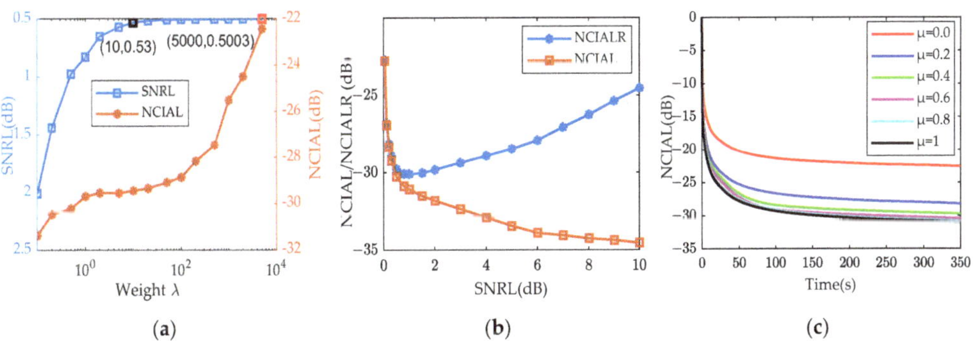

Figure 4. Influences of the weight λ and the SNR loss μ on the NCIAL suppression. (**a**) Influence of the weighting factor λ; (**b**) Influence of the SNR loss μ; (**c**) NCIAL evolution curves with respect to time.

The NCIAL versus the SNRL is further analyzed in the following by setting $\lambda = 100$ and $\mu = [0.0\ 0.1\ 0.2\ 0.4\ 0.6\ 0.8\ 1\ 1.5\ 2\ 3\ 4\ 5\ 6\ 7\ 8\ 9\ 10]$. Figure 4b demonstrates the optimization results of the NCIAL and NCIAL ratio (NCIALR, the sum of NCIAL (dB) and SNRL (dB) representing the suppression of NCIAL relative to the peak value of pulse compression) versus SNRL. It is not difficult to find that the NCIAL can be further suppressed as the SNRL increases. However, the NCIALR cannot be further suppressed when the SNRL is greater than 1.5 dB. This is because although the NCIAL suppression performance is improved as the SNRL increases, the peak value of pulse compression decreases faster than the NCIAL decreases when the SNRL exceeds a certain level. Thus, some conclusions can be drawn according to the simulations for improving the practical application significance of our algorithm, e.g., in high SNR situations, SNRL can reach its minimum point (it represents the SNRL value corresponding to the lowest NCIALR) to improve orthogonality and reduce the pulse compression sidelobe of the waveforms. When in low-SNR situations, SNRL should be less than the minimum point.

The corresponding NCIAL evolution curves with respect to time under different μ are presented in Figure 4c, as can be noted that even a slight SNR loss can also make our algorithm achieve good performance in a short time.

4.2. Doppler Effect Analysis

Example 4: In this subsection, we compare the performance of UOCCSs designed by different algorithms on Doppler resilience. Figure 5 shows the AFs of CCS designed by the MM-CCS algorithm [34], the L-BFGS algorithm [35], and our algorithm. The subfigures in Figure 5a–f shows the zero-Doppler cuts of the AFs. For all algorithms, the waveforms are optimized by 10,000 iterations or 600 s. Here, N = 64, K = 16, the normalized DFR $f \in [-0.2, 0.2]$, $\mu = 0.5$, and $\lambda = 10$ are used.

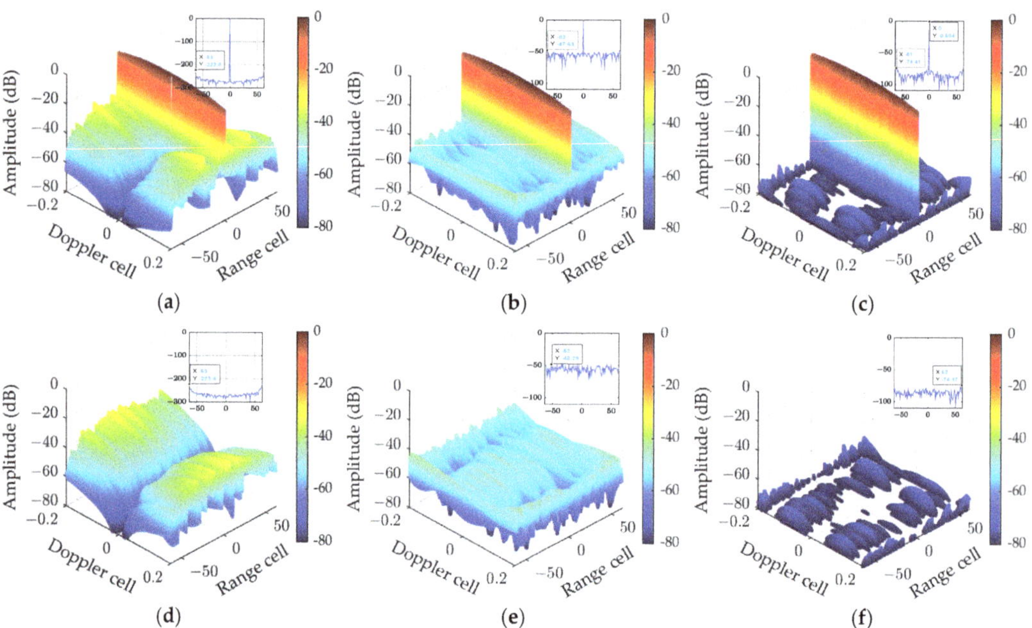

Figure 5. AFs of waveforms obtained by different algorithms. (**a**) Auto-AF by [34]; (**b**) Auto-AF by [35]; (**c**) Auto-AF by ours; (**d**) Cross-AF by [34]; (**e**) Cross-AF by [35]; (**f**) Cross-AF by ours.

It can be observed that the UOCCS along with the receiving filter designed by our algorithm realizes a stable low PSL for both the Auto-AF and the Cross-AF within the given normalized DFR, but this is not the case for both the MM-CCS algorithm and the L-BFGS algorithm. Specifically, the mean normalized sidelobes are about -47.53 dB for the MM-CCS, -56.52 dB for the L-BFGS, and -81.07 dB for our algorithm. The MM-CCS sequence has good ambiguity function properties only in a narrow DFR range. Meanwhile, the actual SNR loss of 0.504 dB of the designed CCS using our algorithm is almost equal to the predefined value (0.5 dB), and the sidelobes and orthogonality of the CCS do not fluctuate significantly in the normalized DFR $f \in [-0.2, 0.2]$. The reason for this is that the objective function in (21) is an integration sum of AFs with different discrete Doppler frequency shift f_q, the sidelobes, and the orthogonality for each discrete frequency shift f_q are optimized equally.

5. Experimental Validation

To verify the performance of the UOCCS waveforms and receiving filters designed by the proposed algorithm more rigorously, practical experiments are carried out using the hardware system as shown in Figure 6a, and the performance of our algorithm is compared with that of the MM-CCS [34] and the L-BFGS algorithm [35]. The block diagram of the experimental system is shown in Figure 6b. The hardware parameters are listed in Table 2. Specifically, in the transmitted part, the designed H and V polarization waveforms are obtained by first generating the base-band waveforms using two 14-bit digital-to-analog converter (DAC) with 2.5 GHz sampling rate, which are then up-converted to Ku band and amplified. Then, the transmitted signals are directly fed into the H and V polarization ports of the orthogonal mode transducer (OMT) of polarization isolation below −50 dB, respectively, to form the simultaneously polarized radar signal. Another OMT is connected to the former OMT through a section of circular waveguide and its two output ports are connected to a two-channel receiver of 200 MHz bandwidth before a 60 dB attenuator is applied. In the two-channel receiver, the attenuated signal is down-converted to in-phase and quadrature-phase (IQ) base-band signals and finally digitized by a 12-bit two-channel analog-to-digital converter (ADC) of 500 MHz sampling rate. We should point out that different Doppler shifts have been modulated into the originally transmitted H and V signals for evaluating the performances of different waveforms on the Doppler tolerance.

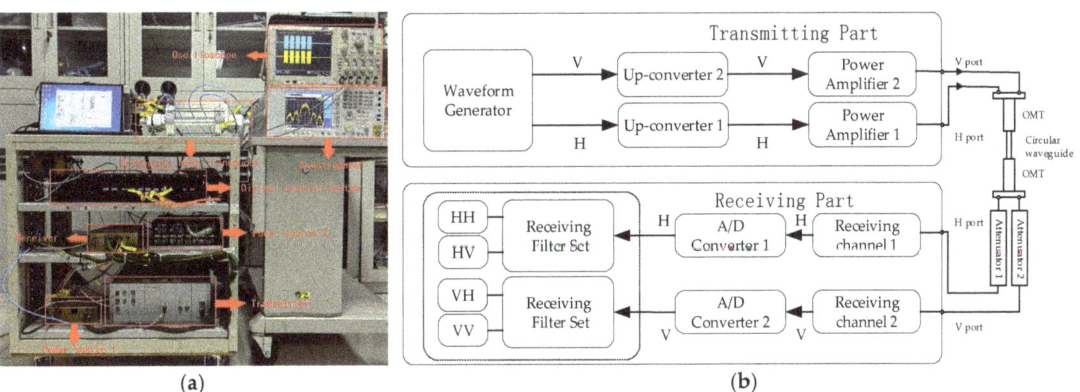

Figure 6. Experiment hardware system, (**a**) Photograph; (**b**) Block diagram.

Table 2. Parameters of the experimental system.

Parameter	Quantity
Carrier Frequency f_c	13.58 GHz
Waveform bandwidth B	40 MHz
Pulse duration T_p	1.6 μs
Pulse PRI T_r	2.1 μs
CPI	10 μs
Code length N	64
Pulse number K	5
A/D sampling rate	500 MHz
ADC resolution	12 bit

In the practical experiment, another set of UOCCS waveforms and receiving filters with $N = 64$ and $K = 5$ by the MM-CCS algorithm [34], the L-BFGS algorithm [35] and our algorithm are implemented into the hardware, whose bandwidth B, pulse duration $T_p(N/B)$, PRI (T_r) as well as the CPI are, respectively, set as 40 MHz, 1.6 μs, 2.1 μs, and 10 μs. The SNR loss $\mu = 0.5$ dB is considered for our algorithm. Figure 7 shows the simulated

AFs before implementation. Figure 8 presents the in-phase components of the sampled waveforms from our waveforms at the receiver end and the corresponding spectrums. Figure 9 presents the AFs of practically implemented waveforms through the transmitter–receiver chain.

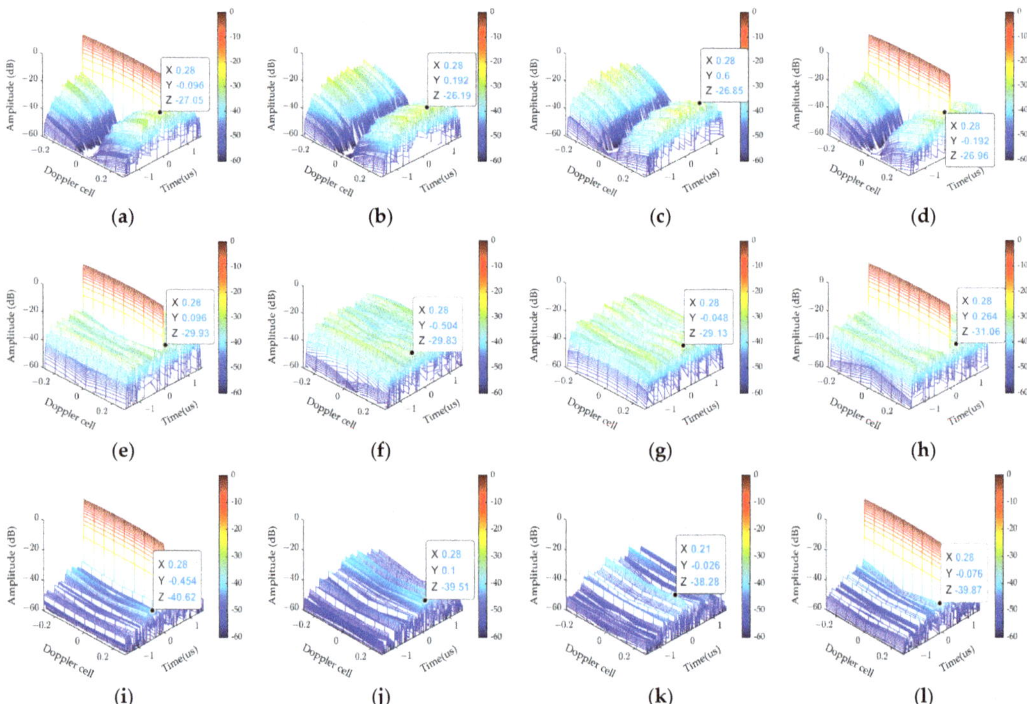

Figure 7. AFs of simulated waveforms from different polarimetric channels before implementation. (**a**) HH by [34]; (**b**) HV by [34]; (**c**) VH for [34]; (**d**) VV by [34]; (**e**) HH by [35]; (**f**) HV by [35]; (**g**) VH by [35]; (**h**) VV by [35]; (**i**) HH by ours; (**j**) HV by ours; (**k**) VH by ours; (**l**) VV by ours.

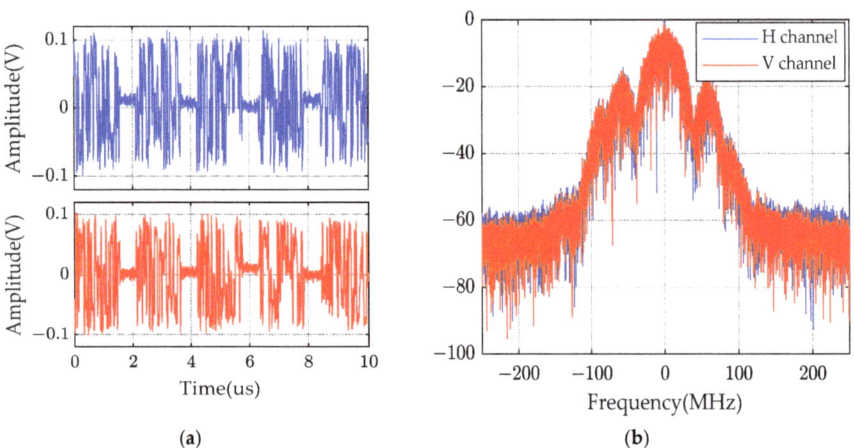

Figure 8. Sampled waveforms by A/D convertor from our waveforms. (**a**) In-phase components in the time domain; (**b**) Spectrums of the sampled signals.

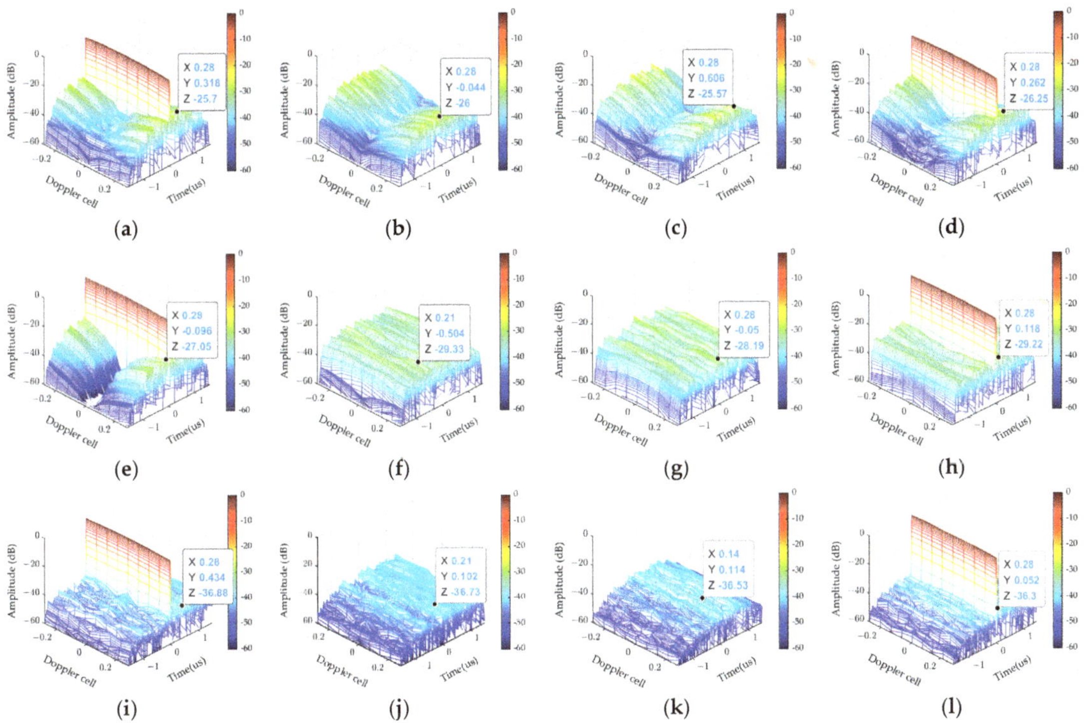

Figure 9. AFs of practically implemented waveforms through the transmitter–receiver chain. (**a**) HH by [34]; (**b**) HV by [34]; (**c**) VH for [34]; (**d**) VV by [34]; (**e**) HH by [35]; (**f**) HV by [35]; (**g**) VH by [35]; (**h**) VV by [35]; (**i**) HH by ours; (**j**) HV by ours; (**k**) VH by ours; (**l**) VV by ours.

Both Figures 7 and 9 demonstrate that the waveforms designed by our method outperform others when the Doppler frequencies are considered although the waveform of [34] performs the best when the Doppler frequency is zero. If we compare Figures 7 and 9, it is easy to note that the real implemented results of all waveforms are obviously not as good as the simulated, and this is mainly due to the following factors: (1) the filter (combination of several filters both in the transmitter channel and in the receiver channel) is applied in hardware but not in the simulation. (2) the IQ imbalance of the transmitter–receiver chain, and the nonlinear effect. To manifest the first factor, another simulation experiment is conducted with a filter applied, which is derived from measurements. To compare the performances, we list the maximum normalized sidelobes (MNSLs) of all waveforms in Table 3, where Simulated 1 denotes the results without filter, while Simulated 2 denotes the results with the filter applied. One can observe the following facts from Table 3: (1) Although the waveform of [34] performs the best when $f_q = 0$, its MNSL is affected by the filter mostly, and when $f_q > 0.07$, the performances of HH, HV, VH, and VV degrade dramatically. (2) Let us take the HH case for example, for the waveform of [34], the measured MNSL of HH increases from -39.41 to -25.7 as f_q increases from 0 to 0.28. For the waveform of [35], the measured MNSL increases from -31.81 to -29.28, while the measured MNSL of our waveform increases from -37.54 to just -36.88. (3) The same trend is observed for HV, VH, and VV polarizations. (4) The degradation of performances by imperfect factors of hardware is less than 2 dB for all waveforms. All in all, our waveforms have the best Doppler tolerance performance.

Table 3. Comparison of the MNSLs (in dB) versus f_q.

		Waveform of [34]				Waveform of [35]				Our Waveform			
		HH	HV	VH	VV	HH	HV	VH	VV	HH	HV	VH	VV
$f_q = 0$	Simulated 1	−61.23	−60.19	−60.19	−60.65	−34.10	−33.56	−33.56	−34.20	−42.91	−42.29	−41.56	−42.87
	Simulated 2	−40.81	−39.85	−41.02	−40.96	−32.17	−33.20	−32.28	−32.98	−38.11	−38.46	−39.29	−38.68
	Measured	−39.41	−39.32	−38.41	−39.17	−31.81	−31.71	−31.44	−31.96	−37.54	−37.13	−38.17	−37.76
$f_q = 0.07$	Simulated 1	−38.83	−37.74	−38.06	−38.31	−32.66	−32.70	−32.78	−32.73	−42.04	−41.72	−40.10	−43.17
	Simulated 2	−35.80	−35.36	−34.02	−37.27	−32.01	−32.45	−32.03	−32.66	−38.41	−39.45	−38.84	−39.62
	Measured	−34.17	−35.55	−33.78	−36.34	−31.01	−30.88	−31.76	−32.05	−37.90	−38.13	−36.88	−38.73
$f_q = 0.14$	Simulated 1	−32.89	−31.70	−31.10	−32.60	−31.18	−30.93	−31.32	−32.45	−41.65	−41.34	−39.16	−42.55
	Simulated 2	−32.19	−30.56	−29.72	−33.04	−30.90	−30.47	−30.36	−32.29	−38.61	−38.73	−37.59	−37.53
	Measured	−30.04	−30.64	−30.39	−32.47	−30.64	−30.35	−30.02	−30.19	−38.29	−37.82	−36.53	−37.16
$f_q = 0.21$	Simulated 1	−29.58	−28.37	−27.80	−29.25	−30.00	−30.11	−30.2	−33.25	−41.27	−40.39	−38.28	−41.13
	Simulated 2	−29.18	−28.06	−27.26	−29.51	−29.13	−29.74	−29.18	−32.73	−38.42	−38.97	−38.42	−37.96
	Measured	−27.90	−28.0	−27.19	−29.47	−29.54	−29.33	−29.42	−30.43	−37.36	−36.73	−37.33	−36.81
$f_q = 0.28$	Simulated 1	−27.05	−26.19	−26.85	−26.96	−29.93	−29.83	−29.13	−31.06	−40.62	−39.51	−38.96	−39.87
	Simulated 2	−26.84	−26.30	−25.56	−27.12	−28.98	−29.56	−28.65	−30.91	−38.35	−38.54	−38.59	−37.49
	Measured	−25.70	−26.00	−25.57	−26.25	−29.28	−29.87	−28.19	−29.22	−36.88	−37.08	−37.89	−36.30

We must emphasize that the digitization errors of DAC and ADC have been considered in the simulation. We also need to further explain the two effects of the filter on the AFs: (1) the signal band is reduced; however the large signal bandwidth is very helpful for obtaining low PSL and ISL. (2) the phase of the waveform can be influenced by the filter resulting in waveform distortion [51,52].

6. Conclusions

The problem for the joint design of a pair of UOCCSs and receiving filters with high Doppler tolerance for SPR is focused on under the mismatch constraint. A nonconvex objective function is constructed based on AF, which is optimized with respect both to the CIAL and the orthogonality within the defined DFR by adopting an alternatively iterative scheme implemented via the MM method. The main steps, i.e., the computation of Toeplitz matrix-vector multiplication terms in the proposed algorithms, are implemented via the FFT with high computational efficiency. Moreover, SQUAREM is adopted to accelerate the algorithm. Compared with the representative MM-CCS and L-BFGS algorithms, the proposed algorithm achieves a significant improvement in the optimized performance and the convergence speed. Simulations are carried out to demonstrate that the CIAL can be significantly suppressed at the cost of appropriate SNR loss by jointly designing the waveforms and the receiving filters. Moreover, a practical experiment based on the transmitter–receiver hardware chain is conducted to validate the design, and the same superior performance of our design has been demonstrated when compared with other designs.

The unimodular constraint on the transmitted CCS is currently considered in this work. Some more general constraints (such as the discrete phase constraint [53], or peak-to-average ratio [54]) may also be applicable. We plan to try some other reformulations on the joint problem with more general constraints applied and more suitable algorithms developed to incorporate into the optimization framework in the future. At last, we should point out that the proposed framework can be extended to design other waveforms for MIMO radar and multichannel communication systems as well, where orthogonal sequences are highly desired.

Author Contributions: Conceptualization, Y.C., Y.Z. and D.L.; methodology, Y.C. and Y.Z.; validation, Y.C. and J.Y.; formal analysis, Y.C. and D.L.; writing—original draft preparation, Y.C.; writing—review and editing, Y.C. and Y.Z. and J.Y. All authors have read and agreed to the published version of the manuscript.

Funding: This work was funded by the National Natural Science Foundation of China under Grant 61971402.

Data Availability Statement: Not applicable.

Acknowledgments: The authors acknowledge the equipment and technical support provided by the National Space Science Center, Chinese Academy of Sciences. We also appreciate the anonymous reviewers.

Appendix A

1. Derivation of Equation (30):

$$
\begin{aligned}
\mathrm{Re}\left(\mathrm{vec}\left(\mathbf{P}^{\dagger}_{m\tilde{m}q}\right)^{\dagger}\mathbf{Q}_{m\tilde{m}q}\,\mathrm{vec}\left(\mathbf{P}^{(l)\dagger}_{m\tilde{m}q}\right)\right)
&= \mathrm{Re}\left(\sum_{n=1-L}^{L-1} w_{m,\tilde{m}}(n,f_q)\,\mathrm{vec}\left(\mathbf{P}^{\dagger}_{m\tilde{m}q}\right)^{\dagger}\mathrm{vec}(\mathbf{T}_n)\mathrm{vec}(\mathbf{T}_n)^{\dagger}\mathrm{vec}\left(\mathbf{P}^{(l)\dagger}_{m\tilde{m}q}\right)\right)\\
&= \mathrm{Re}\left(\sum_{n=1-L}^{L-1} w_{m,\tilde{m}}(n,f_q)\,\mathrm{Tr}\left(\mathbf{T}_n\mathbf{P}_{m\tilde{m}q}\right)\mathrm{Tr}\left(\mathbf{P}^{(l)\dagger}_{m\tilde{m}q}\mathbf{T}_n^{\dagger}\right)\right)\\
&= \mathrm{Re}\left(\sum_{n=1-L}^{L-1} w_{m,\tilde{m}}(n,f_q)\,\mathrm{Tr}\left(\mathbf{T}_n\mathbf{u}_{mq}\mathbf{v}_{\tilde{m}}^{\dagger}\right)\mathrm{Tr}\left(\mathbf{v}_{\tilde{m}}^{(l)}\mathbf{u}_{mq}^{\dagger}\mathbf{T}_n^{\dagger}\right)\right)\\
&= \mathrm{Re}\left(\sum_{n=1-L}^{L-1} w_{m,\tilde{m}}(n,f_q)\,\mathbf{v}_{\tilde{m}}^{\dagger}\mathrm{Tr}\left(\mathbf{v}_{\tilde{m}}^{(l)}\mathbf{u}_{mq}^{\dagger}\mathbf{T}_n^{\dagger}\right)\mathbf{T}_n\mathbf{u}_{mq}\right)\\
&= \mathrm{Re}\left(\sum_{n=1-L}^{L-1} w_{m,\tilde{m}}(n,f_q)\,\mathbf{v}_{\tilde{m}}^{\dagger}\mathbf{u}_{mq}^{\dagger}\mathbf{T}_n^{\dagger}\mathbf{v}_{\tilde{m}}^{(l)}\mathbf{T}_n\mathbf{u}_{mq}\right)\\
&= \mathrm{Re}\left(\mathbf{v}_{\tilde{m}}^{\dagger}\left(\sum_{n=1-L}^{L-1} w_{m,\tilde{m}}(n,f_q)\left(\mathbf{v}_{\tilde{m}}^{\dagger(l)}\mathbf{T}_n\mathbf{u}_{mq}\right)^{*}\mathbf{T}_n\right)\mathbf{u}_{mq}\right)\\
&= \mathrm{Re}\left(\mathbf{v}_{\tilde{m}}^{\dagger}\left(\sum_{n=1-L}^{L-1} w_{m,\tilde{m}}(n,f_q)\left(\Lambda_{\mathbf{u}_{m},\mathbf{v}_{\tilde{m}}^{(l)}}(n,f_q)\right)^{*}\mathbf{T}_n\right)\mathbf{u}_{mq}\right)\\
&= \mathrm{Re}\left(\mathbf{v}_{\tilde{m}}^{\dagger}\mathbf{R}_{m\tilde{m}^{(l)}q}\mathbf{u}_{mq}\right)
\end{aligned}
\tag{A1}
$$

where

$$
\mathbf{R}_{m\tilde{m}^{(l)}q} = \sum_{n=1-L}^{L-1} w_{m,\tilde{m}}(n,f_q)\left(\Lambda_{\mathbf{u}_{m},\mathbf{v}_{\tilde{m}}^{(l)}}(n,f_q)\right)^{*}\mathbf{T}_n
\tag{A2}
$$

2. Derivation of Equation (36):

$$
\begin{aligned}
\mathrm{Re}\left(\mathrm{vec}\left(\mathbf{P}^{\dagger}_{m\tilde{m}q}\right)^{\dagger}\mathbf{Q}_{m\tilde{m}q}\,\mathrm{vec}\left(\mathbf{P}^{(l)\dagger}_{m\tilde{m}q}\right)\right)
&= \mathrm{Re}\left(\sum_{n=1-L}^{L-1} w_{m,\tilde{m}}(n,f_q)\,\mathrm{vec}\left(\mathbf{P}^{\dagger}_{m\tilde{m}q}\right)^{\dagger}\mathrm{vec}(\mathbf{T}_n)\mathrm{vec}(\mathbf{T}_n)^{\dagger}\mathrm{vec}\left(\mathbf{P}^{(l)\dagger}_{m\tilde{m}q}\right)\right)\\
&= \mathrm{Re}\left(\sum_{n=1-L}^{L-1} w_{m,\tilde{m}}(n,f_q)\,\mathrm{Tr}\left(\mathbf{T}_n\mathbf{P}_{m\tilde{m}q}\right)\mathrm{Tr}\left(\mathbf{P}^{(l)\dagger}_{m\tilde{m}q}\mathbf{T}_n^{\dagger}\right)\right)\\
&= \mathrm{Re}\left(\sum_{n=1-L}^{L-1} w_{m,\tilde{m}}(n,f_q)\,\mathrm{Tr}\left(\mathbf{T}_n\mathbf{u}_{mq}\mathbf{v}_{\tilde{m}}^{\dagger}\right)\mathrm{Tr}\left(\mathbf{v}_{\tilde{m}}\mathbf{u}_{mq}^{(l)\dagger}\mathbf{T}_n^{\dagger}\right)\right)\\
&= \mathrm{Re}\left(\sum_{n=1-L}^{L-1} w_{m,\tilde{m}}(n,f_q)\,\mathbf{v}_{\tilde{m}}^{\dagger}\mathrm{Tr}\left(\mathbf{v}_{\tilde{m}}\mathbf{u}_{mq}^{(l)\dagger}\mathbf{T}_n^{\dagger}\right)\mathbf{T}_n\mathbf{u}_{mq}\right)\\
&= \mathrm{Re}\left(\sum_{n=1-L}^{L-1} w_{m,\tilde{m}}(n,f_q)\,\mathbf{v}_{\tilde{m}}^{\dagger}\mathbf{u}_{mq}^{(l)\dagger}\mathbf{T}_n^{\dagger}\mathbf{v}_{\tilde{m}}\mathbf{T}_n\mathbf{u}_{mq}\right)\\
&= \mathrm{Re}\left(\mathbf{u}_{mq}^{\dagger}\left(\sum_{n=1-L}^{L-1} w_{m,\tilde{m}}(n,f_q)\,\mathbf{T}_n\left(\mathbf{v}_{\tilde{m}}^{\dagger}\mathbf{T}_n\mathbf{u}_{mq}^{(l)}\right)\right)\mathbf{v}_{\tilde{m}}\right)\\
&= \mathrm{Re}\left(\mathbf{u}_{mq}^{\dagger}\left(\sum_{n=1-L}^{L-1} w_{m,\tilde{m}}(n,f_q)\,A_{\mathbf{u}_{m}^{(l)},\mathbf{v}_{\tilde{m}}}(n,f_q)\mathbf{T}_n\right)\mathbf{v}_{\tilde{m}}\right)\\
&= \mathrm{Re}\left(\mathbf{u}_{mq}^{\dagger}\mathbf{R}_{m^{(l)}\tilde{m}q}\mathbf{v}_{\tilde{m}}\right)
\end{aligned}
\tag{A3}
$$

where

$$
\mathbf{R}_{m^{(l)}\tilde{m}q} = \sum_{n=1-L}^{L-1} w_{m,\tilde{m}}(n,f_q)\left(A_{\mathbf{u}_{m}^{(l)},\mathbf{v}_{\tilde{m}}}(n,f_q)\right)\mathbf{T}_n
\tag{A4}
$$

References

1. Lee, J.S.; Pottier, E. *Polarimetric Radar Imaging: From Basics to Applications*; CRC: Boca Raton, FL, USA, 2009.
2. Hurtado, M.; Nehorai, A. Polarimetric detection of targets in heavy inhomogeneous clutter. *IEEE Trans. Signal Process.* **2008**, *56*, 1349–1361. [CrossRef]
3. Liang, L.; Zhang, Y.; Li, D. A novel method for polarization orientation angle estimation over steep terrain and comparison of deorientation algorithms. *IEEE Trans. Geosci. Remote Sens.* **2020**, *59*, 4790–4801. [CrossRef]
4. Liang, L.; Zhang, Y.; Li, D. Fast Huynen–Euler Decomposition and its Application in Disaster Monitoring. *IEEE J. Sel. Top. Appl. Earth Obs. Remote Sens.* **2021**, *14*, 4231–4243. [CrossRef]
5. Giuli, D.; Fossi, M.; Facheris, L. Radar target scattering matrix measurement through orthogonal signals. *IEE Proc. F Radar Signal Process.* **1993**, *140*, 233–242. [CrossRef]
6. Li, C.; Li, Y.; Yang, Y.; Pang, C.; Wang, X. Moving target's scattering matrix estimation with a polarimetric radar. *IEEE Trans. Geosci. Remote Sens.* **2020**, *58*, 5540–5551. [CrossRef]
7. Wang, F.; Yin, J.; Pang, C.; Li, Y.; Wang, X. A unified framework of Doppler resilient sequences design for simultaneous polarimetric radars. *IEEE Trans. Geosci. Remote Sens.* **2022**, *60*, 5109615. [CrossRef]
8. Wang, X. Study on Wideband Polarization Information Processing. Ph.D. Thesis, Graduate School of National University of Defense Technology, Changsha, China, 1999.
9. Wang, X.; Li, Y.; Dai, H.; Chang, Y.; Dai, D.; Ma, L.; Liu, Y. Research on instantaneous polarization radar system and external experiment. *Chinese Sci. Bull.* **2010**, *55*, 1560–1567. [CrossRef]
10. Wang, F.; Pang, C.; Li, Y.; Wang, X. Algorithms for designing unimodular sequences with high Doppler tolerance for simultaneous fully polarimetric radar. *Sensors* **2018**, *18*, 905. [CrossRef]
11. Chen, Z.; Liang, J.; Song, K.; Yang, Y.; Deng, X. On designing good doppler tolerance waveform with low PSL of ambiguity function. *Signal Process.* **2023**, *210*, 109075. [CrossRef]
12. Hu, C.; Chen, Z.; Dong, X.; Cui, C. Multistatic geosynchronous SAR resolution analysis and grating lobe suppression based on array spatial ambiguity function. *IEEE Trans. Geosci. Remote Sens.* **2020**, *58*, 6020–6038. [CrossRef]
13. Cui, G.; Fu, Y.; Yu, X.; Li, J. Local ambiguity function shaping via unimodular sequence design. *IEEE Signal Process. Lett.* **2017**, *24*, 977–981. [CrossRef]
14. Ye, Z.; Zhou, Z.; Fan, P.; Liu, Z.; Lei, X.; Tang, X. Low ambiguity zone: Theoretical bounds and Doppler-resilient sequence design in integrated sensing and communication systems. *IEEE J. Sel. Areas Commun.* **2022**, *40*, 1809–1822. [CrossRef]
15. He, H.; Stoica, P.; Li, J. Designing unimodular sequence sets with good correlations-Including an application to MIMO radar. *IEEE Trans. Signal Process.* **2009**, *57*, 4391–4405. [CrossRef]
16. Li, C.; Yang, Y.; Li, Y.; Wang, X. Frequency calibration for the scattering matrix of the simultaneous polarimetric radar. *IEEE Geosci. Remote Sens. Lett.* **2018**, *15*, 429–433. [CrossRef]
17. Zheng, Z.; Zhang, Y.; Peng, X.; Xie, H.; Chen, J.; Mo, J.; Sui, Y. MIMO Radar Waveform Design for Multipath Exploitation Using Deep Learning. *Remote Sens.* **2023**, *15*, 2747. [CrossRef]
18. Li, C.; Yang, Y.; Pang, C.; Li, Y.; Wang, X. A simultaneous polarimetric measurement scheme using one amplitude modulation waveform. *IEEE Geosci. Remote Sens. Lett.* **2019**, *17*, 72–76. [CrossRef]
19. Liu, B. Orthogonal discrete frequency-coding waveform set design with minimized autocorrelation sidelobes. *IEEE Trans. Aerosp. Electron. Syst.* **2009**, *45*, 1650–1657. [CrossRef]
20. Yu, X.; Alhujaili, K.; Cui, G.; Monga, V. MIMO radar waveform design in the presence of multiple targets and practical constraints. *IEEE Trans. Signal Process.* **2020**, *68*, 1974–1989. [CrossRef]
21. Zhang, Y.; Liao, G.; Xu, J.; Lan, L. Mainlobe Deceptive Jammer Suppression Based on Quadratic Phase Coding in FDA-MIMO Radar. *Remote Sens.* **2022**, *14*, 5831. [CrossRef]
22. Liu, T.; Sun, J.; Wang, G.; Lu, Y. A Multi-Objective Quantum Genetic Algorithm for MIMO Radar Waveform Design. *Remote Sens.* **2022**, *14*, 2387. [CrossRef]
23. Stoica, P.; He, H.; Li, J. New algorithms for designing unimodular sequences with good correlation properties. *IEEE Trans. Signal Process.* **2009**, *57*, 1415–1425. [CrossRef]
24. Song, J.; Babu, P.; Palomar, D.P. Sequence design to minimize the weighted integrated and peak sidelobe levels. *IEEE Trans. Signal Process.* **2015**, *64*, 2051–2064. [CrossRef]
25. Feng, X.; Zhao, Z.; Li, F.; Cui, W.; Zhao, Y. Radar Phase-Coded Waveform Design with Local Low Range Sidelobes Based on Particle Swarm-Assisted Projection Optimization. *Remote Sens.* **2022**, *14*, 4186. [CrossRef]
26. He, H.; Stoica, P.; Li, J. On Aperiodic-Correlation Bounds. *IEEE Signal Process. Lett.* **2010**, *17*, 253–256. [CrossRef]
27. Pezeshki, A.; Calderbank, A.R.; Moran, W.; Howard, S.D. Doppler Resilient Golay Complementary Waveforms. *IEEE Trans. Inf. Theory* **2008**, *54*, 4254–4266. [CrossRef]
28. Li, H.; Liu, Y.; Liao, G.; Chen, Y. Joint Radar and Communications Waveform Design Based on Complementary Sequence Sets. *Remote Sens.* **2023**, *15*, 645. [CrossRef]
29. Chen, H.H.; Yeh, Y.C.; Bi, Q.; Jamalipour, A. On a MIMO-based open wireless architecture: Space-time complementary coding. *IEEE Commun. Mag.* **2007**, *45*, 104–112. [CrossRef]
30. Song, Y.; Wang, Y.; Xie, J.; Yang, Y.; Tian, B.; Xu, S. Ultra-Low Sidelobe Waveforms Design for LPI Radar Based on Joint Complementary Phase-Coding and Optimized Discrete Frequency-Coding. *Remote Sens.* **2022**, *14*, 2592. [CrossRef]

31. Levanon, N.; Cohen, I.; Itkin, P. Complementary pair radar waveforms–evaluating and mitigating some drawbacks. *IEEE Aerosp. Electron. Syst. Mag.* **2017**, *32*, 40–50. [CrossRef]
32. Aubry, A.; De Maio, A.; Jiang, B.; Zhang, S. Ambiguity function shaping for cognitive radar via complex quartic optimization. *IEEE Trans. Signal Process.* **2013**, *61*, 5603–5619. [CrossRef]
33. Tang, J.; Zhang, N.; Ma, Z.; Tang, B. Construction of Doppler Resilient Complete Complementary Code in MIMO Radar. *IEEE Trans. Signal Process.* **2014**, *62*, 4704–4712. [CrossRef]
34. Song, J.; Babu, P.; Palomar, D.P. Sequence set design with good correlation properties via majorization-minimization. *IEEE Trans. Signal Process.* **2016**, *64*, 2866–2879. [CrossRef]
35. Wang, F.; Pang, C.; Wu, H.; Li, Y.; Wang, X. Designing Constant Modulus Complete Complementary Sequence With High Doppler Tolerance for Simultaneous Polarimetric Radar. *IEEE Signal Process. Lett.* **2019**, *26*, 1837–1841. [CrossRef]
36. Xu, L.; Zhou, S.; Liu, H. Simultaneous optimization of radar waveform and mismatched filter with range and delay-Doppler sidelobes suppression. *Digit. Signal Process.* **2018**, *83*, 346–358. [CrossRef]
37. Wang, Y.C.; Dong, L.; Xue, X.; Yi, K.-C. On the Design of Constant Modulus Sequences with Low Correlation Sidelobes Levels. *IEEE Commun. Lett.* **2012**, *16*, 462–465. [CrossRef]
38. Wu, L.; Babu, P.; Palomar, D.P. Transmit waveform/receive filter design for MIMO radar with multiple waveform constraints. *IEEE Trans. Signal Process.* **2017**, *66*, 1526–1540. [CrossRef]
39. Beauchamp, R.M.; Tanelli, S.; Peral, E.; Chandrasekar, V. Pulse compression waveform and filter optimization for spaceborne cloud and precipitation radar. *IEEE Trans. Geosci. Remote Sens.* **2016**, *55*, 915–931. [CrossRef]
40. Wang, F.; Xia, X.-G.; Pang, C.; Cheng, X.; Li, Y.; Wang, X. Joint Design Methods of Unimodular Sequences and Receiving Filters with Good Correlation Properties and Doppler Tolerance. *IEEE Trans. Geosci. Remote Sens.* **2022**, *61*, 5100214. [CrossRef]
41. Zhou, K.; Quan, S.; Li, D.; Liu, T.; He, F.; Su, Y. Waveform and filter joint design method for pulse compression sidelobe reduction. *IEEE Trans. Geosci. Remote Sens.* **2022**, *60*, 5107615. [CrossRef]
42. Wang, F.; Feng, S.; Yin, J.; Pang, C.; Li, Y.; Wang, X. Unimodular sequence and receiving filter design for local ambiguity function shaping. *IEEE Trans. Geosci. Remote Sens.* **2022**, *60*, 5113012. [CrossRef]
43. Xiong, W.; Hu, J.; Zhong, K.; Sun, Y.; Xiao, X.; Zhu, G. MIMO Radar Transmit Waveform Design for Beampattern Matching via Complex Circle Optimization. *Remote Sens.* **2023**, *15*, 633. [CrossRef]
44. Zhou, K.; Li, D.; Quan, S.; Liu, T.; Su, Y.; He, F. SAR waveform and mismatched filter design for countering interrupted-sampling repeater jamming. *IEEE Trans. Geosci. Remote Sens.* **2022**, *60*, 5214514. [CrossRef]
45. Yang, J.; Zhou, J.; Liu, L.; Li, Y. A novel strategy of pareto-optimal solution searching in multi-objective particle swarm optimization (MOPSO). *Comput. Math. Appl.* **2009**, *57*, 1995–2000. [CrossRef]
46. Everett, H. Generalized Lagrange multiplier method for solving problems of optimum allocation of resources. *Oper. Res.* **1963**, *11*, 399–417. [CrossRef]
47. Raydan, M.; Svaiter, B.F. Relaxed steepest descent and Cauchy-Barzilai-Borwein method. *Comput. Optim. Appl.* **2002**, *21*, 155–167. [CrossRef]
48. Varadhan, R.; Roland, C. Simple and globally convergent methods for accelerating the convergence of any EM algorithm. *Scand. J. Stat.* **2008**, *35*, 335–353. [CrossRef]
49. Roughgarden, T. *Beyond the Worst-Case Analysis of Algorithms*; Cambridge University Press: Cambridge, UK, 2021.
50. He, H.; Li, J.; Stoica, P. *Waveform Design for Active Sensing Systems: A Computational Approach*; Cambridge University Press: Cambridge, UK, 2012.
51. Chen, R.; Cantrell, B. Highly bandlimited radar signals. In Proceedings of the 2002 IEEE Radar Conference, Long Beach, CA, USA, 25 April 2002.
52. Liu, J.; Zhang, Y.; Dong, X. Linear-FM random radar waveform compressed by dechirping method. *IET Radar Sonar Navig.* **2019**, *13*, 1107–1115. [CrossRef]
53. Chen, Y.; Lin, R.; Cheng, Y.; Li, J. Joint Design of Periodic Binary Probing Sequences and Receive Filters for PMCW Radar. *IEEE Trans. Signal Process.* **2022**, *70*, 5996–6010. [CrossRef]
54. De Maio, A.; Huang, Y.; Piezzo, M.; Zhang, S.; Farina, A. Design of optimized radar codes with a peak to average power ratio constraint. *IEEE Trans. Signal Process.* **2011**, *59*, 2683–2697. [CrossRef]

 *remote sensing*

Communication

Enhanced Radar Detection in the Presence of Specular Reflection Using a Single Transmitting Antenna and Three Receiving Antennas

Yong Yang * and Xue-Song Wang

School of Electronic Science, National University of Defense Technology, Changsha 410073, China
* Correspondence: youngtfvc@163.com

Abstract: Radar target echoes undergo fading in the presence of specular reflection, which is adverse to radar detection. To address this problem, this paper proposes a radar detection method that uses a single transmitting antenna and three receiving antennas. The proposed method uses the maximum absolute value of the difference in the radar received signal power among the three receiving antennas as the test statistic. First, the target echo in the presence of specular reflection is analyzed. Then, selection of the required number of radar antennas and the heights at which they must be situated are discussed. Subsequently, analytical expressions of the radar detection probability and the false alarm probability are derived. Finally, simulation results are presented, which show that the proposed method improves radar detection performance in the presence of specular reflection.

Keywords: radar detection; specular reflection; multiple antennas; detection probability

Citation: Yang, Y.; Wang, X.-S. Enhanced Radar Detection in the Presence of Specular Reflection Using a Single Transmitting Antenna and Three Receiving Antennas. *Remote Sens.* **2023**, *15*, 3204. https://doi.org/10.3390/rs15123204

Academic Editors: Stefano Tebaldini and Deodato Tapete

Received: 21 March 2023
Revised: 8 June 2023
Accepted: 18 June 2023
Published: 20 June 2023

1. Introduction

Multipath interference affects the radar detection of low altitude targets over the sea. In multipaths, the directly arriving target echo and the sea surface-reflected target echo overlap with each other. The resulting combined multipath returns combine either constructively or destructively with the direct-path signal, which produces a stronger or weaker total received signal at random. In most cases, multipath attenuates the radar received signal intensity. It is disadvantageous to radar detection [1–5]. Therefore, overcoming the negative effect of multipath is a key issue for radar detection of low altitude targets over the sea.

Multipath scattering from a rough surface involves two components: specular reflection and diffuse reflection. In most cases, specular reflection dominates multipath scattering, so many scholars only consider specular reflection for multipath scattering [6,7]. In the specular reflection case, the radar detection performance can be improved if prior knowledge of the radar-target environment is known [8]. However, the target position usually varies with time, and such prior knowledge is difficult to obtain. Without prior knowledge, radar systems often employ frequency diversity to overcome multipath interference [9–11]. For example, [10] developed adaptive orthogonal frequency-division multiplexing (OFDM) signals for moving target detection in multipaths. In [11], an order statistics-based detection method was proposed to improve the radar target detection performance in multipaths based on frequency diversity. In fact, multipath returns have different propagation ways. This implies that the amplitude or the phase of the multipath returns are different at some fixed location in the space. Thus, spatial diversity can be used to overcome the negative effect the multipath [12,13]. For example, MIMO radar is proposed to detect the target in the presence of multipath [14]. However, MIMO radar is difficult to be implemented in reality. In addition, array antennas and multiple subapertures (which are collectively referred to as multiple antennas) are widely used for radar tracking in multipaths [15–18]. Since multiple antennas can be used for radar tracking in this way, it is likely that they can also be used for radar detection in multipaths. However, to our knowledge, there has been little previous work on radar detection in multipaths using multiple antennas. Whether radar detection performance in multipaths can be enhanced by using multiple antennas is an interesting question. There are some problems that need to be addressed when using multiple antennas for radar detection in multipaths:

(1)   How many antennas should be used in the system?
(2)   Where should the antenna heights be set?
(3)   What test statistic should be used?

In this paper, we propose using a single transmitting antenna and three receiving antennas to improve radar detection performance in the presence of specular reflection. The maximum absolute value of the difference in the radar-received signal power among the three antennas is used as the test statistic. The advantages of using three receiving antennas are demonstrated, and the requirements for setting the radar antenna heights are discussed. Mathematical expressions for the radar detection probability and the false alarm probability are derived. Simulation results are given to validate the proposed method.

The rest of this paper is organized as follows. Radar target echoes in the presence of specular reflection are analyzed in Section 2. In Section 3, the selection of the number of antennas and the antenna heights are discussed. Mathematical derivations of the radar detection probability and the false alarm probability are presented in Section 4. The simulation results that demonstrate the validity of the proposed method are given in Section 5. Finally, Section 6 concludes the paper.

2. Radar Target Echo in the Presence of Specular Reflection

Assuming that radar transmits and receives signals using the same single antenna, a schematic diagram of the radar specular reflection is shown in Figure 1. In the presence of specular reflection, four different paths contribute to the radar-received target echoes: direct-direct (ABA), direct-reflected (ABOA), reflected-direct (AOBA) and reflected-reflected (AOBOA) paths [19]. Then, the received target echoes can be expressed by

$$
\begin{aligned}
s_t = A \exp(j\varphi) \Bigg[1 + \sqrt{\tfrac{G_r(\theta_r)}{G_r(\theta_d)}} \rho_s \exp(j\phi) + \\
\sqrt{\tfrac{G_t(\theta_r)}{G_t(\theta_d)}} \rho_s \exp(j\phi) + \sqrt{\tfrac{G_t(\theta_r)G_r(\theta_r)}{G_t(\theta_d)G_r(\theta_d)}} \rho_s^2 \exp(j2\phi) \Bigg]
\end{aligned}
\tag{1}
$$

where A and φ are the amplitude and phase of the directly arriving target echo, respectively; $G_t(\theta)$ and $G_r(\theta)$ are the radar transmitting and receiving antenna gains at an angle θ, respectively; θ_d and θ_r are the elevation angles of the direct and reflected paths, respectively; ρ_s is the amplitude of the specular reflection coefficient; $\phi = \phi_l + \phi_\rho$, where ϕ_ρ is the phase of the specular reflection coefficient, which is a constant; $\phi_l = \frac{2\pi}{\lambda}(l_r + l_t - R)$, where λ is the wavelength; and R, l_r, and l_t are the lengths of paths AB, AO, and BO, respectively, as shown in Figure 1. Clearly, ϕ and ϕ_l are both sensitive to the radar height h_r and to the target height h_t.

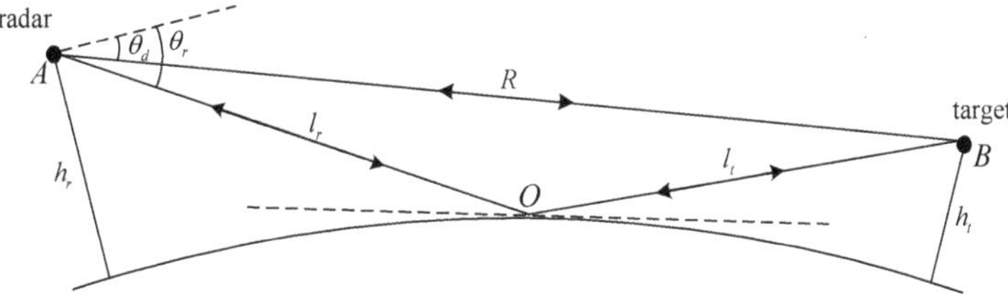

Figure 1. Schematic diagram of radar specular reflection.

In the case when the radar operates at low grazing angles, $\theta_r \approx \theta_d$, $G_t(\theta_r) \approx G_t(\theta_d)$, and $G_r(\theta_r) \approx G_r(\theta_d)$. Thus, Equation (1) can be simplified to

$$s_t = A \exp(\mathrm{j}\varphi)[1 + \rho_s \exp(\mathrm{j}\phi)]^2, \tag{2}$$

From (2), it can be seen that the target echo power varies as the ϕ varies. The ϕ varies with changes in the radar antenna height, target height and target distance. Therefore, the target echo power varies with changes in the radar antenna height and the target location. To demonstrate this conclusion indirectly, Figure 2 presents ϕ_l under various radar antenna heights and target locations, where the radar wavelength is 0.03 m.

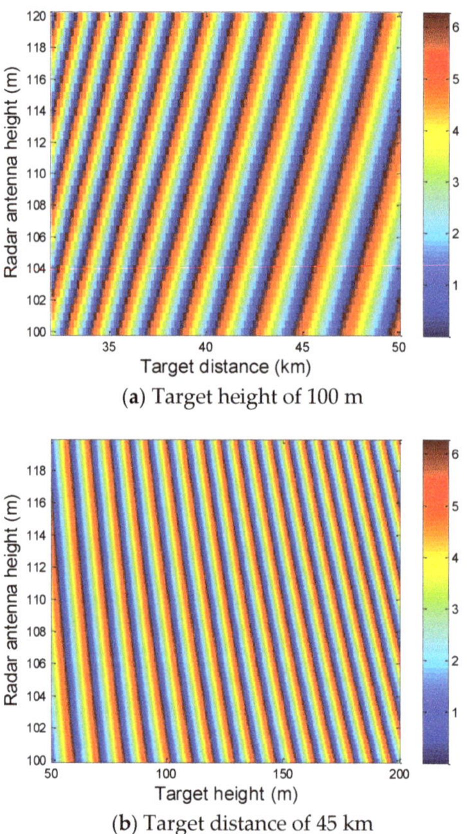

(**a**) Target height of 100 m

(**b**) Target distance of 45 km

Figure 2. ϕ_l versus radar antenna height and target location.

In Figure 2, ϕ_l varies with changes in the radar antenna height, target height and target distance. Therefore, ϕ also varies with the radar antenna height, target height and target distance, which induces the target echo power to vary in the same manner. Thus, we can conclude that the target echo powers that are received by antennas at different heights are also likely different. However, the clutter mean powers that are received by the antennas at the different heights are almost identical if the height difference between antennas is within several meters. Because the ground surface or sea surface represents an area target, the clutter mean power is mainly related to the distance between the radar antenna and the reflected surface. The height differences between the multiple antennas are much smaller than the distance between the radar and the reflected surface, which will not induce the clutter mean power difference in antennas. Overall, the target echo powers received by each of the multiple antennas may be different, whereas the clutter mean powers received by the same multiple antennas are almost identical. Thus, we can use multipath antennas at different heights and use the differences in the signal powers received by the multiple antennas as test statistics to decide whether a target exists or not. In the following section, we first discuss how to select the antenna number and their heights.

3. Selection of Antenna Numbers and Heights

Assuming that the radar uses a single transmitting antenna and multiple receiving antennas and that these antennas are set at different heights. To better detect the target, the differences among the target echo powers received by the multiple antennas needs to be obvious at all times. However, the target location is unknown beforehand, and this location changes with time, which may cause the differences among the target echo powers received by the multiple antennas to be obvious when the target is at certain locations and small when the target is at other locations. Therefore, selection of the number of antennas and their heights is vital to ensure that there is at least a difference among the target echo powers received by the multiple antennas, which is always obvious for any target location.

3.1. Selection of Antenna Number

Assuming that the radar uses a single transmitting antenna and two receiving antennas. Antenna one both transmits and receives signals, while antenna two only receives signals. The simulated target echo powers that are received by the two antennas are shown in Figure 3, where "ant 1" and "ant 2" denote antenna one and antenna two, respectively. In the simulations, the two antenna heights are randomly set at 200 m and 210 m. The radar transmitted power is 50 kW, the radar wavelength is 0.03 m, the maximum antenna gain is 43 dB, the half power beam width is $4°$, and the target height is 50 m. The target maneuver is not considered in this paper. Please refer to [20] for maneuvering target detection.

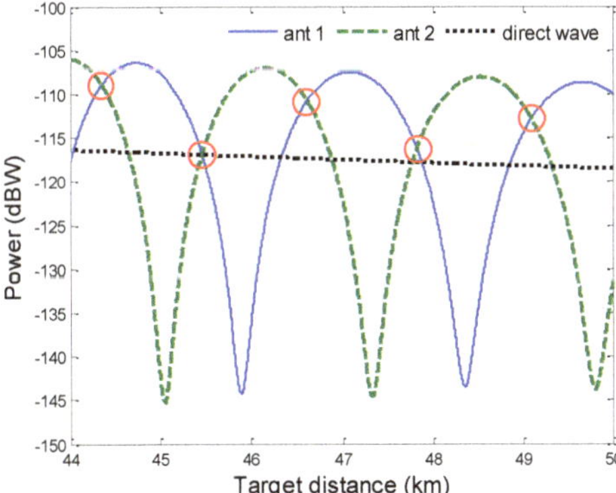

Figure 3. Received target echo powers received by two antennas with different heights.

Figure 3 shows that the target echo powers that are received by the two antennas are obviously different in most cases. However, regardless of the height difference between the two antennas, there are always some target locations at which the difference between the received target echo powers of the two antennas is small, such as the locations that are labeled with circles in Figure 3. The radar detection probability will be low at these locations if we use the difference between the received target echo powers of the two antennas as the test statistic. To address this problem, we attempt to add one receiving antenna and use a single transmitting antenna and three receiving antennas.

Denoting the target echo powers received by the three antennas as z_1, z_2 and z_3. The differences between the target echo powers are $|z_1 - z_2|$, $|z_1 - z_3|$ and $|z_2 - z_3|$. Then, we choose the maximum value of the above differences as the test statistic. Thus, while

$|z_1 - z_2|$ may be small at certain locations, either $|z_1 - z_3|$ or $|z_2 - z_3|$ may be apparent at these locations. To demonstrate this prediction, Figure 4 presents the simulated target echo powers that were received by three antennas set at different heights, where "ant 3" denotes antenna three, which only receives signals, and the heights of the three antennas are set at 200 m, 206 m and 212 m. The other parameters are the same as those used in Figure 3.

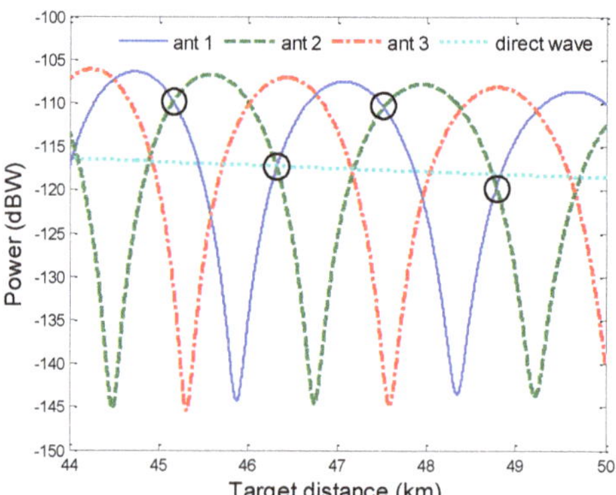

Figure 4. Target echo powers received by three antennas with different heights.

Figure 4 shows that the differences among the target echo powers that are received by the three antennas are apparent in most cases. Despite the fact that $|z_1 - z_2|$ is small in the locations labeled by circles, $|z_1 - z_3|$ or $|z_2 - z_3|$ is apparent at these locations. This finding verifies our prediction above and illustrates the value of using the third antenna. Next, we will discuss how to set the heights of the three antennas.

3.2. Setting of Antenna Heights

From Figure 2, it can be seen that the target echo power is sensitive to the radar antenna height. In addition, Figures 3 and 4 show that the differences among the target echo powers of the multiple antennas are related to the antenna heights. Therefore, setting the antenna height carefully is important to ensure that there is always an apparent difference in the target echo powers of multiple antennas for any target location. In general, there are four antenna height setting requirements:

(1) The antenna height must be set high enough to ensure that the target is within the radar line of sight region when considering the earth's curvature.
(2) The distance between the transmitting and receiving antennas should be sufficient to maintain a fixed antenna isolation degree.
(3) The clutter mean powers received by the three antennas are approximately equal to each other.
(4) There is always an apparent difference value for the target echo powers of the multiple antennas for any target location.

To meet the first requirement, the minimum radar antenna height can be calculated by referring to the work of [21]. The second and third requirements are also both easily satisfied if the distances between the three antenna heights are moderate. The fourth requirement is the most important and is the most difficult to satisfy. As Figures 2 and 3 indicate, the target echo power varies with changes in the target height and the target distance, regardless of the antenna height. Thus, the maximum target echo power difference for the three antennas varies with changes in target location, and it may be small or apparent as the target location changes. For this reason, there is no optimal height setting available for the three antennas that will satisfy the fourth requirement. Without loss of generality, we set the three antenna heights to 200 m, 206 m and 212 m. A schematic diagram of the configuration of the three antennas is shown in Figure 5. Because the three antennas are identical, it can be assumed that antenna 1 randomly transmits and receives signals, while antenna 2 and antenna 3 receive signals only. In the next step, the detection probability and the false alarm probability for the radar using the three antennas will be derived.

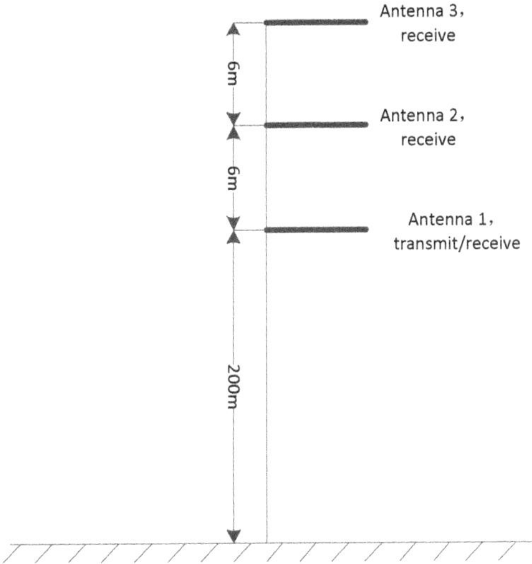

Figure 5. Schematic diagram of the configuration of the three antennas.

4. Detection and False Alarm Probabilities

In the presence of specular reflection, the signal received by the ith antenna can be written as

$$x_i = \begin{cases} c_i + n_i & , \ H_0 \\ s_i[1 + \rho_s \exp(j\phi_i)]^2 + c_i + n_i & , \ H_1 \end{cases} \quad i = 1,2,3, \tag{3}$$

where H_0 and H_1 denote that the target is absent and present, respectively, the subscript i denotes the ith antenna, s_i denotes the directly arriving target echo, c_i represents the complex Gaussian distributed clutter with zero mean and variance σ_c^2, n_i is the complex Gaussian distributed thermal noise with zero mean and variance σ_n^2, and the clutter and the thermal noise are independent of each other.

Expanding (3) gives the received signal of the ith antenna under H_1 as

$$\begin{aligned} x_i &= s_i[1 + \rho_s \exp(j\phi_i)]^2 + c_i + n_i \\ &= s_x a_x - a_y s_y + c_x + n_x + j(a_y s_x + a_x s_y + c_y + n_y) \end{aligned} \tag{4}$$

where $a_x = 1 + 2\rho_s \cos\phi + \rho_s^2 \cos 2\phi$, $a_y = \rho_s^2 \sin 2\phi + 2\rho_s \sin\phi$, $s_i = s_x + js_y$, $c_i = c_x + jc_y$, and $n_i = n_x + jn_y$.

For the Swerling I fluctuation target, its mean power is denoted by P_s. Then, based on (4), the real and imaginary parts of x_i can be derived, and both are found to have Gaussian distributions with zero means and variances in $\left(a_x^2 + a_y^2\right)P_s/2 + \sigma_c^2 + \sigma_n^2$. Because $z_i = |x_i|^2$, the probability density function (PDF) of z_i under H_1 can be obtained as

$$\begin{aligned} f(z_i|H_1) &= \frac{1}{\left(a_x^2 + a_y^2\right)P_s + P_n + P_c} \cdot \\ &\quad \exp\left[-\frac{z_i}{\left(a_x^2 + a_y^2\right)P_s + P_n + P_c}\right], i = 1,2,3 \end{aligned} \tag{5}$$

where $P_n = 2\sigma_n^2$ and $P_c = 2\sigma_c^2$.

From (5), the PDF of $z_{12} = |z_1 - z_2|$ under H_1 can be derived by [22]

$$
\begin{aligned}
f(z_{12}|H_1) &= \int_0^\infty f_{z_1}(z_{12} + z_2|H_1)f(z_2|H_1)\mathrm{d}z_2 + \\
&\quad \int_{z_{12}}^\infty f_{z_1}(z_2 - z_{12}|H_1)f(z_2|H_1)\mathrm{d}z_2 \\
&= \frac{1}{\left(a_x^2 + a_y^2\right)P_s + P_n + P_c} \cdot \\
&\quad \exp\left[-\frac{z_{12}}{\left(a_x^2 + a_y^2\right)P_s + P_n + P_c}\right]
\end{aligned}
\tag{6}
$$

where $f_{z_1}(z_{12} + z_2|H_1)$ means the PDF of z_1 under H_1 with independent variable z_1 substituted by $z_{12} + z_2$.

Similarly, the probability density functions of z_{13} and z_{23} are the same as that of z_{12}. Choosing the test statistic as

$$
L = \max(z_{12}, z_{13}, z_{23}),
\tag{7}
$$

The radar detection probability can be calculated by [see the Appendix A]

$$
\begin{aligned}
P_d &= \Pr[\max(z_{12}, z_{13}, z_{23}) > \eta|H_1] \\
&= 1 - \Pr[|z_1 - z_2| < \eta \cap |z_1 - z_3| < \eta \cap |z_2 - z_3| < \eta], \\
&= 1 - \exp\left(-\frac{3\eta}{\chi}\right)\left[\exp\left(\frac{\eta}{\chi}\right) - 1\right]^3
\end{aligned}
\tag{8}
$$

where η is the detection threshold, and $\chi = \left(a_x^2 + a_y^2\right)P_s + P_n + P_c$.

Setting $P_s = 0$ in (8) then yields the radar false alarm probability as

$$
P_f = 1 - \exp\left(-\frac{3\eta}{P_n + P_c}\right)\left[\exp\left(\frac{\eta}{P_n + P_c}\right) - 1\right]^3,
\tag{9}
$$

5. Simulation Results and Analysis

In this section, the detection performance of the radar using a single transmitting antenna and three receiving antennas in the presence of specular reflection is presented by simulation. We compare the detection performance with that of the radar employing a single antenna and that of the radar employing a single transmitting antenna and two receiving antennas.

For the radar employing a single transmitting antenna and two receiving antennas, the test statistic is z_{12}, and the radar detection probability is given by

$$
P_d = \int_\eta^\infty f(z_{12})\mathrm{d}z_{12} = \exp\left[-\frac{\eta}{\left(a_x^2 + a_y^2\right)P_s + P_n + P_c}\right],
\tag{10}
$$

By letting $P_s = 0$ in (10), we acquire the false alarm probability for the radar employing a single transmitting antenna and two receiving antennas as

$$
P_f = \exp\left(-\frac{\eta}{P_n + P_c}\right),
\tag{11}
$$

The detection probability and the false alarm probability for the radar using a single antenna are the same as (10) and (11), respectively. Therefore, in the presence of specular reflection, the detection performance of the radar using a single transmitting antenna and two receiving antennas is the same as that of the single antenna radar.

In the simulations, the clutter mean power $P_c = 10$, the thermal noise mean power $P_n = 1$, and $\rho_s = 0.9$. Figure 6 shows the false alarm probability versus detection threshold, where the notation "three antennas" means radar employing a single transmitting antenna and three receiving antennas, and the notation "single antenna" denotes the single antenna radar.

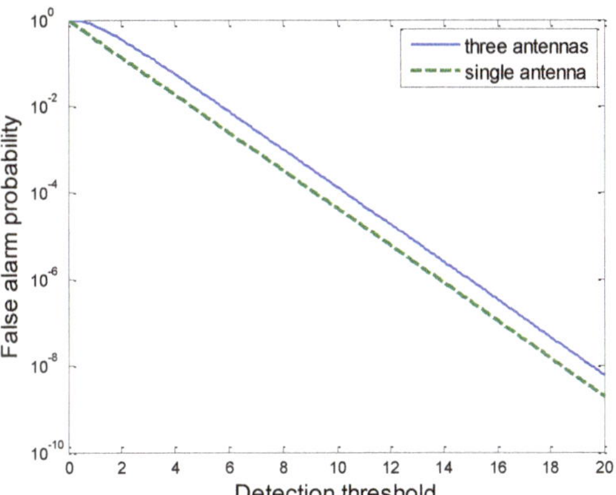

Figure 6. False alarm probability versus detection threshold.

The detection threshold can be obtained by interpretation according to Figure 6 when the false alarm probability is fixed. Figure 6 illustrates that the detection threshold for the radar with the single transmitting antenna and three receiving antennas is higher than that for the single antenna radar to maintain a constant false alarm probability.

The theoretical and simulated detection probabilities for the radar employing a single transmitting antenna and three receiving antennas in the presence of specular reflection are shown in Figure 7, where the Monte Carlo simulation times are 10,000, $P_f = 10^{-3}$, and $\phi_l = \pi$. In addition, the corresponding detection probabilities of the radar using a single antenna in the presence and absence of specular reflection are also presented for comparison. In Figure 7, the notations "with" and "without" denote the presence and absence of specular reflection, respectively.

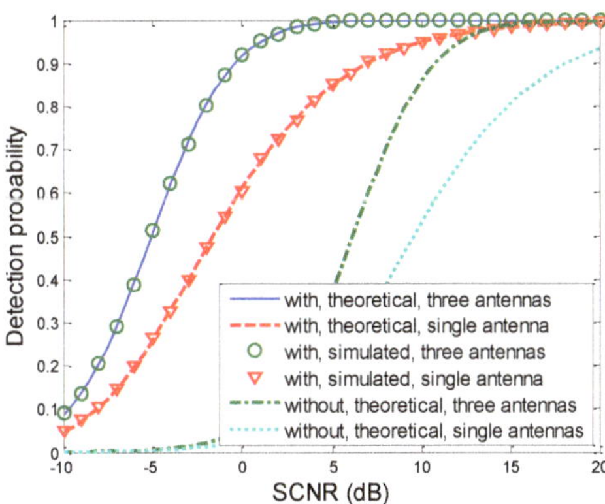

Figure 7. Theoretical and simulated detection probabilities for radar using a single transmitting antenna and three receiving antennas.

Figure 7 shows that the simulated radar detection probability agrees well with the theoretical radar detection probability, which demonstrates the correctness of the theoretical derivation in Section 4. In addition, Figure 7 illustrates that the detection performance of the radar using a single transmitting antenna and three receiving antennas is better than that of the radar with a single antenna.

Because the target echo power is sensitive to ϕ_l and ϕ_l varies with changes in the target location, we present the radar detection probabilities under various ϕ_l of the first antenna in Figure 8. The corresponding detection probabilities of the single antenna radar under the same ϕ_l values are presented in Figure 8 for comparison. Figure 8 shows that radar detection performance in the presence of specular reflection varies with respect to changes in ϕ_l. However, the detection probability for the radar with a single transmitting antenna and three receiving antennas is always higher than that for the single antenna radar, which verifies the effectiveness of the proposed method.

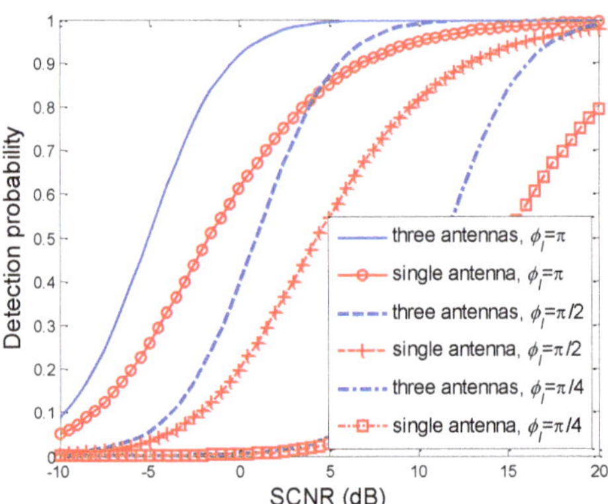

Figure 8. Radar detection probabilities under various ϕ_l.

6. Conclusions

In this paper, radar utilizing a single transmitting antenna and three receiving antennas is proposed for target detection in the presence of specular reflection. This method takes advantage of space diversity to overcome the passive effects of specular reflection on radar detection performance. Based on the characteristic that the target echo powers for antennas set at different heights in the presence of specular reflection are different, the method takes the maximum absolute value of the differences in the received signal powers among the three receiving antennas as the test statistic. Analytical expressions of the radar detection probability and the false alarm probability are obtained. Simulation results show that radar detection performance in the presence of specular reflection is enhanced when using the proposed method.

The impact of the number of receiving antenna is not analyzed concretely in this paper. How to choose the number of the receiving antenna and how to set their height need further investigation. In addition, validation of the proposed method using experimental data is also needed to be studied in the future.

Author Contributions: Conceptualization, Y.Y. and X.-S.W.; methodology, Y.Y. and X.-S.W.; software, Y.Y.; validation, Y.Y. and X.-S.W.; writing—review and editing, Y.Y.; project administration, Y.Y.; funding acquisition, Y.Y. All authors have read and agreed to the published version of the manuscript.

Funding: This work is supported by the National Natural Science Foundation of China under Grant 62171447.

Data Availability Statement: Data sharing is not applicable to this article.

Conflicts of Interest: The authors declare no conflict of interest.

Appendix A

Because z_{12}, z_{13} and z_{23} are similar to each other, we have

$$
\begin{aligned}
\Pr\ & [\max(z_{12}, z_{13}, z_{23}) < \eta] \\
=\ & \Pr[z_{12} < \eta, z_{12} > z_{13}, z_{12} > z_{23}] + \\
& \Pr[z_{13} < \eta, z_{13} > z_{12}, z_{13} > z_{23}] + \\
& \Pr[z_{23} < \eta, z_{23} > z_{12}, z_{23} > z_{13}] \\
=\ & 3 \cdot \Pr[z_{12} < \eta, z_{12} > z_{13}, z_{12} > z_{23}] \\
=\ & 3 \cdot \Pr[z_{12} < \eta, z_{12} > \max(z_{13}, z_{23})]
\end{aligned}
\tag{A1}
$$

Using the notation $t = \max(z_{13}, z_{23})$, (A1) can then be given by

$$
\Pr[\max(z_{12}, z_{13}, z_{23}) < \eta] = 3 \int_0^\eta \int_t^\eta f(z_{12}) \mathrm{d}z_{12} \cdot f(t) \mathrm{d}t,
\tag{A2}
$$

where η is the detection threshold, and

$$
f(t) = F_{z_{13}}(t) f_{z_{23}}(t) + f_{z_{13}}(t) F_{z_{23}}(t),
\tag{A3}
$$

where $f_{z_{23}}(t) = f_{z_{13}}(t) = \frac{1}{\chi} \exp\left[-\frac{t}{\chi}\right]$, $\chi = \left(a_x^2 + a_y^2\right) P_s + P_n + P_c$, and $F_{z_{13}}(t) = F_{z_{23}}(t) = \Pr[z_{23} < t] = \int_0^t f(z_{23}) \mathrm{d}z_{23} = 1 - \exp\left[-\frac{t}{\chi}\right]$. Then, $f(t)$ can be further simplified to

$$
f(t) = 2 F_{z_{13}}(t) f_{z_{23}}(t) = \frac{2}{\chi} \exp\left(-\frac{t}{\chi}\right) - \frac{2}{\chi} \exp\left(-\frac{2t}{\chi}\right),
\tag{A4}
$$

Substituting (A4) and (6) into (A2) yields

$$
\Pr[\max(z_{12}, z_{13}, z_{23}) < \eta] = \exp\left(-\frac{3\eta}{\chi}\right) \left[\exp\left(\frac{\eta}{\chi}\right) - 1\right]^3,
\tag{A5}
$$

References

1. Jang, Y.; Lim, H.; Yoon, D. Multipath effect on radar detection of nonfluctuating targets. *IEEE Trans. Aerosp. Electron. Syst.* **2015**, *51*, 792–795. [CrossRef]
2. Yang, Y.; Feng, D.J.; Wang, X.S.; Xiao, S.P. Effects of K distributed sea clutter and multipath on radar detection of low altitude sea surface targets. *IET Radar Sonar Navig.* **2014**, *8*, 757–766. [CrossRef]
3. Haspert, K.; Tuley, M. Comparison of predicted and measured multipath impulse responses. *IEEE Trans. Aerosp. Electron. Syst.* **2011**, *47*, 1696–1709. [CrossRef]
4. Teti, J.G. Wide-band airborne radar operating considerations for low-altitude surveillance in the presence of specular multipath. *IEEE Trans. Antennas Propag.* **2000**, *48*, 176–191. [CrossRef]
5. Aubry, A.; De Maio, A.; Foglia, G.; Orlando, D. Diffuse multipath exploitation for adaptive radar detection. *IEEE Trans. Signal Process.* **2015**, *63*, 1268–1281. [CrossRef]
6. Barton, D.K. Low-angle radar tracking. *Proc. IEEE* **1974**, *62*, 687–704. [CrossRef]
7. Blair, W.D.; Brandt-Pearce, M. Statistics of Monopulse Measurements of Rayleigh Targets in the Presence of Specular and Diffuse Multipath. In Proceedings of the 2001 IEEE Radar Conference (Cat. No.01CH37200), Atlanta, GA, USA, 3 May 2001; pp. 369–375.
8. Hayvaci, H.T.; De Maio, A.; Erricolo, D. Improved detection probability of a radar target in the presence of multipath with prior knowledge of the environment. *IET Radar Sonar Navig.* **2013**, *7*, 36–46. [CrossRef]
9. Wilson, S.L.; Carlson, B.D. Radar detection in multipath. *IEE Proc. Radar Sonar Navig.* **1999**, *146*, 45–54. [CrossRef]
10. Sen, S.; Nehorai, A. Adaptive OFDM radar for target detection in multipath scenarios. *IEEE Trans. Signal Process.* **2011**, *59*, 78–90. [CrossRef]
11. Cao, Y.; Wang, S.; Wang, Y.; Zhou, S. Target detection for low angle radar based on multi-frequency order-statistics. *J. Syst. Eng. Electron.* **2015**, *26*, 267–273. [CrossRef]
12. Krolik, J.L.; Farrell, J.; Steinhardt, A. Exploiting Multipath Propagation for GMTI in Urban Environments. In Proceedings of the IEEE Radar Conference, Verona, NY, USA, 24–27 April 2006; pp. 65–68.
13. Durek, J. *Multipath Exploitation Radar Industry Day*; DARPA: Herndon, VA, USA, 2009.
14. Shi, J.; Hu, G.; Zhou, H. Entropy-based multipath detection model for MIMO radar. *J. Syst. Eng. Electron.* **2017**, *28*, 51–57. [CrossRef]

15. Xu, Z.H.; Xiao, S.P.; Xiong, Z.Y. *Low Angle Tracking Techniques for Array Radar*; Science Press: Beijing, China, 2014.
16. Gordon, W.B. Improved three subaperture method for elevation angle estimation. *IEEE Trans. Aerosp. Electron. Syst.* **1983**, *19*, 114–122. [CrossRef]
17. Cantrell, B.H.; Gordon, W.B.; Trunk, G.V. Maximum likelihood elevation angle estimates of radar targets using subapertures. *IEEE Trans. Aerosp. Electron. Syst.* **1981**, *AES-17*, 213–221. [CrossRef]
18. Chen, B.X.; Hu, T.J.; Zheng, Z.L.; Wang, F.; Zhang, S.H. Method of altitude measurement based on beam split in VHF radar and its application. *Acta Electron. Sin.* **2007**, *35*, 1021–1025.
19. Daeipour, E.; Blair, W.D.; Bar-Shalom, Y. Bias compensation and tracking with monopulse radars in the presence of multi path. *IEEE Trans. Aerosp. Electron. Syst.* **1997**, *33*, 863–882. [CrossRef]
20. Xia, L.; Gao, H.; Liang, L.; Lu, T.; Feng, B. Radar maneuvering target detection based on product scale zoom discrete chirp fourier transform. *Remote Sens.* **2023**, *15*, 1792. [CrossRef]
21. Sinha, A.; Bar-Shalom, Y.; Blair, W.D.; Kirubarajan, T. Radar measurement extraction in the presence of sea-surface multipath. *IEEE Trans. Aerosp. Electron. Syst.* **2003**, *39*, 550–567. [CrossRef]
22. Papoulis, A.; Pillai, S.U. *Probability, Random and Stochastic Processes*; McGraw-Hill: New York, NY, USA, 2001.

 *remote sensing*

Article

Target Parameter Estimation Algorithm Based on Real-Valued HOSVD for Bistatic FDA-MIMO Radar

Yuehao Guo [1], Xianpeng Wang [1,*], Jinmei Shi [2], Lu Sun [3] and Xiang Lan [1]

1 School of Information and Communication Engineering, Hainan University, Haikou 570228, China
2 College of Information Engineering, Hainan Vocational University of Science and Technology, Haikou 571158, China
3 Department of Communication Engineering, Institute of Information Science Technology, Dalian Maritime University, Dalian 116026, China
* Correspondence: wxpeng2016@hainanu.edu.cn

Abstract: Since there is a frequency offset between each adjacent antenna of FDA radar, there exists angle-range two-dimensional dependence in the transmitter. For bistatic FDA-multiple input multiple output (MIMO) radar, range direction of departure (DOD) direction of arrival (DOA) information is coupled in transmitting the steering vector. How to decouple the three information has become the focus of research. Aiming at the issue of target parameter estimation of bistatic FDA-MIMO radar, a real-valued parameter estimation algorithm based on high-order-singular value decomposition (HOSVD) is developed. Firstly, for decoupling DOD and range in transmitter, it is necessary to divide the transmitter into subarrays. Then, the forward–backward averaging and unitary transformation techniques are utilized to convert complex-valued data into real-valued data. The signal subspace is obtained by HOSVD, and the two-dimensional spatial spectral function is constructed. Secondly, the dimension of spatial spectrum is reduced by the Lagrange algorithm, so that it is only related to DOA, and the DOA estimation is obtained. Then the frequency increment between subarrays is used to decouple the DOD and range information, and eliminate the phase ambiguity at the same time. Finally, the DOD and range estimation automatically matched with DOA estimation are obtained. The proposed algorithm uses the multidimensional structure of high-dimensional data to promote performance. Meanwhile, the proposed real-valued tensor-based method can effectively cut down the computing time. Simulation results verify the high efficiency of the developed method.

Keywords: bistatic FDA-MIMO radar; unitary transformation technique; HOSVD; DOA-DOD-range estimation

 check for
updates

Citation: Guo, Y.; Wang, X.; Shi, J.; Sun, L.; Lan, X. Target Parameter Estimation Algorithm Based on Real-Valued HOSVD for Bistatic FDA-MIMO Radar. *Remote Sens.* **2023**, *15*, 1192. https://doi.org/10.3390/rs15051192

Academic Editor: Okan Yurduseven

Received: 18 January 2023
Revised: 11 February 2023
Accepted: 13 February 2023
Published: 21 February 2023

1. Introduction

Multiple input multiple output (MIMO) radar was developed in 2004, which can make up for the drawbacks of phased array (PA) [1–3]. Different from PA radar, the transmitted waveforms of MIMO radar are orthogonal to each other [4–6]. When the receiver completes the matched filtering, it can produce a large number of virtual array elements, which can dramatically promote radar performance [7–9]. However, the advantages of MIMO radar in range estimation are not prominent. Therefore, in 2006, Antonik et al. proposed frequency diversity array (FDA) [10,11]. With the development of array signal processing, some scholars developed FDA-MIMO radar [12,13]. There are two categories of FDA-MIMO: statistical radar [14,15] and collocated radar [16,17]. In this paper, the bistatic FDA-MIMO in collocated radar is taken as the research object.

FDA-MIMO radar adds frequency increment in transmitter antennas, the transmitter waveform is affected by both range and angle [18,19]. The transmitting waveform has two-dimensional dependence on range and angle [20,21]. Therefore, FDA-MIMO can estimate angle and range concurrently [22,23]. Moreover, since FDA-MIMO radar adds the information of range dimension, the degree of freedom (DOF) is increased [24,25].

Thus, FDA-MIMO radar can achieve more tasks in different environments [26]. Therefore, FDA-MIMO radar can be applied to achieve parameter estimation.

Since FDA-MIMO can provide more DOFs of the system, it can provide more target parameter information. Therefore, many parameter estimation studies for FDA-MIMO have been performed. In [27], an algorithm based on the multiple signal classification (MUSIC) method to achieve target angle-range is proposed. Since the estimation method requires two spectral peak searches to obtain angle and range, the estimation method has a lot of operational redundancy. In order to cut down the operation time, a two-stage algorithm via rotation invariance technique (ESPRIT) approach is developed to obtain the angle-range estimation [28]. The angle and range estimated by this algorithm are automatically matched. In [29], in order to cut down more operation time. Liu et al. constructed the unitary matrix to transform the subspace into real-valued data, further reducing the operational redundancy. However, the precision of this approach will be declined in low signal-to-noise ratio (SNR). Therefore, a tensor-based FDA-MIMO radar algorithm is proposed [30], this algorithm realizes spatial spectrum dimensionality reduction through Lagrange multiplier method, thus reducing the computational complexity of spectral peak search part, and achieving decoupling of angle and range. Xu et al. proposed an algorithm based on high-order-singular value decomposition (HOSVD), which retains the multi-dimensional structure of data and can obtain superior result. Although the above algorithms can realize the estimation of FDA-MIMO, the accuracy of estimation will sharply decline in low SNR and snapshot. However, in bistatic FDA-MIMO radar, these algorithms will fail because of the coupling problem between the DOD and the range of the target. At present, the research of target parameter estimation for bistatic FDA-MIMO radar is still insufficient, and the existing algorithms are mostly carried out for monostatic FDA-MIMO radar.

In this paper, a real-valued parameter estimation approach based on HOSVD is developed. This algorithm solves the problem of three-dimensional direction of arrival (DOA)-direction of departure (DOD)-range estimation in bistatic FDA-MIMO radar. Firstly, for eliminating the coupling between DOD and range, the transmitter is divided into several subarrays. Then a three-dimensional tensor data model is constructed. The original tensor is converted into a real-valued tensor by unitary transformation technique. The HOSVD algorithm is employed to obtain signal subspace. The spatial spectrum function is constructed through obtained subspace. Then the one-dimensional spatial spectrum only related to DOA information is obtained by Lagrange algorithm. The DOA is estimated by one-dimensional spatial spectral. Utilizing the constructed transmitting subarray, the DOD and range information are decoupled and the periodic ambiguity of phase is eliminated. Finally, the automatically matched DOD and range are estimated. The developed approach preserves the multidimensional structure. Furthermore, the matched DOA, DOD and range estimation can be obtained through the reduced-dimension MUSIC algorithm. The presented method not only has high precision, but also cut down the computational complexity.

In summary, the contributions of the developed algorithm are summarized as:

(1) The developed approach can achieve joint DOA-DOD-range estimation for bistatic FDA-MIMO. The tensor signal subspace is obtained by HOSVD method. The original structure of the received data is preserved, which can greatly improve the estimation accuracy;

(2) The proposed approach is a real-valued operation, and it utilizes the reduced-dimension MUSIC algorithm, which estimates DOA by utilizing one-dimensional spatial spectrum. It greatly reduces operational redundancy while ensuring the performance advantages;

(3) The presented method eliminates the coupling of DOD information and range information by subarray division of transmitter. Accurate DOD and range estimations are achieved.

The notations are presented in Table 1.

Table 1. Notations.

Notation	Definition
$(\cdot)^*$	conjugate
$(\cdot)^T$	transpose
$(\cdot)^H$	conjugate-transpose
$(\cdot)^{-1}$	inverse
$(\cdot)^\dagger$	pseudo-inverse
$\otimes$	Kronecker product
$\odot$	Khatri–Rao product
$\circ$	outer product
$\sqcup_n$	the concatenation along the n-th mode
I_Q	$Q \times Q$ identity matrix
0_Q	$Q \times Q$ zero matrix
$\lfloor \cdot \rfloor$	floor operator
$(\cdot)!$	factorial
$diag(\cdot)$	diagonalization of matrix
$angle(\cdot)$	the phase of array elements

2. Signal Model

The bistatic FDA-MIMO radar is taken as the research object. From Figure 1, we can know that there are M transmitting antennas and N receiving antennas, all of which are uniform linear arrays (ULA) with half wavelength spacing. d_t and d_r are antenna spacing. On the basis of the definition of FDA-MIMO radar, there is a frequency increment between antennas at transmitter. The carrier frequency of the m-th antenna is given by [31]

$$f_m = f_0 + (m-1)\Delta f, \quad m = 1, 2, \cdots, M, \tag{1}$$

where f_0 represents carrier frequency. Δf stands for frequency increment. The signal transmitted by m-th transmitting antenna can be defined as

$$s_m(t) = \sqrt{\frac{E}{M}} \psi_m(t) e^{j2\pi f_m t},$$

$$0 \le t \le T, \quad m = 1, 2, \cdots, M, \tag{2}$$

where E stands for energy, $\psi_m(t)$ represents transmitting waveform. T stands for delay. Since transmitting waveforms are orthogonal to each other, the following expression is given

$$\int_T \psi_m(t)\psi_n^*(t-\tau) e^{j2\pi(m-n)\Delta f t} \mathrm{d}t = \begin{cases} 0, & m \neq n, \forall \tau \\ 1, & m = n, \forall \tau. \end{cases} \tag{3}$$

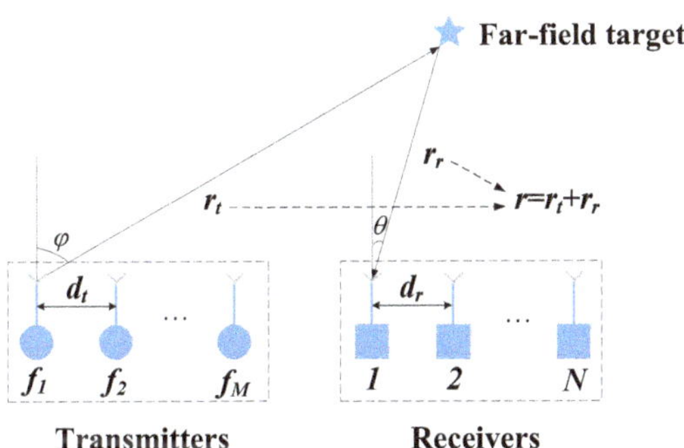

Figure 1. Bistatic FDA-MIMO radar.

Assume K targets in the far-field. DOA, DOD, and range of k-th target are written as θ_k, φ_k, and r_k. The data of the k-th target received by n-th receiving antenna can be defined as [19,32]

$$y_n(t) = \rho_k \sum_{m=1}^{M} \psi_m\left(t - \tau_{m,k}^t - \tau_{n,k}^r\right) e^{j2\pi f_m(t - \tau_{m,k}^t - \tau_{n,k}^r)}, \tag{4}$$

where ρ_k is complex-valued reflection coefficient of the k-th target. $\tau_{m,k}^t$ and $\tau_{n,k}^r$ stand for the time delay, which can be given by

$$\begin{aligned} \tau_{m,k}^t &= \left(r_{1,k}^t - (m-1)d_t \sin(\theta_k)\right)/c, \\ \tau_{n,k}^r &= \left(r_{1,k}^r - (n-1)d_r \sin(\varphi_k)\right)/c, \end{aligned} \tag{5}$$

where $c = 3 \times 10^8$ m/s.

The received snapshot is given by [33]

$$\mathbf{X} = \sum_{k=1}^{K} \xi_k \mathbf{a}_R(\theta_k) \mathbf{a}_T^T(r_k, \varphi_k) + \mathbf{F}, \tag{6}$$

$$\xi_k = \rho_k e^{j2\pi f_0 r_k/c}, \tag{7}$$

where $\mathbf{F}$ represents noise vector.

Then $\mathbf{a}_T(r_k, \varphi_k)$ is defined as [34]

$$\mathbf{a}_T(r_k, \varphi_k) = \mathbf{r}(r_k) \odot \mathbf{d}(\varphi_k) \in \mathbb{C}^{M \times 1}, \tag{8}$$

$$\mathbf{r}(r_k) = \left[1, e^{-j2\pi \Delta f r_k/c}, \ldots, e^{-j2\pi \Delta f(M-1)r_k/c}\right]^T \in \mathbb{C}^{M \times 1}, \tag{9}$$

$$\mathbf{d}(\varphi_k) = \left[1, e^{j2\pi d_t f_0 \sin \varphi_k/c}, \ldots, e^{j2\pi d_t f_0(M-1)\sin \varphi_k/c}\right]^T \in \mathbb{C}^{M \times 1}. \tag{10}$$

From Equation (8), we can know that the range and DOD information are coupled with each other. However, the DOA information of target is only related to $\mathbf{a}_R(\theta_k)$, which is given by

$$\mathbf{a}_R(\theta_k) = \left[1, e^{j2\pi d_r f_0 \sin \theta_k/c}, \ldots, e^{j2\pi d_r f_0(N-1)\sin \theta_k/c}\right]^T \in \mathbb{C}^{N \times 1}. \tag{11}$$

Therefore, the DOA estimation can be obtained from $\mathbf{a}_R(\theta_k)$. However, to obtain DOD and range estimation, $\mathbf{a}_T(r_k, \varphi_k)$ needs to be decoupled. Therefore, the transmitter is converted to P subarrays. Since each subarray is independent, the frequency increment of subarrays is unequal. The subarray $\mathbf{a}_{TS}(r_k, \varphi_k)$ can be written as

$$\mathbf{a}_{TS}(r_k, \varphi_k) = \begin{bmatrix} \mathbf{a}_{TS}^1(r_k, \varphi_k) \\ \mathbf{a}_{TS}^2(r_k, \varphi_k) \\ \vdots \\ \mathbf{a}_{TS}^P(r_k, \varphi_k) \end{bmatrix}. \tag{12}$$

The frequency of mth element in pth ($p = 1, 2, \cdots, P$) subarray is defined as $f_{TS}^{m,p}$, which can be given by

$$f_{TS}^{m,p} = f_{TS}^{1,p} + (m-1)\Delta f_p, m = 1, 2, \cdots, M_{TS}^p, \tag{13}$$

where Δf_p stands for frequency increment of pth subarray.

Figure 2 is subarray model. It can be seen that the transmitter is converted to P subarrays. Therefore, the steering vector of the pth subarray is given by

$$a_{TS}^p(r_k, \varphi_k) = e^{j\frac{2\pi}{c}\left(\left(\sum_{q=1}^{p-1} M_{TS}^q\right)d_t f_1 \sin(\varphi_k) + (f_1 - f_{TS}^{1,p} r_k)\right)} \times \begin{bmatrix} 1 \\ e^{j\frac{2\pi}{c}\left(d_t f_1 \sin(\varphi_k) - \Delta f_p r_k\right)} \\ \vdots \\ e^{j\left(M_{TS}^p - 1\right)\frac{2\pi}{c}\left(d_t f_1 \sin(\varphi_k) - \Delta f_p r_k\right)} \end{bmatrix}. \tag{14}$$

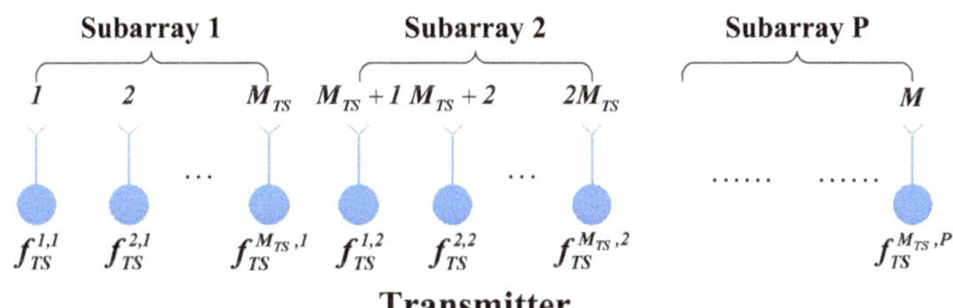

Figure 2. Subarray model.

In this paper, the number of antennas in each subarray is equal, that is, $M_{TS}^p = M/P, (p = 1, 2, \cdots, P)$. In addition, the frequency of the last antenna in the former subarray is identical to that of first element in latter subarray. In other words, $f_{TS}^{M_{TS}^p,p} = f_{TS}^{1,p+1}$, where $f_{TS}^{1,1} = f_1, \Delta f_1 < \Delta f_2 \cdots < \Delta f_P$.

Equation (6) can be converted to tensor form, which can be given by

$$\mathcal{X}(n, m, l) = \sum_{k=1}^{K} A_R(n, k) \circ A_{TS}(m, k) \circ S(l, k) + F_{n,m,l}, \tag{15}$$

where $A_T(m, k)$ is (m, k)-th element of A_T, $A_R(n, k)$ denotes (n, k)-th element of A_R, $S(l, k)$ represents (l, k)-th element of S. $S = [\xi_1, \xi_2, \ldots, \xi_k]^T$. $F_{n,m,l}$ stands for noise tensor, while l denotes number of snapshots.

3. The Proposed Method

3.1. Real-Valued Signal Subspace Estimation

For changing $\mathcal{X}$ into a centro-Hermitian tensor, forward–backward averaging technique is used

$$\mathcal{Z} = [\mathcal{X} \sqcup_3 (\mathcal{X}^* \times_1 \Gamma_N \times_2 \Gamma_M \times_3 \Gamma_L)]. \tag{16}$$

where the element on the anti-diagonal of Γ_n is 1, and the other elements are 0. Through the unitary transformation, a new real-valued tensor is obtained

$$\mathcal{Z}_{real} = \mathcal{Z} \times_1 \mathbf{E}_N^H \times_2 \mathbf{E}_M^H \times_3 \mathbf{E}_{2L}^H \tag{17}$$

where the unitary matrices are defined as

$$\mathbf{E}_K = \frac{1}{\sqrt{2}} \begin{bmatrix} \mathbf{I}_K & j\mathbf{I}_K \\ \Gamma_K & -j\Gamma_K \end{bmatrix} \tag{18}$$

$$\mathbf{E}_{2K+1} = \frac{1}{\sqrt{2}} \begin{bmatrix} \mathbf{I}_K & 0 & j\mathbf{I}_K \\ \mathbf{0}_{1\times K} & \sqrt{2} & \mathbf{0}_{1\times K} \\ \Gamma_K & 0 & -j\Gamma_K \end{bmatrix} \tag{19}$$

Firstly, HOSVD algorithm is employed for $\mathcal{Z}_{real}$, which can be written as

$$\mathcal{Z}_{real} = \mathcal{G}_{real} \times_1 \mathbf{U}_1 \times_2 \mathbf{U}_2 \times_3 \mathbf{U}_3, \tag{20}$$

where $\mathcal{G}_{real} \in \mathbb{C}^{M \times N \times L}$ is core tensor, $U_1 \in \mathbb{C}^{M \times M}$, $U_2 \in \mathbb{C}^{N \times N}$ and $U_3 \in \mathbb{C}^{L \times L}$ are composed of left singular of the mode-n tensor unfolding of $\mathcal{Z}_{real}$, respectively. In other words, $[\mathcal{Z}_{real}]_{(n)} = U_n \Lambda_n V_n^{\mathrm{H}} (n = 1, 2, 3)$. Then the truncated HOSVD is employed to obtain signal subspace estimation, which is given by

$$\mathcal{Z}_s = \mathcal{G}_s \times_1 U_{s1} \times_2 U_{s2}, \tag{21}$$

where $\mathcal{G}_s = \mathcal{Z}_{real} \times_1 U_{s1}^{\mathrm{H}} \times_2 U_{s2}^{\mathrm{H}} \times_3 U_{s3}^{\mathrm{H}}$ is the truncated core tensor. $U_{sn}(n = 1, 2, 3)$ is composed of the column vector of U_n.

Substituting $\mathcal{G}_s$ into Equation (21), Equation (21) is given by

$$\mathcal{Z}_s = \mathcal{Z}_{real} \times_1 (U_{s1} U_{s1}^{\mathrm{H}}) \times_2 (U_{s2} U_{s2}^{\mathrm{H}}) \times_3 U_{s3}^{\mathrm{H}}. \tag{22}$$

According to [35,36], the properties of the mode product is given by

$$\begin{cases} \mathcal{J} \times_i W \times_j G = \mathcal{J} \times_j W \times_i G, j \neq i \\ \mathcal{J} \times_i W \times_i G = \mathcal{J} \times_i (WG), j = i \end{cases} \tag{23}$$

$$[\mathcal{J} \times_1 W_1 \times_2 W_2 \times \cdots \times_I W_I]_{(i)} =$$
$$W_i \cdot [\mathcal{J}]_{(i)} \cdot [W_I \otimes \ldots \otimes W_{i+1} \otimes W_{i-1} \otimes \ldots \otimes W_2 \otimes W_1]^{\mathrm{T}} \tag{24}$$

where $\mathcal{J}$ is a tensor, W and G are matrices.

According to Equations (23) and (24), the signal subspace U_s is obtained by mode-3 unfolding of $\mathcal{Z}_s$.

$$U_s = [\mathcal{Z}_s]_{(3)}^{\mathrm{T}} = (U_{s2} U_{s2}^{\mathrm{H}} \otimes U_{s1} U_{s1}^{\mathrm{H}}) [\mathcal{Z}_{real}]_{(3)}^{\mathrm{T}} U_{s3}^*, \tag{25}$$

where $[\mathcal{Z}_{real}]_{(3)} = U_3 \Lambda_3 V_3^{\mathrm{H}}$. In addition, $[\mathcal{Z}_{real}]_{(3)}^{\mathrm{T}} \approx V_{s3}^* \Lambda_{s3} U_{s3}^{\mathrm{T}}$. Therefore, Equation (25) can be rewritten as

$$U_s = (U_{s2} U_{s2}^{\mathrm{H}} \otimes U_{s1} U_{s1}^{\mathrm{H}}) V_{s3}^* \Lambda_{s3}. \tag{26}$$

Therefore, the tensor-based signal subspace estimation has been obtained.

3.2. DOA Estimation

According to MUSIC algorithm [37], the noise subspace is achieved by orthogonal projection, which can be expressed as [38]

$$U_{noise} U_{noise}^{\mathrm{H}} = I_{NM} - U_o U_o{}^{\mathrm{H}}, \tag{27}$$

where U_o is the orthogonal basis of U_s.

The spectrum function is given by

$$f(\theta, \varphi, r) = \frac{1}{[\mathbf{E}_N^H a_R(\theta) \otimes \mathbf{E}_M^H a_{TS}(r_k, \varphi_k)]^{\mathrm{H}} [I_{NM} - U_o U_o{}^{\mathrm{H}}][\mathbf{E}_N^H a_R(\theta) \otimes \mathbf{E}_M^H a_{TS}(r_k, \varphi_k)]}. \tag{28}$$

From Equation (28), we can know that parameter estimation can be obtained by three-dimensional spectral peak searching. In order to cut down the computing redundancy, the Lagrange multiplier approach is employed to cut down the dimension of the spectrum function [39].

On the basis of the Kronecker product [40], the expression of $\mathbf{E}_N^H a_R(\theta) \otimes \mathbf{E}_M^H a_{TS}(r, \varphi)$ can be simplified, which is given by

$$\begin{aligned} \mathbf{E}_N^H a_R(\theta) \otimes \mathbf{E}_M^H a_{TS}(r, \varphi) &= [\mathbf{E}_N^H a_R(\theta) I_1] \otimes [\mathbf{E}_M^H a_{TS}(r, \varphi)] \\ &= [\hat{a}_R(\theta) I_1] \otimes [\mathbf{E}_M^H a_{TS}(r, \varphi)] \\ &= [\hat{a}_R(\theta) \otimes \mathbf{E}_M^H][I_1 a_{TS}(r, \varphi)] \\ &= [\hat{a}_R(\theta) \otimes \mathbf{E}_M^H] a_{TS}(r, \varphi). \end{aligned} \tag{29}$$

where $\hat{a}_R(\theta) = E_N^H a_R(\theta)$. Then according to Equation (28), $F(\theta, \varphi, r)$ is defined as

$$
\begin{aligned}
F(\theta, \varphi, r) &= [E_N^H a_R(\theta) \otimes E_M^H a_{TS}(r, \varphi)]^H U_{orth}[E_N^H a_R(\theta) \otimes E_M^H a_{TS}(r, \varphi)] \\
&= a_{TS}(r, \varphi)^H [\hat{a}_R(\theta) \otimes E_M^H]^H U_{orth}[\hat{a}_R(\theta) \otimes E_M^H] a_{TS}(r, \varphi) \\
&= a_{TS}(r, \varphi)^H F(\theta) a_{TS}(r, \varphi),
\end{aligned}
\tag{30}
$$

where $F(\theta) = [\hat{a}_R(\theta) \otimes E_M^H]^H U_{orth}[\hat{a}_R(\theta) \otimes E_M^H]$, $U_{orth} = U_{noise} U_{noise}^H$.

The quadratic optimization problem needs to be considered in Equation (30). That is, the following constraints are considered [39], it is given by

$$
\begin{aligned}
e_o^H a_{TS}(r, \varphi) &= 1 \\
\Rightarrow a_{TS}(r, \varphi) &= (e_o^H)^{-1},
\end{aligned}
\tag{31}
$$

where $e_o^H = [1, 0, \cdots, 0]^T \in \mathbb{C}^{M \times 1}$.

For solving the extreme value issue in Equation (30), Lagrange multiplier approach is employed. The Lagrange function of Equation (30) is given by

$$
L(\theta, \varphi, r) = a_{TS}(r, \varphi)^H F(\theta) a_{TS}(r, \varphi) + \lambda(e_o^H a_{TS}(r, \varphi) - 1),
\tag{32}
$$

where λ is Lagrangian multiplier.

The partial derivative of Equation (32) is given by

$$
\begin{aligned}
\frac{\partial L(\theta, \varphi, r)}{\partial a_{TS}(r, \varphi)} &= 2F(\theta) a_{TS}(r, \varphi) - \lambda e_o = 0 \\
\Rightarrow a_{TS}(r, \varphi) &= \frac{\lambda}{2} F(\theta)^{-1} e_o.
\end{aligned}
\tag{33}
$$

Substituting Equation (31) into Equation (33), we can obtain

$$
\begin{aligned}
\frac{\lambda}{2} F(\theta)^{-1} e_o &= a_{TS}(r, \varphi) = (e_o^H)^{-1} \\
\Rightarrow \lambda &= \frac{2}{e_o^H F(\theta)^{-1} e_o}.
\end{aligned}
\tag{34}
$$

Therefore, $a_{TS}(r, \varphi)$ is given by

$$
a_{TS}(r, \varphi) = \frac{F(\theta)^{-1} e_o}{e_o^H F(\theta)^{-1} e_o}.
\tag{35}
$$

Finally, the DOA estimation can be obtained through Equations (28) and (35).

$$
\begin{aligned}
\hat{\theta} &= \arg\max \quad f(\theta, \varphi, r) \\
&= \arg\min \quad a_{TS}(r, \varphi)^H F(\theta) a_{TS}(r, \varphi) \\
&= \arg\min \quad e_o^{-1} F(\theta) e_o^{-H} \\
&= \arg\max \quad e_o^H F(\theta)^{-1} e_o.
\end{aligned}
\tag{36}
$$

The result of DOA estimation is to search the reduced dimension MUSIC spectral peak. The first K largest peaks obtained are the DOAs of corresponding K targets.

3.3. DOD and Range Estimation

By using the DOA estimation and combining with Equation (35), the estimation of the transmit steering vector $\hat{\mathbf{a}}_T(r_k, \varphi_k)$ is achieved. The scale ambiguity of $\hat{\mathbf{a}}_T(r_k, \varphi_k)$ should be eliminated by normalization process. Therefore, the phase of $\hat{\mathbf{a}}_T(r_k, \varphi_k)$ is expressed as

$$\mathbf{\Phi}_T^p = diag \begin{bmatrix} e^{j\frac{2\pi}{c}[d_t f_1 \sin \varphi_1 - \Delta f_1 r_1]} \\ e^{j\frac{2\pi}{c}[d_t f_1 \sin \varphi_2 - \Delta f_2 r_2]} \\ \vdots \\ e^{j\frac{2\pi}{c}[d_t f_1 \sin \varphi_k - \Delta f_K r_k]} \end{bmatrix}. \tag{37}$$

The phase of the kth diagonal element in $\mathbf{\Phi}_T^p$ corresponds to kth target, which is

$$\begin{cases} \frac{2\pi}{c} d_t f_1 \sin(\varphi_k) - \frac{2\pi}{c} \Delta f_1 r_k = angle(\boldsymbol{\phi}_{TS}^{k,1}) - 2z_1 \pi \\ \frac{2\pi}{c} d_t f_1 \sin(\varphi_k) - \frac{2\pi}{c} \Delta f_2 r_k = angle(\boldsymbol{\phi}_{TS}^{k,2}) - 2z_2 \pi \\ \qquad\qquad \vdots \\ \frac{2\pi}{c} d_t f_1 \sin(\varphi_k) - \frac{2\pi}{c} \Delta f_K r_k = angle(\boldsymbol{\phi}_{TS}^{k,P}) - 2z_P \pi, \end{cases} \tag{38}$$

where $\boldsymbol{\phi}_{TS}^{k,p}$ corresponds to the phase of the kth target in pth subarray, $z_i \in \mathbb{Z}, i = 1, 2, \cdots, P$. The value of z_i cannot be determined due to the phase period ambiguity of $angle(\boldsymbol{\phi}_{TS}^{k,p})$. Since $\Delta f_1 < \Delta f_2 < \cdots < \Delta f_P$, subtract the subformula in Equation (38), which is given by

$$\begin{cases} \frac{2\pi}{c}(\Delta f_2 - \Delta f_1) r_k = angle(\boldsymbol{\phi}_{TS}^{k,1}) - angle(\boldsymbol{\phi}_{TS}^{k,2}) + 2\pi(z_2 - z_1) \\ \frac{2\pi}{c}(\Delta f_3 - \Delta f_2) r_k = angle(\boldsymbol{\phi}_{TS}^{k,2}) - angle(\boldsymbol{\phi}_{TS}^{k,3}) + 2\pi(z_3 - z_2) \\ \qquad\qquad \vdots \\ \frac{2\pi}{c}(\Delta f_P - \Delta f_{P-1}) r_k = angle(\boldsymbol{\phi}_{TS}^{k,P-1}) - angle(\boldsymbol{\phi}_{TS}^{k,P}) + 2\pi(z_P - z_{P-1}). \end{cases} \tag{39}$$

It can be seen from Equation (39) that only range information is included in Equation (39). However, there is still phase period ambiguity in Equation (39). In order to eliminate phase period ambiguity, we define

$$\mathbf{h}_k = \begin{bmatrix} h_{k,1} \\ h_{k,2} \\ \vdots \\ h_{k,P-1} \end{bmatrix} = \begin{bmatrix} \frac{\Delta f_2 - \Delta f_1}{c} \\ \frac{\Delta f_3 - \Delta f_2}{c} \\ \vdots \\ \frac{\Delta f_P - \Delta f_{P-1}}{c} \end{bmatrix}, \tag{40}$$

$$\zeta_{k,i} = angle(\boldsymbol{\phi}_{TS}^{k,i}) - angle(\boldsymbol{\phi}_{TS}^{k,i+1}) + (z_{i+1} - z_i), (i = 1, 2, ..., P-1). \tag{41}$$

Then Equation (39) can be rewritten as

$$\begin{cases} h_{k,1} * r_k = \zeta_{k,1} \\ h_{k,2} * r_k = \zeta_{k,2} \\ \qquad \vdots \\ h_{k,P-1} * r_k = \zeta_{k,P-1} \end{cases} \Rightarrow h_k r_k = \zeta_k, \tag{42}$$

where $\boldsymbol{\zeta}_k = [\zeta_{k,1}, \zeta_{k,2}, \cdots, \zeta_{k,P-1}]^{\mathrm{T}}$.

Equation (42) can be simplified as

$$2\pi\frac{\Delta f_{TS}^{i+1}-\Delta f_{TS}^{i}}{c}r_k$$
$$=angle(\boldsymbol{\phi}_{TS}^{k,i})-angle(\boldsymbol{\phi}_{TS}^{k,i+1})+2\pi(z_{i+1}-z_i). \tag{43}$$

For determining the range parameter, the following conditions need to be met

$$0<2\pi\frac{\Delta f_{TS}^{i+1}-\Delta f_{TS}^{i}}{c}r_k\leqslant 2\pi. \tag{44}$$

Since the range r_k is positive and $\Delta f_{TS}^1<\Delta f_{TS}^2\cdots<\Delta f_{TS}^P$, $2\pi\frac{\Delta f_{TS}^{i+1}-\Delta f_{TS}^{i}}{c}r_k>0$. When $2\pi\frac{\Delta f_{TS}^{i+1}-\Delta f_{TS}^{i}}{c}r_k\leqslant 2\pi$, $(z_{i+1}-z_i)$ has a unique solution, which can be expressed as

$$z_{i+1}-z_i=\begin{cases}1, & angle(\boldsymbol{\phi}_{TS}^{k,i})<angle(\boldsymbol{\phi}_{TS}^{k,i+1})\\0, & angle(\boldsymbol{\phi}_{TS}^{k,i})>angle(\boldsymbol{\phi}_{TS}^{k,i+1}).\end{cases} \tag{45}$$

According to Equation (44), the range r_k can be written as

$$r_k\leqslant\frac{c}{\Delta f_{TS}^{i+1}-\Delta f_{TS}^{i}}. \tag{46}$$

Combining Equations (42) and (46), the range r_k can be expressed as

$$r_k\leqslant\frac{c}{max(\Delta f_{TS}^{i+1}-\Delta f_{TS}^{i})}. \tag{47}$$

Therefore, the effective range of radar estimation is affected by radar frequency.

By substituting Equation (45) into Equation (42) and employing the least square (LS) approach, the range estimation $\hat{r}_k$ can be obtained, which is given by

$$\hat{r}_k=h_k^\dagger\zeta_k, \tag{48}$$

where the $\hat{r}_k$ is the range estimation of kth target.

By substituting $\hat{r}_k$ into Equation (39), we can obtain

$$\begin{cases}\frac{2\pi}{c}d_tf_1\sin(\varphi_k)=angle(\boldsymbol{\phi}_{TS}^{k,1})-2z_1\pi+\frac{2\pi}{c}\Delta f_{TS}^1 r_k\\\frac{2\pi}{c}d_tf_1\sin(\varphi_k)=angle(\boldsymbol{\phi}_{TS}^{k,2})-2z_2\pi+\frac{2\pi}{c}\Delta f_{TS}^2 r_k\\\qquad\qquad\vdots\\\frac{2\pi}{c}d_tf_1\sin(\varphi_k)=angle(\boldsymbol{\phi}_{TS}^{k,P})-2z_P\pi+\frac{2\pi}{c}\Delta f_{TS}^P r_k.\end{cases} \tag{49}$$

Since $2d_tf_1/c\leqslant 1$, the parameter $z_i(i=1,2,\cdots,P)$ can be obtained, which is expressed as

$$z_i=\left\lfloor\frac{angle(\boldsymbol{\phi}_{TS}^{k,i})-\frac{2\pi}{c}\Delta f_{TS}^i\hat{r}_k+\pi}{2\pi}\right\rfloor,i=1,2,\cdots,P. \tag{50}$$

To simplify Equation (49), we define

$$h_\varphi=\begin{bmatrix}\frac{2\pi}{c}f_1d_t\\\frac{2\pi}{c}f_1d_t\\\vdots\\\frac{2\pi}{c}f_1d_t\end{bmatrix}\in\mathbb{C}^{P\times 1}, \tag{51}$$

$$\boldsymbol{\psi}_\varphi = \begin{bmatrix} angle(\boldsymbol{\phi}_{TS}^{k,1}) - 2z_1\pi + \frac{2\pi}{c}\Delta f_{TS}^1 r_k \\ angle(\boldsymbol{\phi}_{TS}^{k,2}) - 2z_2\pi + \frac{2\pi}{c}\Delta f_{TS}^2 r_k \\ \vdots \\ angle(\boldsymbol{\phi}_{TS}^{k,P}) - 2z_P\pi + \frac{2\pi}{c}\Delta f_{TS}^P r_k \end{bmatrix} \in \mathbb{C}^{P\times 1}. \tag{52}$$

Then Equation (49) can be rewritten as

$$\boldsymbol{h}_\varphi \sin\varphi_k = \boldsymbol{\psi}_\varphi. \tag{53}$$

Utilizing the total LS approach, the DOD $\hat{\varphi}_k$ of kth target can be estimated, which is given by

$$\hat{\varphi}_k = \arcsin\left[\left(\boldsymbol{h}_\varphi^{\mathrm{T}}\boldsymbol{h}_\varphi\right)^{-1}\boldsymbol{h}_\varphi^{\mathrm{T}}\boldsymbol{\psi}_\varphi\right]. \tag{54}$$

The DOA, DOD, and range obtained by the proposed approach are automatically matched without additional matching process.

4. Algorithm Analysis

4.1. Algorithm Summary

The developed method for bistatic FDA-MIMO radar can be simplified in Algorithm 1.

Algorithm 1 Target parameter estimation algorithm based on real-valued HOSVD for bistatic FDA-MIMO radar.

1: $\boldsymbol{a}_T(r_k, \varphi_k)$ is converted to subarray by Equation (12),
2: Signal can be converted to tensor form $\boldsymbol{\mathcal{X}}$,
3: Construct the real-valued tensor $\boldsymbol{\mathcal{Z}}_{real}$ by Equation (17),
4: Calculate $\boldsymbol{U}_s$ by Equation (26),
5: Construct reduced dimensional spatial spectrum function by Equation (30),
6: $\hat{\theta}$ is achieved by Equation (36),
7: Decouple DOD and range information by Equation (39),
8: Eliminate phase period ambiguity by Equation (40),
9: $\hat{r}_k$ is obtained by Equation (48),
10: $\hat{\varphi}_k$ is estimated by Equation (54).

4.2. Computational Complexity

In order to highlight the advantages of the developed approach, the computational complexity of our method is given by

(1) The computational complexity of HOSVD for $\boldsymbol{\mathcal{X}} \in \mathbb{C}^{M\times N\times L}$ is $O(1/4MNL(M+N+L))$ in Equation (20);
(2) The computational complexity of signal subspace estimation is $O(KLMN)$ in Equation (26);
(3) The computational complexity of dimensionality reduction for three-dimensional spatial spectrum in Equation (30) is $O(M^2N^2K^2(MN+K^2))$;
(4) The computational complexity of spatial spectrum search for DOA estimation in Equation (36) is $O(d_\theta(MK)!(MK-1))$, where d_θ is the DOA search time;
(5) Computing DOD and range requires $O(\sum\limits_{i=1}^{P}(N(M_{TS}-1)(2M_{TS}NK+4K^2)+K^3))$.

The computational complexity of this process is relatively small, so it can be ignored.

In brief, the computational complexity of the developed approach is $O\{KMNL+M^2N^2K^2(MN+K^2)+d_\theta(MK)!(MK-1)+1/4MNL(M+N+L)\}$.

For proving the superiority of our method, the computational complexity of the developed approach is compared with that of ESPRIT [28], Tensor-ESPRIT [30], RD-MUSIC [20], and MUSIC [27]. Table 2 shows the computational complexity comparison.

Table 2. Computational complexity comparison.

Method	Computational Complexity
Proposed	$O\{KMNL + M^2N^2K^2(MN + K^2) + d_\theta(MK)!(MK - 1)$ $+1/4MNL(M + N + L)\}$
ESPRIT	$O\{(2MN)^2L + (2MN)^3 + 4(5MN - 2M - 2N)K^2$ $+K^3(L + M) + MNK^2 + 31K^3\}$
Tensor-ESPRIT	$O\{2(MNL)^3 + MNL(M + N + L) + MLK(N + K)$ $+K^3(L + M) + MNK^2 + 31K^3\}$
RD-MUSIC	$O\{4KMNL + M^2N^2K^2(MN + K^2) + d_\theta(MK)!(MK - 1)$ $+MNL(M + N + L)\}$
MUSIC	$O\{LM^2N^2 + 4(MN)^3 + 4d_\theta d_\varphi d_r MN(MN + 1)\}$

5. Simulation Results

Several numerical simulations are developed to prove the superiority of our method in parameter estimation. ESPRIT [28], tensor ESPRIT [30], and RD-MUSIC [20] are employed for performance comparison. In addition, Cramér–Rao bound (CRB) [41,42] is employed to evaluate the precision of the developed algorithm. In this part, bistatic FDA-MIMO radar with $M = 18$ transmitter antennas and $N = 18$ receiver antennas is considered. In the simulation, it is supposed that there are three uncorrelated targets at $(\theta_1, \varphi_1, r_1) = (-50°, -15°, 30,000 \text{ m})$, $(\theta_2, \varphi_2, r_2) = (0°, 20°, 60,000 \text{ m})$, and $(\theta_3, \varphi_3, r_3) = (30°, -10°, 0 \text{ m})$. To assess the precision of our method, the root mean square error (RMSE) is employed, where the RMSEs of angle and range estimation can be given by

$$\text{RMSE}_{\theta,\varphi} = \frac{1}{K} \sum_{k=1}^{K} \sqrt{\frac{1}{\zeta} \sum_{i=1}^{\zeta} \left\{ (\tilde{\theta}_{k,i} - \theta_k)^2 + (\tilde{\varphi}_{k,i} - \varphi_k)^2 \right\}}, \tag{55}$$

$$\text{RMSE}_r = \frac{1}{K} \sum_{k=1}^{K} \sqrt{\frac{1}{\zeta} \sum_{i=1}^{\zeta} (\tilde{r}_{k,i} - r_k)^2}, \tag{56}$$

where the results of the i-th Monte Carlo of θ_k, φ_k and r_k are $\tilde{\theta}_{k,i}$, $\tilde{\varphi}_{k,i}$ and $\tilde{r}_{k,i}$. ζ is the total Monte Carlo times, $\zeta = 500$.

Figures 3 and 4 prove the effectiveness of the developed algorithm, where $\text{SNR} = 20$ dB and $L = 200$. Figure 3 demonstrates the spatial spectrum of DOA. It can be seen that DOA estimation is accurately achieved through spatial spectrum searching. The spatial spectrum of the proposed algorithm is clearer and more accurate than that of MUSIC algorithm, and has more sharp peaks. We can know from Figure 4 that our method can accurately obtain the DOA, DOD, and range information of three targets. It can testify the availability of our method.

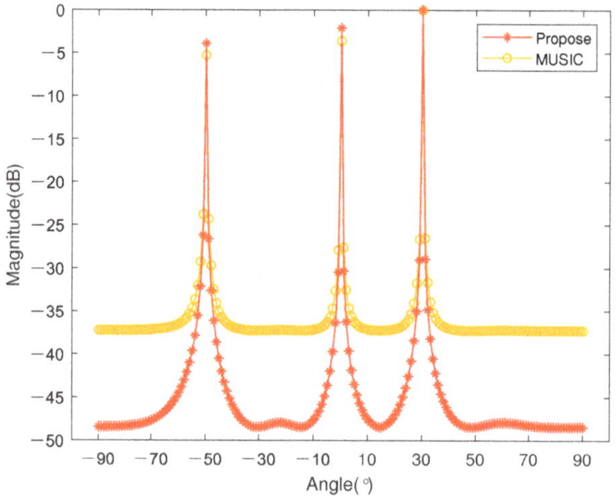

Figure 3. The spatial spectrum, SNR = 20 dB, and $L = 200$.

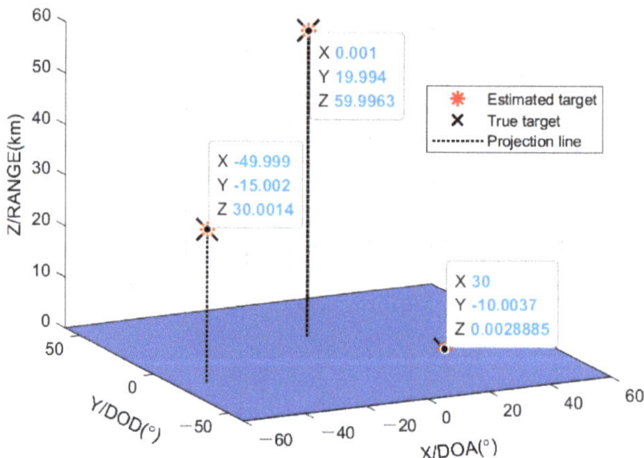

Figure 4. Estimation results, SNR = 20 dB, $L = 200$.

Figures 5 and 6 show the comparisons of the computational complexity with the number of array elements and the number of snapshots, respectively. It is known from Figure 5 that the computational complexity of all algorithms increases with the increase in the number of array elements. The computational complexity of our algorithm is close to that of Tensor-ESPRIT and RD-MUSIC, and far lower than that of MUSIC. Figure 6 shows that the computational complexity of our method is lower than that of RD-MUSIC and slightly higher than that of Tensor-ESPRIT. In general, since our method is a real-valued operation, the computational complexity of our method is lower than that of RD-MUSIC, which is close to that of Tensor-ESPRIT algorithm.

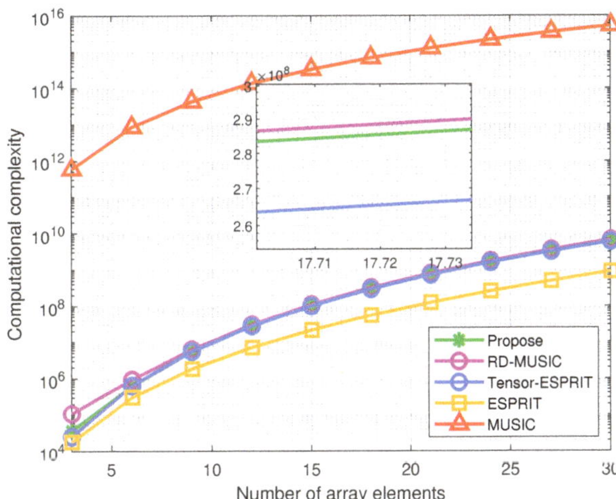

Figure 5. Computational complexity comparison versus the number of array elements.

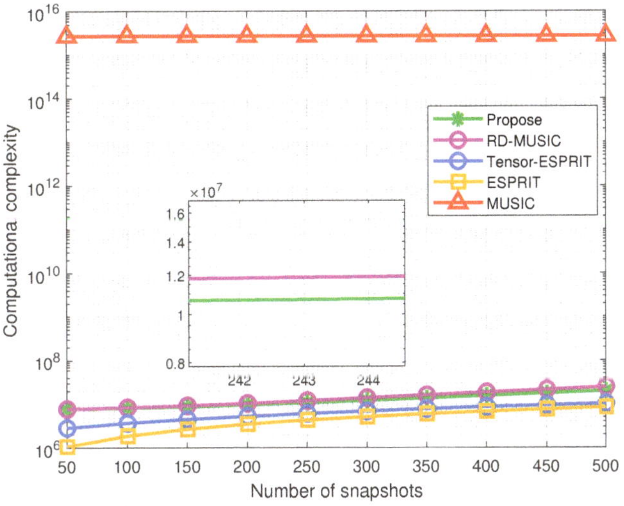

Figure 6. Computational complexity comparison versus the number of snapshots.

For exploring the impact of different SNR on the precision of the algorithm, the first group of comparative tests is proposed, where $L = 50$. Figures 7–9 show the comparison of DOA-DOD-range estimations of different methods. From this set of comparative experiments, we can see that our method has superior precision than other algorithms in DOA, DOD, and range estimation, and nearer to the CRB. This is because the developed algorithm takes advantage of the multi-dimensional structure of data, and can obtain accurate estimations through spectral peak searching. On the one hand, the accuracy of our method in DOA and DOD estimation is very close to that of the RD-MUSIC, but the computational redundancy is reduced. On the other hand, in terms of range estimation, since the subarray division can more accurately decouple the range information and DOD information, more accurate range estimation can be obtained. Moreover, the accuracy of tensor ESPRIT is more accurate than that of ESPRIT due to the use of multi-dimensional structure characteristics.

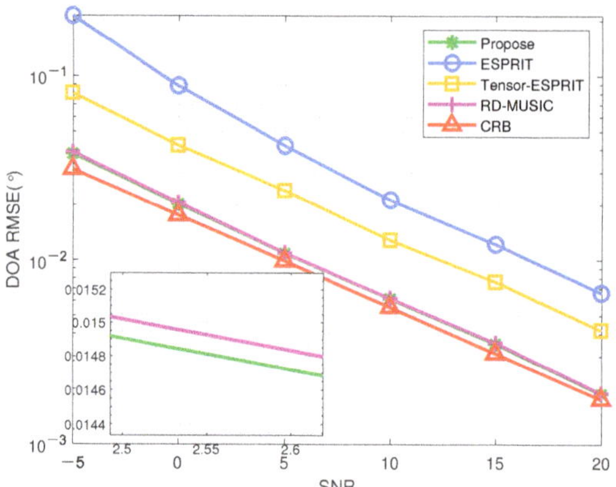

Figure 7. DOA estimation comparison versus SNR, $L = 50$.

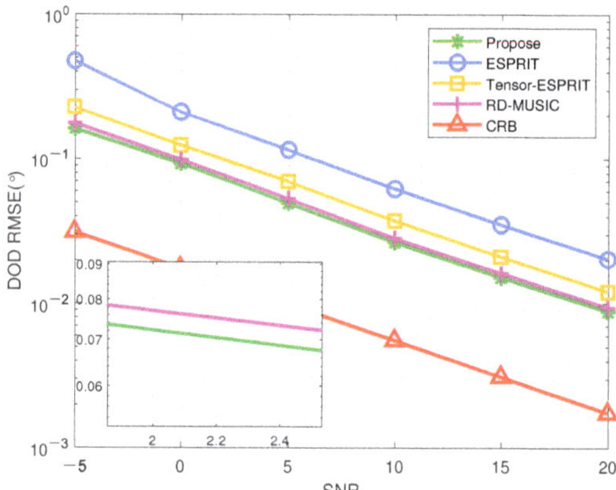

Figure 8. DOD estimation comparison versus SNR, $L = 50$.

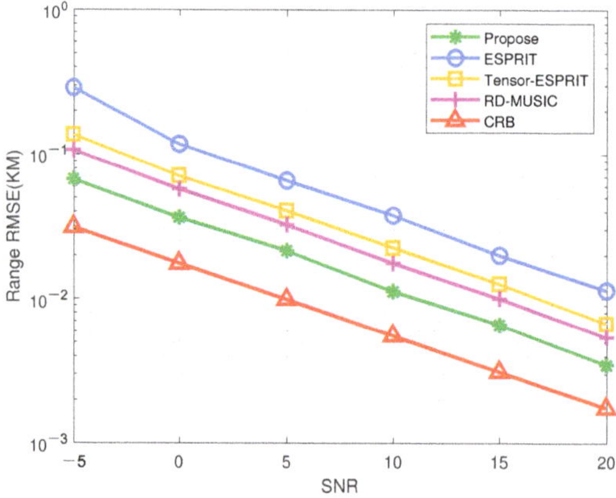

Figure 9. Range estimation comparison versus SNR, $L = 50$.

The second group of experiments confirms the advantage of our algorithm under different snapshots, where SNR = 5 dB. From Figures 10–12 we can know that DOA-DOD-range estimation accuracy of the developed algorithm are higher to the other algorithms. Moreover, the estimation accuracy of the developed algorithm is more stable in the case of low snapshots. It is because our algorithm utilizes high-dimensional data to improve the performance under small snapshots. In terms of DOA and DOD estimation, the performance of our method is slightly better than that of the RD-MUSIC. However, in range estimation, the accuracy of our method is higher to the RD-MUSIC because the subarray division solves the coupling problem of the transmitter. Moreover, the precision of tensor ESPRIT algorithm is obviously higher than ESPRIT algorithm in small snapshots. With the increase in snapshot, the precision gap between the two methods will gradually decrease. That is to say, the accuracy of the two algorithms is very close in high snapshots. Only the precision of our method is nearer to CRB curve. This demonstrates the superiority of our method.

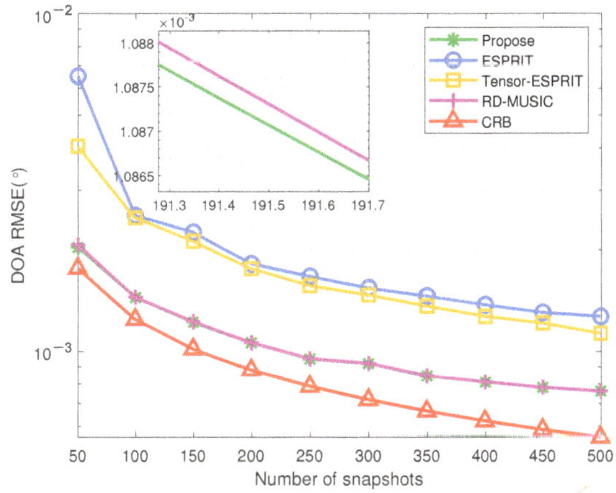

Figure 10. DOA estimation comparison versus the number of snapshots, SNR = 5 dB.

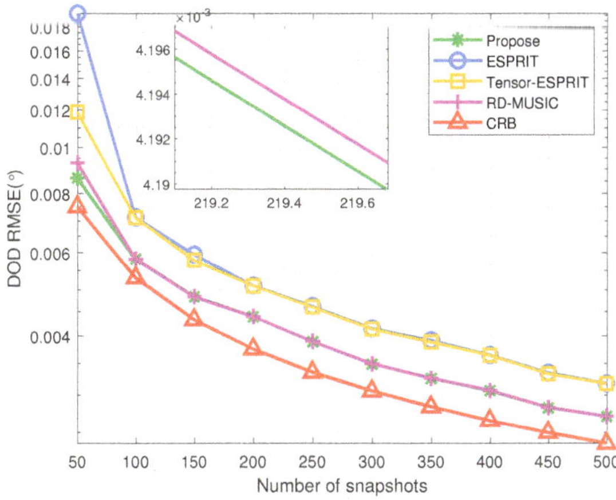

Figure 11. DOD estimation comparison versus the number of snapshots, SNR = 5 dB.

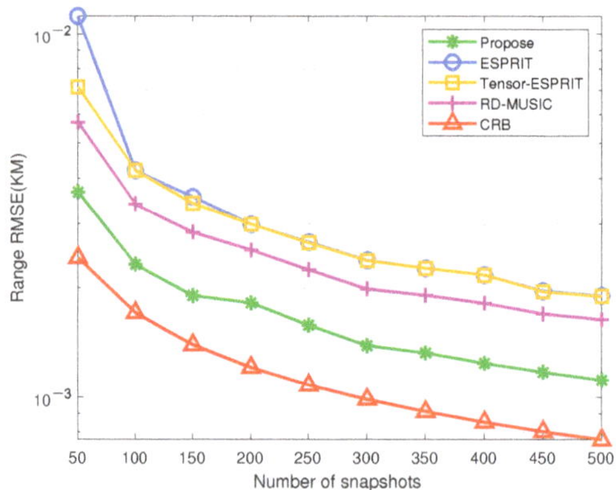

Figure 12. Range estimation comparison versus the number of snapshots, SNR = 5 dB.

Another criterion utilized to access the precision of the developed algorithm is probability of the successful detection (PSD), where PSD is given by

$$\mathrm{PSD} = \frac{D}{\zeta} \times 100\%, \tag{57}$$

where D is the number of times of obtaining the correct estimation result. If the angle error is lower than $0.1°$ and the range error is lower than 0.1 km, it is the correct estimation result.

For further exploring the estimation accuracy of our method, the concept of PSD is introduced to design the third group of experiments, where $L = 50$. It can be seen from Figures 13–15 that in DOA estimation, DOD estimation and range estimation, the PSD of the developed method is higher than the other methods at the same SNR. This means that the estimation accuracy of the developed algorithm is the highest at the same SNR. In addition, when SNR = 0 dB, PSD of our algorithm can reach 100% in DOA estimation. It shows that our method can obtain more accurate results at low SNR. It is worth noting that the proposed method has less advantages than the RD-MUSIC algorithm in angle estimation, but has greater advantages in range estimation. In conclusion, the estimation accuracy of the developed approach is higher than the other methods.

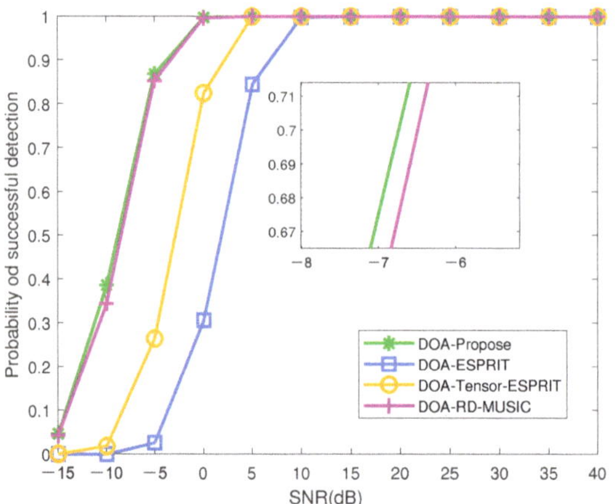

Figure 13. PSD of DOA estimation versus SNR, $L = 50$.

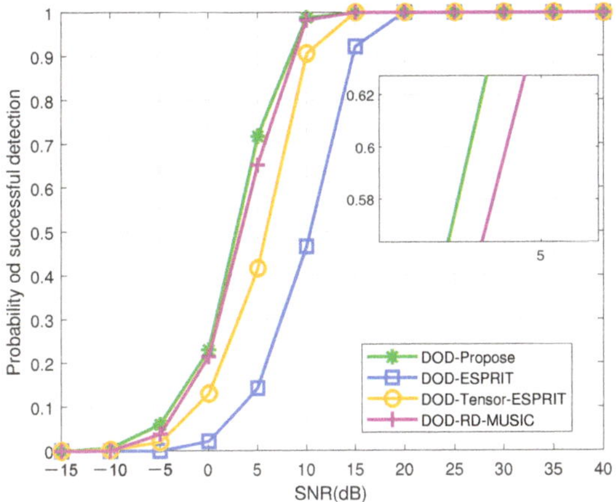

Figure 14. PSD of DOD estimation versus SNR, $L = 50$.

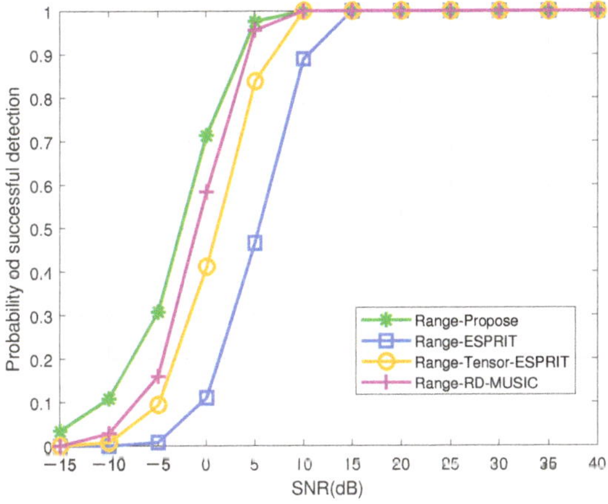

Figure 15. PSD of range estimation versus SNR, $L = 50$.

6. Conclusions

In this paper, a joint DOA-DOD-range estimation algorithm with low computational complexity based on real-valued HOSVD is developed for bistatic FDA-MIMO radar. The developed algorithm can realize high-precision DOA-DOD-range estimation with low computational complexity. By employing unitary transformation technique, our method converts data into real-valued data. Then our algorithm utilizes HOSVD to estimate the signal subspace and construct a two-dimensional spatial spectral function. Then, the spatial spectral function is transformed into one-dimension by Lagrange algorithm, and the DOA estimation is obtained. The proposed approach divides the transmitting array into several subarrays, which decouples the DOD and range information. Finally, the decoupled phase is employed to achieve DOD and range estimations. Our method preserves the original multidimensional structure of the data and effectively improves the estimation precision. Numerical simulations demonstrate the preponderance of our method.

Author Contributions: Conceptualization, X.W. and L.S.; methodology, Y.G. and X.W.; writing—original draft preparation, Y.G.; writing—review and editing, X.W. and J.S.; supervision, X.W.; funding acquisition, X.W. and X.L. All authors have read and agreed to the published version of the manuscript.

Funding: This work was supported by Natural Science Foundation of Hainan Province (620RC555), the National Natural Science Foundation of China (No. 61861015, 61961013, 62101088), Radar Signal Processing National Defense Science and Technology Key Laboratory Fund (6142401200101).

Data Availability Statement: Not applicable.

Conflicts of Interest: The authors declare no conflict of interest.

References

1. Fishler, E.; Haimovich, A.; Blum, R.; Chizhik, D.; Cimini, L.; Valenzuela, R. MIMO radar: An idea whose time has come. In Proceedings of the 2004 IEEE Radar Conference (IEEE Cat. No. 04CH37509), Philadelphia, PA, USA, 29 April 2004; pp. 71–78.
2. Fishler, E.; Haimovich, A.; Blum, R.; Cimini, R.; Chizhik, D.; Valenzuela, R. Performance of MIMO radar systems: Advantages of angular diversity. In Proceedings of the Conference Record of the Thirty-Eighth Asilomar Conference on Signals, Systems and Computers, Pacific Grove, CA, USA, 7–10 November 2004; Volume 1, pp. 305–309.
3. Wang, X.; Yang, L.T.; Meng, D.; Dong, M.; Ota, K.; Wang, H. Multi-UAV cooperative localization for marine targets based on weighted subspace fitting in SAGIN environment. *IEEE Internet Things J.* **2021**, *9*, 5708–5718. [CrossRef]
4. Cong, J.; Wang, X.; Lan, X.; Huang, M.; Wan, L. Fast target localization method for FMCW MIMO radar via VDSR neural network. *Remote Sens.* **2021**, *13*, 1956. [CrossRef]
5. Wang, X.; Wan, L.; Huang, M.; Shen, C.; Han, Z.; Zhu, T. Low-complexity channel estimation for circular and noncircular signals in virtual MIMO vehicle communication systems. *IEEE Trans. Veh. Technol.* **2020**, *69*, 3916–3928. [CrossRef]
6. Liu, W.; Xie, W.; Liu, J.; Wang, Y. Adaptive double subspace signal detection in Gaussian background—Part I: Homogeneous environments. *IEEE Trans. Signal Process.* **2014**, *62*, 2345–2357. [CrossRef]
7. Jin, M.; Liao, G.; Li, J. Joint DOD and DOA estimation for bistatic MIMO radar. *Signal Process.* **2009**, *89*, 244–251. [CrossRef]
8. Guo, Y.; Wang, X.; Wang, W.; Huang, M.; Shen, C.; Cao, C.; Bi, G. Tensor-based angle estimation approach for strictly noncircular sources with unknown mutual coupling in bistatic MIMO radar. *Sensors* **2018**, *18*, 2788. [CrossRef]
9. Wen, F.; Shi, J.; Zhang, Z. Generalized spatial smoothing in bistatic EMVS-MIMO radar. *Signal Process.* **2022**, *193*, 108406. [CrossRef]
10. Liu, Q.; Wang, X.; Huang, M.; Lan, X.; Sun, L. DOA and Range Estimation for FDA-MIMO Radar with Sparse Bayesian Learning. *Remote Sens.* **2021**, *13*, 2553. [CrossRef]
11. Guo, Y.; Wang, X.; Shi, J.; Lan, X.; Wan, L. Tensor-Based Target Parameter Estimation Algorithm for FDA-MIMO Radar with Array Gain-Phase Error. *Remote Sens.* **2022**, *14*, 1405. [CrossRef]
12. Zhuang, L.; Liu, X. Application of frequency diversity to suppress grating lobes in coherent MIMO radar with separated subapertures. *EURASIP J. Adv. Signal Process.* **2009**, *2009*, 481792. [CrossRef]
13. Antonik, P.; Wicks, M.C.; Griffiths, H.D.; Baker, C.J. Frequency diverse array radars. In Proceedings of the 2006 IEEE Conference on Radar, Verona, NY, USA, 24–27 April 2006; 3p.
14. Fishler, E.; Haimovich, A.; Blum, R.S.; Cimini, L.J.; Chizhik, D.; Valenzuela, R.A. Spatial diversity in radars—Models and detection performance. *IEEE Trans. Signal Process.* **2006**, *54*, 823–838. [CrossRef]
15. Haimovich, A.M.; Blum, R.S.; Cimini, L.J. MIMO radar with widely separated antennas. *IEEE Signal Process. Mag.* **2007**, *25*, 116–129. [CrossRef]
16. Li, J.; Stoica, P. MIMO radar with colocated antennas. *IEEE Signal Process. Mag.* **2007**, *24*, 106–114. [CrossRef]
17. Li, X.; Cheng, T.; Su, Y.; Peng, H. Joint time-space resource allocation and waveform selection for the collocated MIMO radar in multiple targets tracking. *Signal Process.* **2020**, *176*, 107650. [CrossRef]
18. Antonik, P.; Wicks, M.C.; Griffiths, H.D.; Baker, C.J. Multi-mission multi-mode waveform diversity. In Proceedings of the 2006 IEEE Conference on Radar, Verona, NY, USA, 24–27 April 2006; 3p.
19. Xu, J.; Liao, G.; Zhu, S.; Huang, L.; So, H.C. Joint range and angle estimation using MIMO radar with frequency diverse array. *IEEE Trans. Signal Process.* **2015**, *63*, 3396–3410. [CrossRef]
20. Xu, T.; Wang, X.; Huang, M.; Lan, X.; Sun, L. Tensor-Based Reduced-Dimension MUSIC Method for Parameter Estimation in Monostatic FDA-MIMO Radar. *Remote Sens.* **2021**, *13*, 3772. [CrossRef]
21. Chen, H.; Shao, H. Sparse reconstruction based target localization with frequency diverse array MIMO radar. In Proceedings of the 2015 IEEE China Summit and International Conference on Signal and Information Processing (ChinaSIP), Chengdu, China, 12–15 July 2015; pp. 94–98.
22. Wang, Y.; Wang, W.Q.; Shao, H. Frequency diverse array radar Cramér-Rao lower bounds for estimating direction, range, and velocity. *Int. J. Antennas Propag.* **2014**, *2014*, 830869. [CrossRef]
23. Wang, W.Q.; Shao, H. Range-angle localization of targets by a double-pulse frequency diverse array radar. *IEEE J. Sel. Top. Signal Process.* **2013**, *8*, 106–114. [CrossRef]
24. Li, B.; Bai, W.; Zhang, Q.; Zheng, G.; Zhang, M.; Wan, P. The spatially separated polarization sensitive FDA-MIMO radar: A new antenna structure for unambiguous parameter estimation. *MATEC Web Conf.* **2018**, *173*, 02015. [CrossRef]
25. Liu, W.; Liu, J.; Gao, Y.; Wang, G.; Wang, Y.L. Multichannel signal detection in interference and noise when signal mismatch happens. *Signal Process.* **2020**, *166*, 107268. [CrossRef]

26. Guo, Y.; Wang, X.; Lan, X.; Su, T. Traffic Target Location Estimation Based on Tensor Decomposition in Intelligent Transportation System. *IEEE Trans. Intell. Transp. Syst.* **2022**, 1–19. [CrossRef]
27. Xiong, J.; Wang, W.Q.; Gao, K. FDA-MIMO radar range–angle estimation: CRLB, MSE, and resolution analysis. *IEEE Trans. Aerosp. Electron. Syst.* **2017**, *54*, 284–294. [CrossRef]
28. Yan, Y.; Cai, J.; Wang, W.Q. Two-stage ESPRIT for unambiguous angle and range estimation in FDA-MIMO radar. *Digit. Signal Process.* **2019**, *92*, 151–165. [CrossRef]
29. Liu, F.; Wang, X.; Huang, M.; Wan, L.; Wang, H.; Zhang, B. A novel unitary ESPRIT algorithm for monostatic FDA-MIMO radar. *Sensors* **2020**, *20*, 827. [CrossRef] [PubMed]
30. Xu, T.; Wang, X.; Su, T.; Wan, L.; Sun, L. Vehicle Location in Edge Computing Enabling IoTs Based on Bistatic FDA-MIMO Radar. *IEEE Access* **2021**, *9*, 46398–46408. [CrossRef]
31. Yao, A.M.; Rocca, P.; Wu, W.; Massa, A.; Fang, D.G. Synthesis of time-modulated frequency diverse arrays for short-range multi-focusing. *IEEE J. Sel. Top. Signal Process.* **2016**, *11*, 282–294. [CrossRef]
32. Chen, K.; Yang, S.; Chen, Y.; Qu, S.W. Accurate models of time-invariant beampatterns for frequency diverse arrays. *IEEE Trans. Antennas Propag.* **2019**, *67*, 3022–3029. [CrossRef]
33. Xu, J.; Liao, G.; Zhu, S.; So, H.C. Deceptive jamming suppression with frequency diverse MIMO radar. *Signal Process.* **2015**, *113*, 9–17. [CrossRef]
34. Shao, H.; Li, X.; Wang, W.Q.; Xiong, J.; Chen, H. Time-invariant transmit beampattern synthesis via weight design for FDA radar. In Proceedings of the 2016 IEEE Radar Conference (RadarConf), Philadelphia, PA, USA, 2–6 May 2016; pp. 1–4.
35. De Lathauwer, L.; De Moor, B.; Vandewalle, J. A multilinear singular value decomposition. *SIAM J. Matrix Anal. Appl.* **2000**, *21*, 1253–1278. [CrossRef]
36. Kolda, T.G.; Bader, B.W. Tensor decompositions and applications. *SIAM Rev.* **2009**, *51*, 455–500. [CrossRef]
37. Wang, X.; Wang, W.; Liu, J.; Liu, Q.; Wang, B. Tensor-based real-valued subspace approach for angle estimation in bistatic MIMO radar with unknown mutual coupling. *Signal Process.* **2015**, *116*, 152–158. [CrossRef]
38. Cong, J.; Wang, X.; Yan, C.; Yang, L.T.; Dong, M.; Ota, K. CRB weighted source localization method based on deep neural networks in multi-UAV network. *IEEE Internet Things J.* **2022**, *1*. [CrossRef]
39. Zhang, X.; Xu, L.; Xu, L.; Xu, D. Direction of departure (DOD) and direction of arrival (DOA) estimation in MIMO radar with reduced-dimension MUSIC. *IEEE Commun. Lett.* **2010**, *14*, 1161–1163. [CrossRef]
40. Zhang, X. *Matrix Analysis and Applications*; Tsinghua University Press: Beijing, China, 2004.
41. Gui, R.; Wang, W.Q.; Cui, C.; So, H.C. Coherent pulsed-FDA radar receiver design with time-variance consideration: SINR and CRB analysis. *IEEE Trans. Signal Process.* **2017**, *66*, 200–214. [CrossRef]
42. Mao, Z.; Liu, S.; Qin, S.; Huang, Y. Cramér-Rao Bound of Joint DOA-Range Estimation for Coprime Frequency Diverse Arrays. *Remote Sens.* **2022**, *14*, 583. [CrossRef]

remote sensing

Article

An ADMM-qSPICE-Based Sparse DOA Estimation Method for MIMO Radar

Yongwei Zhang [1], Yongchao Zhang [1,2,*], Jiawei Luo [1], Yulin Huang [1,2], Jianan Yan [1], Yin Zhang [1,2] and Jianyu Yang [1]

[1] School of Information and Communication Engineering, University of Electronic Science and Technology of China (UESTC), Chengdu 611731, China; zhyw49@uestc.edu.cn (Y.Z.); jiawei_luo@std.uestc.edu.cn (J.L.); yulinhuang@uestc.edu.cn (Y.H.); yjn@std.uestc.edu.cn (J.Y.); yinzhang@uestc.edu.cn (Y.Z.); jyyang@uestc.edu.cn (J.Y.)

[2] Yangtze Delta Region Institute, University of Electronic Science and Technology of China (UESTC), Quzhou 324003, China

* Correspondence: yongchaozhang@uestc.edu.cn

Abstract: In recent years, sparse direction-of-arrival (DOA) estimation for multiple-input multiple-output (MIMO) radar has attracted extensive attention and been extensively studied, especially the method based on the classic least absolute shrinkage and selection operator (LASSO) estimator. The alternating-direction method of multipliers (ADMM) is an effective method for solving this problem at the cost of introducing an additional user parameter. To avoid introducing an additional user parameter, this paper adopts an equivalent transformation in the form of the generalized SParse Iterative Covariance-based Estimation (qSPICE) cost function to obtain a mean squared minimized form of the cost function. Then, the problem is transformed into a sparse optimization problem in the form of a weighted LASSO. Next, this unconstrained optimization problem is decomposed into three subproblems, which are solved separately to reduce the dimension of each problem and thus reduce the overall computational complexity based on ADMM. Simulation results and measured data indicate that the proposed method significantly outperforms the traditional super-resolution DOA estimation method and ADMM-LASSO method and slightly outperforms qSPICE in terms of resolution and sidelobe suppression capability. In addition, the proposed method has a much lower computational complexity and substantially fewer iterations than qSPICE.

Keywords: multiple-input multiple-output radar; direction-of-arrival estimation; sparse signal processing

Citation: Zhang, Y.; Zhang, Y.; Luo, J.; Huang, Y.; Yan, J.; Zhang, Y.; Yang, J. An ADMM-qSPICE-Based Sparse DOA Estimation Method for MIMO Radar. *Remote Sens.* **2023**, *15*, 446. https://doi.org/10.3390/rs15020446

Academic Editors: Guolong Cui, Yong Yang, Xianxiang Yu and Bin Liao

Received: 20 November 2022
Revised: 7 January 2023
Accepted: 9 January 2023
Published: 11 January 2023

1. Introduction

In recent years, a new type of antenna system, multiple-input multiple-output (MIMO), has been introduced [1,2]. A MIMO antenna system is generally defined as a system that has multiple transmitted linearly independent waveforms and is capable of jointly processing multiple received antenna signals. MIMO radar draws on the idea of MIMO communication [3]. Its mechanism is to adopt transmit waveform diversity technology to improve the angular (spatial) resolution of the radar system while reducing the number of physical channels and antenna aperture, compared with traditional array antennas.

Estimation of the direction-of-arrival (DOA) of multiple targets in noise-polluted received data is the most important task in the practical application of centralized MIMO radar. The most fundamental DOA estimation method is the delay-and-sum (DAS) method, which can be implemented efficiently by fast Fourier transform (FFT) due to the Fourier structure of the orientation matrix. However, the resolution of this method is poor; to improve the positioning accuracy of the target source, many high-precision DOA estimation methods have been developed for traditional single-input multiple-output (SIMO) radar [4–7]. More recently, DOA estimation methods have been introduced into the field of centralized MIMO radar DOA estimation [8–10]. Among these methods, estimation of signal parameters via rotational invariance techniques (ESPRIT) and multiple signal classification (MUSIC) are relatively classic because of their simplicity and high-resolution performance [8,9]. Based on the orthogonality characteristics of the signal subspace and noise subspace, the multiple signal classification (MUSIC) method can lead to better DOA estimation performance than the DAS method with higher resolution. Taking advantage of the rotational invariance of the spatial correlation matrix signal subspace, the (ESPRIT) algorithm was proposed with excellent resolution and search-free advantages. In reference [10], the parallel factor analysis (PARAFAC) algorithm was proposed for a colocated MIMO radar with imperfect waveforms. ESPRIT is a low-dimensional and high-efficiency version of PARAFAC [11], and they have similar resolution performance. However, the resolution of these methods is limited, and noticeable performance degradation occurs when there are a few snapshots or a single snapshot. Recently, deep-learning techniques have been applied to DOA estimation for massive MIMO systems and achieved good system performance [12]. However, this method requires many training samples and the training phase is complex.

To further improve the resolution performance of MIMO radar DOA estimation, refs. [13,14] proposed the iterative adaptive method (IAA) for MIMO radar imaging; tests have demonstrated that the IAA method can function stably with a small number of snapshots or a single snapshot and has better angular resolution and target positioning accuracy. However, due to its high computational complexity, this method is difficult to apply in practical engineering [6]. Taking advantage of the sparsity of the source distribution, sparse learning via iterative minimization (SLIM) was proposed [15] for MIMO radar imaging. The method follows an L_q-norm constraint and thus offers a more accurate estimate. In addition, SLIM has been demonstrated to have a higher angular resolution and lower computational complexity than IAA. Furthermore, a sparse spectrum estimation method with a higher resolution, called SParse Iterative Covariance-based Estimation (SPICE), was proposed based on weighted covariance fitting criteria [16,17]. This method is a semiparametric method, which converges globally and requires no selection of user parameters. In the case of a small number of snapshots, its frequency estimation performance is better than those of IAA and SLIM. However, the method has an L_1-norm penalty (sparse constraint) for both signal and noise, which may result in a singular covariance matrix or many conditions. To avoid this problem, an improved algorithm, named qSPICE, was developed by introducing a L_q-norm constraint on noise changes, where $q \geq 1$ [18]. This method offers better estimation performance than SPICE. However, this method requires solving for the covariance matrix and its inverse, as well as performing the associated matrix multiplication operations to estimate each sampling grid point, which may result in a large computational burden, especially in massive MIMO (m-MIMO) signal processing.

With the increasing demand for higher reliability and higher data rates, novel MIMO antenna technologies are booming. In [19], an optimized algorithm based on semidefinite programming (SDP) and minimum mean squared error (MMSE) is proposed for s cognitive radio (CR) MIMO system. This method is proven to perform better in terms of total transmitted power and signal-to-interference plus noise ratio (SINR). To alleviate the large channel state information (CSI) feedback in m-MIMO system, a robust channel estimation scheme is proposed based on the separation mechanism of the channel matrix [20]. Moreover, some MIMO antenna design strategies and beamforming methods have also been well developed, such as m-MIMO antenna technology for 5G communication base stations [21,22], 10-element MIMO antenna design for new 5G smartphones [23,24], ultra-massive MIMO radar technology for terahertz antenna beamforming [25], etc. However, m-MIMO antenna arrangement requires the integration of a huge number of antennas at the base station and a large number of antennas at the user terminal, which will undoubtedly cause hardware implementation challenges and extremely high signal-processing complexity: whether it is for the spatial diversity and beamforming of MIMO communication systems [26], or directions of arrival (DOA) estimation of MIMO radar systems [17].

To reduce the high signal-processing complexity for m-MIMO systems, various advanced matrix inversion acceleration algorithms, such as the Neumann series (NS) algorithm [27] and the Jacobi algorithm [28], may be used to reduce the computational burden of qSPICE or IAA methods. In [29], the NS algorithm was used for matrix inversion approximation (MIA) for m-MIMO signal processing. This method transforms the matrix inverse problem into a matrix multiplication problem, which is suitable for hardware platforms, but its computational complexity is the same or even higher than those of direct inverse methods (e.g., QR-based methods [30]). Although the Jacobi method reduces the complexity from $O(N^3)$ to $O(LN^2)$, where L represents the number of iterations, it converges slowly [31].

The alternating-direction method of multipliers (ADMM) is a powerful technique for solving massive optimization problems. This method is widely used in various fields, such as compressed sensing (CS) [32], regularization estimation [33], image processing [34], and machine learning [35]. In [36], the least absolute shrinkage and selection operator (LASSO)-ADMM algorithm was proposed for solving L_1-norm constrained optimization problems for m-MIMO signal detection. In [37], ADMM was used to reduce the computational complexity of CS-DOA compared with the traditional interior point method (IPM); this method reduces the computational complexity by dividing the problem into multiple subproblems during the iteration process to reduce the dimension of each problem. In [38], an imprecise augmented Lagrange multiplier (ALM)-ADMM algorithm was proposed for solving a weighted mixed $L_{2,1}$-norm penalty minimization problem to improve the DOA estimation performance for MIMO radar signals with missing elements.

Although the ADMM-based methods discussed above can efficiently solve the array DOA estimation problem, the application of an ADMM method introduces two additional user parameters, i.e., a Lagrangian parameter and a sparse regularization parameter [39]; the simultaneous adjustment of these two parameters is very tricky. To avoid this problem, in the present paper, the basic form of qSPICE is expanded, weighted covariance fitting is performed, and a mean square minimized form of the cost function is obtained by an equivalent transformation of the qSPICE cost function. Then, through the optimization properties, the problem is transformed into a sparse optimization problem in the form of a weighted LASSO. Then, this unconstrained optimization problem is decomposed into three subproblems to reduce the dimension of each problem and thus reduce the computational complexity based on ADMM. Due to the absence of user parameters in qSPICE, the proposed method can eliminate one sparse regularization parameter compared with the traditional ADMM-LASSO method, therefore significantly alleviating the difficulty of parameter selection. Meanwhile, the theoretical computational complexity of the proposed method is one order of magnitude lower than that of the traditional qSPICE method. Simulation results and measured data indicate that the estimation performance of the proposed method is not lower than those of qSPICE and ADMM-LASSO, and it incorporates the advantages of both methods, i.e., low complexity and a single user parameter.

The remainder of this paper is organized as follows. Section 2 reviews the MIMO radar signal model and the generalized SPICE and ADMM-LASSO methods. Then, the proposed ADMM-qSPICE DOA estimation method for MIMO radar is derived in detail. In Section 3, simulation and test results are presented, which demonstrate the effectiveness of the proposed method. In Section 4, the limitations of the proposed method are discussed. Section 5 presents the conclusions of this study.

2. Materials and Methods

2.1. MIMO Signal Model

We consider a centralized MIMO radar system equipped with M transmitting antennas and N receiving antennas. The transmitting and receiving antennas are assumed to be very close to each other such that the targets can be considered to be in the same orientation relative to them in the far field. It is assumed that the transmitting and receiving antennas of MIMO radar incorporate uniform linear arrays (ULAs) and that the MIMO radar system uses M transmitting antennas to transmit M mutually orthogonal waveforms with the same bandwidth and center frequency. Suppose there are K target sources. The DOA of the k-th target source is expressed as $\theta_k (k = 1, 2, \cdots, K)$. Based on the orthogonality of the transmitted signal waveform, the receiving array matches the filtered output; that is, the received data vector $\mathbf{y}(t) \in \mathbb{C}^{MN \times 1}$ can be expressed as [17,40,41]

$$\mathbf{y}(t) = \mathbf{A}\mathbf{s}(t) + \mathbf{e}(t) \tag{1}$$

where $\mathbf{s}(t) = [s_1(t), s_2(t), \cdots, s_K(t)]^T \in \mathbb{C}^{K \times 1}$ represents the target source signal, in which $s_k(t) = \alpha_k \cdot \exp(j\omega_k t)$, with α_k and ω_k representing the reflection coefficient and Doppler angular frequency, respectively, of the k-th azimuth target; t denotes the distance-time variable; $\mathbf{A} = \mathbf{A}_t \odot \mathbf{A}_r = [\mathbf{a}_1, \mathbf{a}_2, \cdots, \mathbf{a}_K] \in \mathbb{C}^{MN \times K}$ is the orientation matrix; $\odot$ represents the Khatri-Rao product; and $\mathbf{e}(t)$ denotes the covariance matrix of zero-mean white Gaussian noise $\sigma^2 \mathbf{I}_{MN}$, with $\mathbf{I}_{MN}$ denoting the $MN \times MN$-dimensional unit matrix. $\mathbf{A}_t$ and $\mathbf{A}_r$ represent the transmitting and receiving orientation matrices, respectively, which can be expressed as

$$\mathbf{A}_t = [\mathbf{a}_t(\theta_1), \mathbf{a}_t(\theta_2), \cdots, \mathbf{a}_t(\theta_K)] \in \mathbb{C}^{M \times K} \tag{2}$$

$$\mathbf{A}_r = [\mathbf{a}_r(\theta_1), \mathbf{a}_r(\theta_2), \cdots, \mathbf{a}_r(\theta_K)] \in \mathbb{C}^{N \times K} \tag{3}$$

where $\mathbf{a}_t(\theta_k)$ and $\mathbf{a}_r(\theta_k)$ represent the transmitting and receiving orientation vectors, respectively, which are expressed as

$$\mathbf{a}_t(\theta_k) = [1, \ \exp(j2\pi d_t \sin\theta_k / \lambda), \cdots, \exp(j2\pi d_t (M-1) \sin\theta_k / \lambda)]^T \tag{4}$$

$$\mathbf{a}_r(\theta_k) = [1, \ \exp(j2\pi d_r \sin\theta_k / \lambda), \cdots, \exp(j2\pi d_r (N-1) \sin\theta_k / \lambda)]^T \tag{5}$$

where d_t and d_r represent the transmitting and receiving element spacings, respectively. The receiving element spacing is $d_r = \lambda/2$, where λ represents the carrier wavelength. The transmitting element spacing is $d_t = M_t \cdot d_r$. According to MIMO theory, the equivalent virtual array of a MIMO system can be regarded as a SIMO ULA antenna with MN elements. Assuming there are L snapshots in total, the received data can be discretized as

$$\mathbf{y}(l) = \mathbf{A}\mathbf{s}(l) + \mathbf{e}(l) \tag{6}$$

where $\mathbf{s}(l) \in \mathbb{C}^{K \times 1}$ is the target signal source to be estimated and $\mathbf{e}(l) \in \mathbb{C}^{MN \times 1}$ denotes noise, for $l = 1, 2 \cdots, L$. To simplify the model, it is assumed that the Doppler frequency of the target $\omega_k = 0$; at this point, the only target information to be estimated is the azimuth angle θ_k and scattering intensity α_k.

For many MIMO radar detection applications, such as airspace surveillance radar, the number of targets is considerably smaller than the number of potential source locations. In such a case, more accurate DOAs with higher resolution can be obtained using the sparsity of $\mathbf{s}$. Below, two sparse signal recovery methods are considered, and the proposed method is derived in detail.

2.2. qSPICE

A classic sparse recovery method for solving model (6) is LASSO, whose cost function can be expressed as [42]

$$\underset{\mathbf{s}}{\text{minimize}} \ \frac{1}{2} \|\mathbf{y} - \mathbf{A}\mathbf{s}\|_2^2 + \mu \|\mathbf{s}\|_1 \tag{7}$$

where μ represents a user-adjustable regularization parameter, which balances the sparsity of the solution with the fitting degree of the signal. To avoid tricky parameter selection problems, an interesting alternative, namely the SPICE algorithm, was proposed in [16].

This method is based on the sparse covariance fitting criterion, and the minimization of the cost function can be expressed as

$$\underset{p}{\text{minimize}} \ \left\| \mathbf{R}^{-1/2}\left(\mathbf{y}\mathbf{y}^H - \mathbf{R}\right) \right\|_F^2 \tag{8}$$

where

$$\mathbf{R} = \mathbf{F}\mathbf{P}\mathbf{F}^H \tag{9}$$

$$\begin{aligned} \mathbf{F} &= \begin{bmatrix} \mathbf{A} & \mathbf{I}_{MN} \end{bmatrix} \\ &\overset{\Delta}{=} [\mathbf{a}_1, \mathbf{a}_2, \cdots, \mathbf{a}_{MN+K}] \end{aligned} \tag{10}$$

$$\mathbf{P} = diag(\mathbf{p}) \tag{11}$$

$$\begin{aligned} \mathbf{p} &= \left[|s_1|^2, |s_2|^2, \cdots, |s_K|^2, \sigma_1^2, \sigma_2^2, \cdots, \sigma_{MN}^2 \right] \\ &\overset{\Delta}{=} [p_1, p_2, \cdots, p_{K+MN}] \end{aligned} \tag{12}$$

in which $\mathbf{R}^{-1/2}$ represents the Hermitian positive-definite root mean square of $\mathbf{R}^{-1}$; $\|\cdot\|_F$ denotes the Frobenius matrix norm; $(\cdot)^H$ is the conjugate transpose; $diag(\cdot)$ denotes a diagonal matrix composed of a specified vector; and σ_i denotes the noise variance of MIMO channel $i(i = 1, 2, \cdots, MN)$. The minimization of (8) is equivalent to

$$\underset{p_k \geq 0}{\text{minimize}} \ \ \mathbf{y}^H \mathbf{R}^{-1} \mathbf{y} + \|\mathbf{W}\mathbf{p}\|_1 \tag{13}$$

where

$$\mathbf{W} = diag([w_1, w_2, \cdots, w_{K+MN}]) \tag{14}$$

$$w_j = \frac{\|\mathbf{a}_k\|^2}{\|\mathbf{y}\|^2}, \ j = 1, 2, \cdots, K + MN \tag{15}$$

The constraint of (13) is a weighted L_1 norm; this constraint tends to make its solution sparse. As expressed in (12), the SPICE algorithm simultaneously penalizes signal and noise. However, the penalty does not distinguish between signal and noise terms. Therefore, the minimized solution (13) will also cause the solution for the noise variance to tend to be sparse; that is, some values will be 0. In such a case, the covariance matrix $\mathbf{R}$ is not of full rank, which leads to two problems: unsatisfactory sparsity and an estimated increase in the model order [18]. To address these problems, a generalized SPICE algorithm, namely the qSPICE algorithm, was proposed based on the second constraint in the improved expression (13):

$$\underset{p_k \geq 0}{\text{minimize}} \ \mathbf{y}^H \mathbf{R}^{-1} \mathbf{y} + \|\mathbf{W}_s \mathbf{p}_s\|_1 + \|\mathbf{W}_n \mathbf{p}_n\|_q \tag{16}$$

where

$$\mathbf{p}_s = [p_1, p_2, \ldots, p_K]^T \tag{17}$$

$$\mathbf{p}_n = [p_{K+1}, p_{K+2}, \ldots, p_{K+MN}]^T \tag{18}$$

$$\mathbf{W}_s = \text{diag}([w_1, w_2, \ldots, w_K]) \tag{19}$$

$$\mathbf{W}_n = \text{diag}([w_{K+1}, w_{K+2}, \ldots, w_{K+MN}]) \tag{20}$$

in which $\|\cdot\|_q$ represents the vector q norm ($q > 0$). The sparsity of the noise variance can be directly controlled by selecting a suitable value for q. When $q = 1$, qSPICE degenerates into the classic SPICE algorithm. Therefore, the robustness of the estimation results can be improved by increasing the noise variance constraint term in Equation (16). However, the computational complexity of this method is too high, and it requires multiple matrix multiplication operations and one $MN \times MN$ matrix inversion operation in each iteration. Its computational complexity is $O\left\{ J\left[K(MN)^2 + (MN)^3 \right] \right\}$, where J represents the number of iterations, which may generally be several hundred according to the authors' experience [16]. Such high complexity is not conducive to practical engineering applications.

2.3. ADMM

There is also an effective divide-and-conquer approach for solving the optimization problem (7), namely the ADMM algorithm [43]. It has both the strong convergence characteristics of the multiplier method and the decomposability of the dual maximization problem. The LASSO problem is rewritten as a separable convex optimization problem with linear constraints (ADMM standard form) as follows:

$$\underset{s,z}{\text{minimize}} \quad \frac{1}{2}\|\mathbf{y} - \mathbf{As}\|_2^2 + \mu\|\mathbf{z}\|_1 \tag{21}$$

subject to

$$\mathbf{s} = \mathbf{z} \tag{22}$$

where $\mathbf{z} \in \mathbb{C}^{K \times 1}$ is an intermediate variable. The constrained optimization problem (21) is equivalent to the unconstrained optimization problem (22).

According to the multiplier method, the augmented Lagrangian function is constructed as

$$L_\rho(\mathbf{s}, \mathbf{z}, \mathbf{u}) = \frac{1}{2}\|\mathbf{y} - \mathbf{As}\|_2^2 + \mu\|\mathbf{z}\|_1 + \mathbf{u}^T(\mathbf{s} - \mathbf{z}) + \frac{\rho}{2}\|\mathbf{s} - \mathbf{z}\|_2^2 \tag{23}$$

where $\mathbf{u} \in \mathbb{C}^{K \times 1}$ is an intermediate variable and ρ is an introduced penalty parameter related to the constraint condition that satisfies $\rho > 0$. According to the ADMM algorithm [36], only one variable is updated at a time, while the other two variables are fixed, and the updating process is repeated alternately. For iterations $j = 1, 2, \cdots, J$, the following equations are applied:

$$\begin{aligned}
\mathbf{s}^{j+1} &= \underset{\mathbf{s}}{\text{argmin}} \ L_\rho\left(\mathbf{s}, \mathbf{z}^j, \mathbf{u}^j\right) \\
\mathbf{z}^{j+1} &= \underset{\mathbf{z}}{\text{argmin}} \ L_\rho\left(\mathbf{s}^{j+1}, \mathbf{z}, \mathbf{u}^j\right) \\
\mathbf{u}^{j+1} &= \mathbf{u} + \left(\mathbf{s}^{j+1} - \mathbf{z}^{j+1}\right)
\end{aligned} \tag{24}$$

According to the equations above, each variable is updated successively through the following three steps:
- $\mathbf{s}$ update:

$$\mathbf{s}^{j+1} = \left(\mathbf{A}^T\mathbf{A} + \rho\mathbf{I}\right)^{-1}\left(\mathbf{A}^T\mathbf{y} + \rho\left(\mathbf{z}^j - \mathbf{u}^j\right)\right) \tag{25}$$

- $\mathbf{z}$ update:

$$\mathbf{z}^{j+1} = Soft\left(\mathbf{u}^j + \mathbf{s}^{j+1}, \frac{\mu}{\rho}\right) \tag{26}$$

where $S_{\lambda/\rho}(\cdot)$ is the soft threshold function, which satisfies

$$Soft\left(\mathbf{u}^j + \mathbf{s}^{j+1}, \frac{\mu}{\rho}\right) = \begin{cases} x - \dfrac{\mu}{\rho} & x > \dfrac{\mu}{\rho} \\ 0 & |x| \leq \dfrac{\mu}{\rho} \\ x + \dfrac{\mu}{\rho} & x < -\dfrac{\mu}{\rho} \end{cases} \tag{27}$$

- **u** update:

$$\mathbf{u}^{j+1} = \mathbf{u}^j + \mathbf{s}^{j+1} - \mathbf{z}^{j+1} \tag{28}$$

According to our experience, the ADMM algorithm can lead to a satisfactory value of (25) after dozens of iterations by iteratively alternating among (25)–(28). Since the original problem is decomposed into three subproblems for solution, the variable dimension of each step of the alternating update is low, and the updating steps of $\mathbf{z}$ and $\mathbf{u}$ only involve vector addition and subtraction operations, resulting in low computational complexity. Notably, the computational complexity of this problem is mainly due to the calculation of the first matrix inverse in Equation (25). This matrix is constant in each iteration, so the matrix inverse must only be computed once in the algorithmic process. The computational complexity of this algorithm is $O(K^3)$.

Compared with the qSPICE algorithm, this method has great advantages in terms of computational complexity, but it introduces a new user parameter, namely the Lagrangian parameter μ, on the basis of the sparse regularization parameter ρ of the original LASSO problem. Adjusting both μ and ρ at the same time is highly unpreferrable, which makes the method difficult to apply in practice.

2.4. Proposed Method

To avoid the above problem, we derive a new DOA estimation method with high resolution, high speed, and a single user parameter using the generalized SPICE problem model through the ADMM solution method.

2.4.1. ADMM-qSPICE

Similar to the derivation in [44,45], and assuming that the variances of the noise terms are consistent, i.e., $\forall i$, $\sigma_i = \sigma$, the optimization problem (16) can be equivalent to a weighted mean square LASSO problem [7,18]

$$\underset{\mathbf{s}}{\text{minimize}} \ \|\mathbf{y} - \mathbf{As}\|_2 + \|\mathbf{Ds}\|_1 \tag{29}$$

where the weight matrix $\mathbf{D}$ satisfies

$$\mathbf{D} = \text{diag}\left(\left[\sqrt{\frac{\|\mathbf{a}_1\|_2^2}{(MN)^q \cdot \|\mathbf{y}\|_2^2}}, \sqrt{\frac{\|\mathbf{a}_2\|_2^2}{(MN)^q \cdot \|\mathbf{y}\|_2^2}}, \cdots, \sqrt{\frac{\|\mathbf{a}_K\|_2^2}{(MN)^q \cdot \|\mathbf{y}\|_2^2}}\right]\right) \tag{30}$$

in which q is the control constraint parameter for noise in the generalized SPICE. We note that $\underset{\mathbf{s}}{\text{minimize}} \ \|\mathbf{y} - \mathbf{As}\|_2 + \|\mathbf{Ds}\|_1$ is a Lagrangian form of an optimization problem that is similar to $\underset{\mathbf{s}}{\text{minimize}} \ \|\mathbf{y} - \mathbf{As}\|_2 \text{s.t.} \ \|\mathbf{Ds}\|_1 \leq \varepsilon$. This problem is equivalent to $\underset{\mathbf{s}}{\text{minimize}} \|\mathbf{y} - \mathbf{As}\|_2^2 \text{s.t.} \ \|\mathbf{D's}\|_1 \leq \varepsilon$, where there is a bidirectional mapping relationship in $\mathbf{D} \mapsto \mathbf{D'}$ [44]. In the simulation and processing of the measured data, we find that it is possible that $\mathbf{D} \approx \mathbf{D}'$, which can lead to excellent DOA results. Therefore, we solve the following weighted LASSO problem instead:

$$\underset{\mathbf{s}}{\text{minimize}} \, \|\mathbf{y} - \mathbf{As}\|_2^2 + \|\mathbf{D}'\mathbf{s}\|_1 \tag{31}$$

We rewrite the above equation as

$$\underset{\mathbf{s}}{\text{minimize}} \, \|\mathbf{y} - \mathbf{As}\|_2^2 + \|\mathbf{z}\|_1 \tag{32}$$

subject to

$$\mathbf{D}'\mathbf{s} - \mathbf{z} = 0 \tag{33}$$

and construct the augmented Lagrange function

$$\tilde{L}_\rho(\mathbf{s}, \mathbf{z}, \mathbf{u}) = \frac{1}{2}\|\mathbf{y} - \mathbf{As}\|_2^2 + \|\mathbf{z}\|_1 + \mathbf{u}^T(\mathbf{D}'\mathbf{s} - \mathbf{z}) + \frac{\rho}{2}\|\mathbf{D}'\mathbf{s} - \mathbf{z}\|_2^2 \tag{34}$$

According to the ADMM algorithm [36], only one variable is updated at a time, while the other two variables are fixed, and the update is repeated alternately. For iterations $j = 1, 2, \cdots, J$, we apply

$$\begin{aligned}
\mathbf{s}^{j+1} &= \underset{\mathbf{s}}{\text{argmin}} \, \tilde{L}_\rho\left(\mathbf{s}, \mathbf{z}^j, \mathbf{u}^j\right) \\
\mathbf{z}^{j+1} &= \underset{\mathbf{z}}{\text{argmin}} \, \tilde{L}_\rho\left(\mathbf{s}^{j+1}, \mathbf{z}, \mathbf{u}^j\right) \\
\mathbf{u}^{j+1} &= \mathbf{u}^j + \rho\left(\mathbf{D}'\mathbf{s}^{j+1} - \mathbf{z}^{j+1}\right)
\end{aligned} \tag{35}$$

According to the equations above, each variable can be updated successively through the following three steps: To simplify the form, we let $\boldsymbol{\xi} = \frac{\mathbf{u}}{\rho}$ and express (34) in scaled form:

$$\tilde{L}_\rho(\mathbf{s}, \mathbf{z}, \boldsymbol{\xi}) = \frac{1}{2}\|\mathbf{y} - \mathbf{As}\|_2^2 + \|\mathbf{z}\|_1 + \frac{\rho}{2}\|\mathbf{D}'\mathbf{s} - \mathbf{z} + \boldsymbol{\xi}\|_2^2 - \frac{\rho}{2}\|\boldsymbol{\xi}\|_2^2 \tag{36}$$

- $\mathbf{s}$ update:

Taking the derivative of (36) with respect to $\mathbf{s}$ and setting it to 0 yields

$$\frac{\partial \tilde{L}_\rho}{\partial \mathbf{s}} = -\mathbf{A}^T(\mathbf{y} - \mathbf{As}) + \frac{\rho}{2} \cdot 2 \cdot \mathbf{D}'^T(\mathbf{D}'\mathbf{s} - \mathbf{z} + \boldsymbol{\xi}) = 0 \tag{37}$$

Therefore, the iteration formula for the variable $\mathbf{s}$ can be written as

$$\mathbf{s}^{j+1} = (\mathbf{A}^T\mathbf{A} + \rho\mathbf{D}'^T\mathbf{D}')^{-1}(\mathbf{A}^T\mathbf{y} + \rho\mathbf{D}'^T(\mathbf{z}^j - \boldsymbol{\xi}^j)) \tag{38}$$

- $\mathbf{z}$ update:

Taking the derivative of (36) with respect to $\mathbf{z}$ yields

$$\begin{aligned}
\frac{\partial \tilde{L}_\rho}{\partial \mathbf{z}} &= \frac{\partial(\|\mathbf{z}\|_1 + \frac{\rho}{2}\|\mathbf{D}'\mathbf{s} - \mathbf{z} + \boldsymbol{\xi}\|_2^2 - \frac{\rho}{2}\|\boldsymbol{\xi}\|_2^2)}{\partial \mathbf{z}} \\
&= \frac{\partial(\frac{\rho}{2}\|\mathbf{z}\|_1 + \|\mathbf{z} - (\mathbf{D}'\mathbf{s} + \boldsymbol{\xi})\|_2^2)}{\partial \mathbf{z}}
\end{aligned} \tag{39}$$

The solution to the above problem can be represented by the soft threshold [46] and expressed as

$$
\mathbf{z}^{j+1} = Soft\left(\boldsymbol{\xi}^j + \mathbf{D}'\mathbf{s}^{j+1}, \frac{1}{\rho}\right) = \begin{cases} \left(\boldsymbol{\xi}^j + \mathbf{D}'\mathbf{s}^{j+1}\right) + \dfrac{1}{\rho}, & \boldsymbol{\xi}^j + \mathbf{D}'\mathbf{s}^{j+1} > \dfrac{1}{\rho} \\[2mm] 0, & \left|\boldsymbol{\xi}^j + \mathbf{D}'\mathbf{s}^{j+1}\right| \leq \dfrac{1}{\rho} \\[2mm] \left(\boldsymbol{\xi}^j + \mathbf{D}'\mathbf{s}^{j+1}\right) - \dfrac{1}{\rho}, & \boldsymbol{\xi}^j + \mathbf{D}'\mathbf{s}^{j+1} < -\dfrac{1}{\rho} \end{cases} \tag{40}
$$

- $\mathbf{u}$ update:

The iterative formula for variable $\mathbf{u}$ can be written as

$$
\mathbf{u}^{j+1} = \mathbf{u}^j + \rho(\mathbf{D}'\mathbf{s}^{j+1} - \mathbf{z}^{j+1}), \quad \mathbf{u}^j = \rho\boldsymbol{\xi}^j \tag{41}
$$

By alternating iteration of Equations (38)–(41), the DOA result $\mathbf{s}$ of the proposed method can be obtained. Algorithm 1 presents the pseudocode of this method. The matrix $\boldsymbol{\Delta}$ and vectors $\boldsymbol{\Phi}$ and $\mathbf{c}$ represent intermediate variables. The iteration termination criterion is $\left\|\mathbf{s}^{j+1} - \mathbf{s}^j\right\| \leq \varepsilon$, where ε is a small positive number. $\mathrm{sign}(\cdot)$ represents the sign function, $\odot$ denotes the Hadamard matrix product, and $\max(\cdot)$ denotes the maximum of two numbers.

Algorithm 1 ADMM-qSPICE

1:   $\mathbf{D}'$ is initialized from (30), and $\boldsymbol{\Delta} = (\mathbf{A}^T\mathbf{A} + \rho\mathbf{D}'^T\mathbf{D}')^{-1}$ is calculated. We set $\boldsymbol{\Phi} = \mathbf{A}^T\mathbf{y}$,
    $j = 1$, $\mathbf{u} = \mathbf{0}$, $\mathbf{z} = \mathbf{0}$, and $\mathbf{s} = \mathbf{0}$.
2:   **While** the termination criterion is not satisfied, **do**
3:      $\mathbf{s}^{j+1} = \boldsymbol{\Delta} \cdot \left(\boldsymbol{\Phi} + \rho\mathbf{D}'^T\left(\mathbf{z}^j - \boldsymbol{\xi}^j\right)\right)$ is updated.
4:      We let $\mathbf{c} = \boldsymbol{\xi}^j + \mathbf{D}'\mathbf{s}^{j+1}$.
5:      $\mathbf{z}^{j+1} = \mathrm{sign}(\mathbf{c}) \odot \max(\mathbf{0}, |\mathbf{c}| - 1/\rho)$ is updated.
6:      $\mathbf{u}^{j+1} = \mathbf{u}^j + \rho(\mathbf{D}'\mathbf{s}^{j+1} - \mathbf{z}^{j+1})$ is updated.
7:      $\boldsymbol{\xi}^{j+1} = \mathbf{u}^{j+1}/\rho$ is updated.
8:      We set $j = j + 1$.
9:   **End while**

2.4.2. Computational Complexity Analysis

As presented in Algorithm 1, the proposed ADMM-qSPICE method can be implemented by iterating (39)–(41) and performing some initialization calculations. We update $\mathbf{s}$ using Equation (38); because its left term is a fixed matrix in each iteration of the loop, it can be calculated beforehand for backup. Similarly, $\boldsymbol{\Phi}$ is calculated in advance during initialization for backup, which can significantly reduce the computational complexity of the algorithm. The following is a complexity analysis of the initialization and iteration processes. In Step 1 of Algorithm 1, initializing $\mathbf{D}'$ requires $(K + 2)MN$ multiplication operations. Calculating $\mathbf{A}^T\mathbf{A}$ requires $(MN)^2K$ multiplication operations. Since $\mathbf{D}'$ is a diagonal matrix, calculating $\mathbf{D}'^T\mathbf{D}'$ only requires K multiplication operations, and inversion of the matrix $K \times K$ has a computational complexity of K^3. Therefore, $\boldsymbol{\Delta}$ is required to initialize $(MN)^2K + K^3 + K$. Initializing $\boldsymbol{\Phi}$ requires KMN multiplication operations. Thus, the computational complexity of the initialization portion is $(MN)^2K + K^3 + (2K + 2)MN + K$. In the iterative process of Algorithm 1, updating $\mathbf{s}$ requires $2K^2 + K$ multiplication operations, while updating $\mathbf{z}$ requires $K^2 + K$ multiplication operations. Updating $\mathbf{u}$ requires $K^2 + 2K$ multiplication operations. We suppose J iterations are required; then, the computational complexity of the iterative process is $J(4K^2 + 4K)$. In summary, the complexity of the algorithm is $K^3 + (MN)^2K + 2KMN + 4JK^2 + (4J + 1)K + 2MN$. The computational complexity of the iterative process of the proposed method is not high, and its main computational load corresponds to the initialization of $\boldsymbol{\Delta}$.

A comparison of the DAS method, IAA method, ADMM-LASSO method, qSPICE method and proposed method in terms of computational complexity is shown in Table 1, where $J_1 \sim J_4$ represent the numbers of iterations for IAA, ADMM-LASSO, qSPICE and proposed method, respectively. Notably, the number of iterations J_1 for IAA is set to a fixed value of 12; J_2, J_3 and J_4 depend on the termination criteria of the corresponding algorithms.

Table 1. Complexity comparison.

Method	Number of Multiplication and Division Operations	Computational Complexity	Computational Time ($MN = 512, K = 256$)
DAS (FFT)	$K\log_2 K$	$O(K\log_2 K)$	4.43×10^{-4} s
IAA [13]	$J_1\left[(MN)^3 + 2(MN)^2 K + MNK\right]$	$O\left(J_1(MN)^3\right)$	4.88 s
ADMM-LASSO [36]	$K^3 + (MN + J_2)K^2 + KMN$	$O(K^3)$	0.03 s
qSPICE [18]	$J_3\left[(MN)^3 + (K+1)(MN)^2 + (K^2 + K + 1)MN + 12K\right]$	$O\left(J_3(MN)^3\right)$	11.41 s
Proposed method	$K^3 + (MN)^2 K + 2KMN + 4J_4 K^2 + (4J_4 + 1)K + 2MN$	$O(K^3)$	0.13 s

3. Results

In this section, the performance of the proposed ADMM-qSPICE method is demonstrated by simulation and testing. First, we conduct simulation tests to verify the DOA estimation accuracy and super-resolution ability of the proposed method. Then, we further evaluate the estimation accuracy and calculation time of the proposed method using two sets of measured data. The methods that are compared in this paper are DAS, IAA, ADMM-LASSO and qSPICE. Since only the angle estimation performance of each method is considered in this study, the target is set to be static and located at different azimuth angles of the same distance element. All our simulations and measurements are performed on a PC workstation equipped with 64-bit MATLAB R2018a, an Intel Core i5-9500 CPU (3.0 GHz) and 16 GB RAM. The root mean square error of angle estimation is used to evaluate the simulation performance of the proposed method, which is defined as

$$\text{RMSE} = 10\log_{10}\sqrt{\frac{1}{P}\sum_{p=1}^{P}|\hat{\theta}_p - \theta_p|^2} \tag{42}$$

where θ_p represents the true value of the target angle of the p-th target grid and $\hat{\theta}_p$ denotes the corresponding estimated value. The signal-to-noise ratio (SNR) is defined as

$$\text{SNR} = 10\log_{10}\left(\frac{P_s}{\delta^2}\right) \tag{43}$$

where P_s represents the signal power and σ^2 is the variance of the additive white Gaussian noise.

3.1. Simulation Results

It is assumed that there are two independent sources with the same radiation power, which are located at azimuth $-5°$ and $0°$, respectively. The azimuth scanning range is $[-90°, 90°]$, and the number of target points is $K = 512$. The SNR is set to SNR $= 10$ dB. We consider a 2-T 4-R MIMO radar system with a carrier frequency of $f_c = 77$ GHz, a receiving element spacing of $d_r = \lambda/2 = 1.9$ mm, and a transmitting element spacing of $d_t = M_t \cdot d_r = 7.6$ mm.

Figure 1a shows the results of 10 Monte Carlo tests with the traditional DAS method. This method can be implemented quickly by FFT, but the azimuth resolution of the DOA results is very low, resulting in an inability to resolve two adjacent targets effectively.

Figure 1b shows the results of 10 Monte Carlo tests with the IAA method. This method significantly outperforms the traditional DAS method in terms of sidelobe suppression, resolution and positioning accuracy, and the two targets are distinguished successfully in all 10 tests. However, this method still has various problems, i.e., unsatisfactory resolution, high sidelobes, and high computational complexity. Figure 1c shows the results of 10 Monte Carlo tests with the LASSO-ADMM method in [36], where the user parameters $\lambda = 1$ and $\rho = 10$. We manually adjust these two parameters based on the quality of the DOA result. This quality specifically expressed as target resolution, 3 dB width, and DOA accuracy. The values of $\lambda = 1$ and $\rho = 10$ were obtained by extensive experiments and are the parameters that we believe are good for the ADMM-LASSO method under the simulation conditions in this paper. This method can distinguish the two adjacent targets well in most cases, and the sidelobe suppression effect is better than that of IAA. However, the method requires the adjustment of two user parameters at the same time in the process of implementation. When any of these parameters change, the target's resolution, 3 dB width, and DOA accuracy decrease. Hence, this is a very tricky problem in engineering applications. Figure 1d shows the results of 10 Monte Carlo tests with the qSPICE method, where $q = 1.5$. This method can better recover the amplitude and position information of the target, and the DOA result is more sparse than that in Figure 1c. However, the computational complexity of this method is quite high, especially for m-MIMO array problems. Figure 1e shows the results of 10 Monte Carlo tests with the proposed method, where $q = 1.5$ and $\rho = 1.2$. The DOA results of the proposed method are better than those of DAS, IAA and LASSO-ADMM and are similar to those of qSPICE. In one of the Monte Carlo tests, qSPICE shows a spike with an amplitude of approximately 0.3 at $-26°$, but the proposed method only has a small peak with an amplitude of less than 0.05. The noise suppression ability of the proposed method clearly may be better than that of qSPICE when $\rho = 1.2$ is reasonably adjusted.

Figure 2 shows the root mean square error (RMSE) and Cramer-Rao bound (CRB) for the angle estimation results of the five methods over the SNR range of -5 dB to 30 dB. The RMSE of the proposed method is considerably lower than those of DAS and LASSO-ADMM, slightly lower than that of IAA, and similar to that of SPICE. In addition, as presented in Table 2, the target width of 3 dB when SNR = 10 dB quantitatively verifies the resolution performance of the proposed method. Moreover, the corresponding RMSE is presented. The proposed method clearly outperforms all the comparison methods above in terms of resolution, and its 3 dB width is the narrowest.

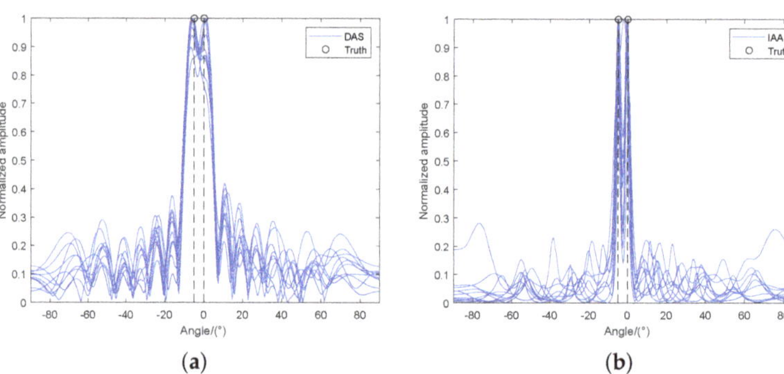

(a)   (b)

Figure 1. *Cont.*

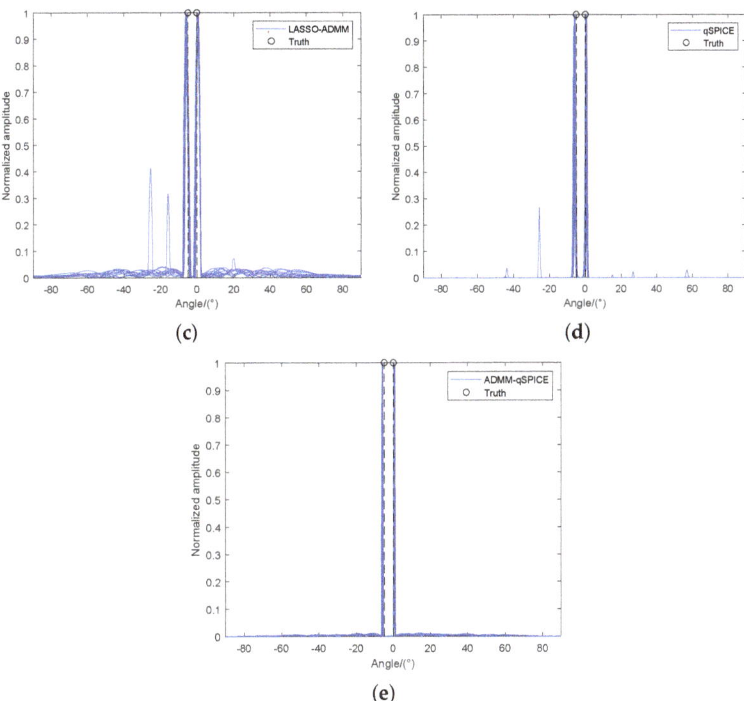

Figure 1. Simulated DOA results. (**a**) DAS method, (**b**) IAA method, (**c**) ADMM−LASSO method ($\lambda = 1$ and $\rho = 10$), (**d**) qSPICE method, and (**e**) proposed method ($\rho = 1.2$).

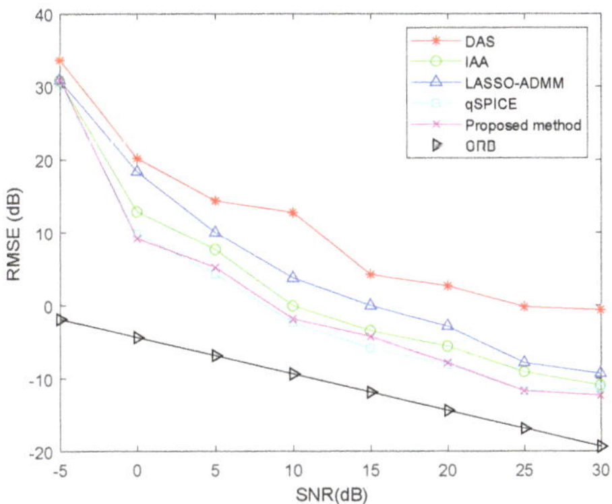

Figure 2. RMSE and CRB of each method.

Table 2. Performance Comparison of Various Methods (SNR = 10 dB).

Method	3 dB Width (Degrees)	RMSE (dB)
DAS	16.80	12.79
IAA	2.26	0.15
ADMM-LASSO	3.22	3.791
qSPICE	1.32	−2.269
ADMM-SPICE	0.98	−1.791

3.2. Measured Results

In this section, a set of MIMO radar measurements are used to verify the performance of the proposed method. An optical image of the original scene captured by an unmanned aerial vehicle (UAV) is shown in Figure 3, where the imaging area contains six vehicles. The MIMO radar parameters used in the test are shown in Table 3. There is a relationship among the carrier frequency, number of array elements and beam width, which jointly affect the azimuth resolution. The pulse width and pulse repetition interval affect the SNR. The bandwidth determines the range resolution.

Table 3. Parameters of the Measured Data.

Parameter	Value
Carrier frequency	77 GHz
Bandwidth	3.75 GHz
Beam width	1.4°
Pulse width	1 ms
Pulse recurrence interval	512 µs
Number of transmitting array elements	12
Number of receiving array elements	16
Range sampling points	261

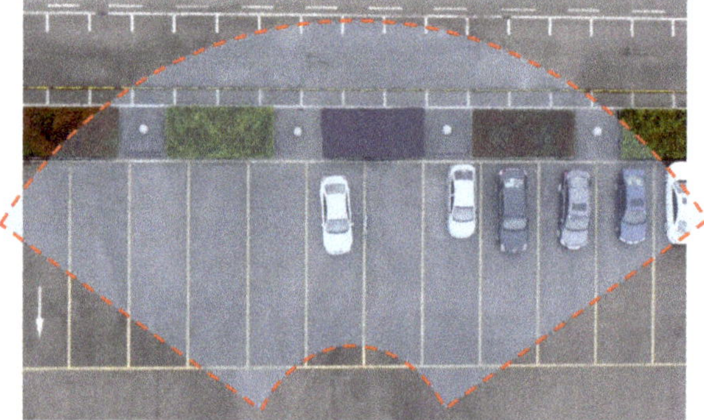

Figure 3. Optical scene.

Figure 4a shows the two-dimensional results of the DAS method, with a processing time of 0.02 s. The echo energy of the middle car is strong; there is a clear front contour, but the sidelobe is quite high. The echo energies of the five vehicles on the right overlap and are blurred because DAS is based on traditional matching filtering, which suffers from low resolution and high sidelobes. Moreover, as can be seen from the Figure 4a, there are strip-shaped strongly scattered objects above the vehicles, which are actually caused by the stone edge of the flower bed. Figure 4b shows the two-dimensional results of the IAA method, for which the processing time is 115.13 s. The IAA method effectively improves the azimuth resolution of the image compared with DAS. The echo sidelobe of the middle car is significantly suppressed, and the vehicles in the dense group on the right can be better distinguished. Figure 4c shows the two-dimensional results of the ADMM-LASSO method, and its processing time is 2.40 s. This method can further improve the azimuth resolution of the target and better suppress the sidelobes compared with IAA. However, the method has two user parameters that must be adjusted, i.e., the sparse regularization parameter μ and Lagrangian parameter ρ. It is difficult to adjust both parameters at the same time in practical applications.

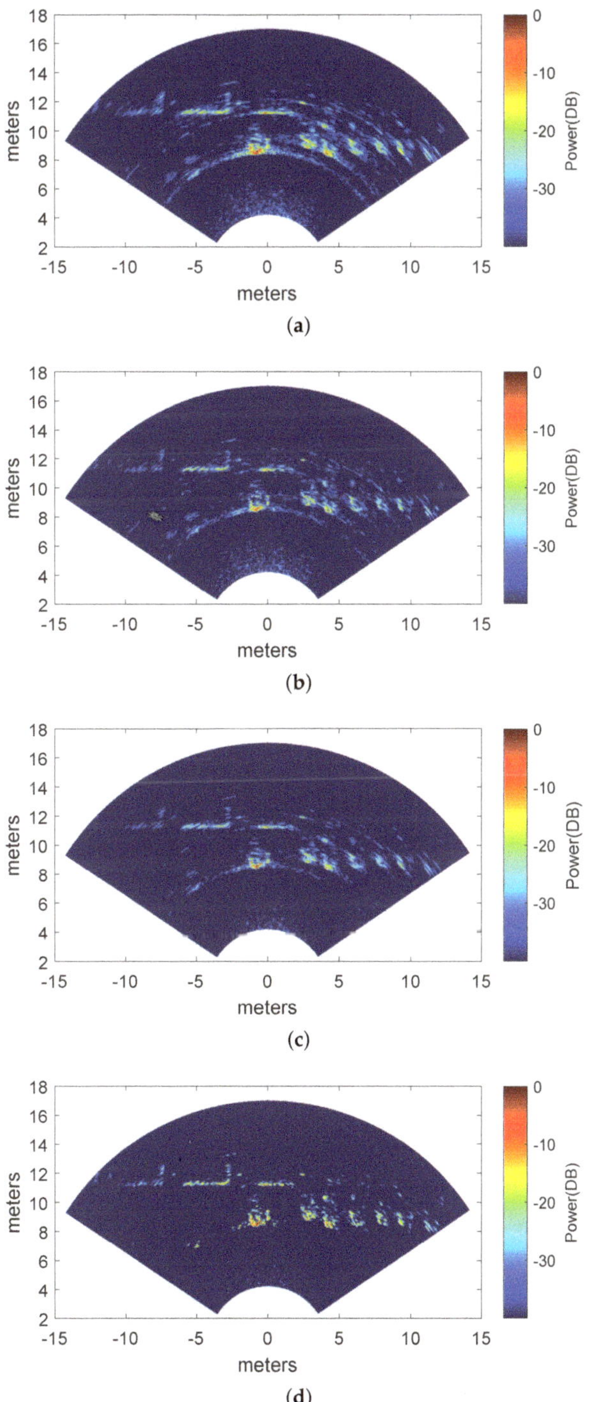

Figure 4. *Cont.*

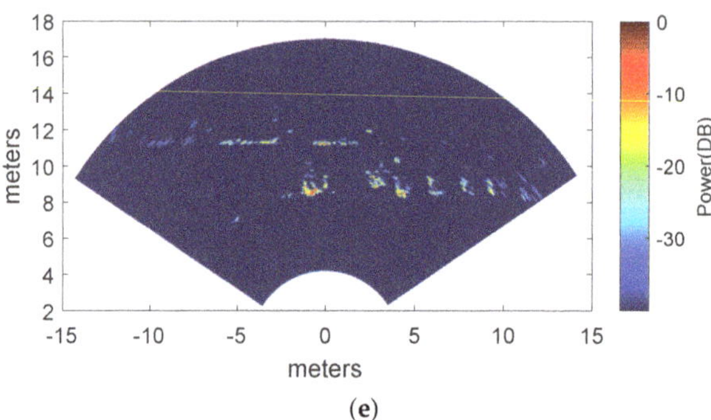

(**e**)

Figure 4. Measured Data. (**a**) Two−dimensional result of the DAS method, (**b**) Two−dimensional results of the IAA method, (**c**) Two−dimensional results of the ADMM−LASSO method, (**d**) Two−dimensional results of the qSPICE method, and (**e**) Two−dimensional results of the ADMM−SPICE method.

Figure 4d shows the two-dimensional results of the qSPICE method, with a processing time of 2704.81 s. As shown in the figure, this method further improves the azimuth resolution of the image compared with ADMM-LASSO. The sidelobes of the car in the center of the scene are significantly suppressed, and the car is more clearly defined. Unfortunately, the computational complexity of this method is too high, and the processing time is too long to be applied to real-time imaging applications with automotive radar, a consequence of the qSPICE method requiring high-dimensional matrix multiplication and inversion operations during the iteration process and requiring thousands of iterations per range element on average. Figure 4e shows the two-dimensional results of the ADMM-qSPICE method, with a processing time of 9.52 s. Compared with ADMM-LASSO, this method not only significantly improves the target resolution but also contains only one user parameter; thus, it is easier to apply in practice. Compared with the qSPICE method, the proposed method further suppresses the vehicle sidelobes. Sparser imaging results and more pronounced vehicle contours enable vehicles to be better detected and positioned. In addition, the computational complexity of this method is much lower than that of qSPICE method, because there is no need to calculate the matrix inverse in the iterations of the method, and the average number of iterations of each range element of the proposed method, i.e., 76, is considerably lower than that of the qSPICE method.

Image entropy (IE) was introduced to quantitatively analyze the two-dimensional target test results of the proposed method, where the lower the image entropy is, the better the image restoration effect [7]. The IE can be explicitly expressed by

$$\text{IE} = \sum_{i=1}^{I} p_i \log_2 p_i \tag{44}$$

where I denotes histogram counts of the two-dimensional image, and p_i represents the probability of each gray level occurring. The image entropy of the processing results above is shown in Table 4 below. DAS clearly has the highest image entropy, which indicates poor image quality. The image entropy of IAA is lower than that of DAS, but the reduction is limited. ADMM-LASSO and qSPICE show significantly reduced image entropy compared with DAS and IAA, indicating that the image quality obtained by these two methods is better. As presented in the last row of Table 4, the image entropy of the proposed method is lower than those of other methods, which indicates that the proposed method has a stronger ability to suppress noise and sidelobes, therefore making the two-dimensional imaging performance on vehicles better and more conducive to vehicle detection and recognition.

Table 4. Image Entropy.

Methods	IE
DAS	4.03
IAA	3.70
ADMM-LASSO	1.55
qSPICE	1.08
ADMM-SPICE	0.97

4. Discussion

4.1. Results Analysis and Limitations

In this paper, an ADMM-qSPICE-based sparse DOA estimation method for MIMO radar was proposed. The proposed method was compared and verified in detail based on complexity analysis results, simulation results, and measured data for point and area targets. As shown in Figures 1 and 4, the ADMM-qSPICE method achieved similar or even better resolution performance than traditional qSPICE. Moreover, the computational complexity of the ADMM-qSPICE method is considerably lower than that of the traditional qSPICE method. However, although the proposed method has the above two advantages, it also has various limitations.

As the ADMM algorithm was used to solve the problem, μ was introduced in the construction of the augmented Lagrangian function, which led to an additional user parameter that needed to be adjusted manually compared with the qSPICE method. Although one parameter is less complex than the two parameters of ADMM-LASSO, it is still difficult for practical applications. we always hope to obtain a user-parameter free method, which can adapt to different SNRs and sparsities. In follow-up efforts, we will study the variation rule of the Lagrangian parameter μ in different scenarios and seek an improved ADMM-SPICE method without user parameters.

Another limitation is that the computational complexity of the method is still large. Its computational complexity mainly comes from the matrix inversion in Equation (38). In the future work, we will further accelerate the calculation of high-dimensional matrix inversion for the proposed method.

4.2. Extended Applications

The proposed method can not only be used for the high accuracy DOA estimation of MIMO radar, but also can be applied to other radar imaging techniques, such as MIMO radar imaging, synthetic aperture radar (SAR) imaging [47] and real aperture radar (RAR) super-resolution imaging processing [6]. This is because, on the one hand, the signal models of the above problems are similar, and they can all be expressed as $\mathbf{y} = \mathbf{As} + \mathbf{e}$. On the other hand, it is because the proposed method has good adaptability, few user parameters, low complexity, and easy implementation under various platforms.

In addition, the proposed method can also be employed to spatial diversity, beamforming, terminal locating and other techniques for m-MIMO antenna communication systems [48]. These techniques may apply to some equipment such as base stations with m-MIMO antennas and smartphones with new 5G MIMO antennas mentioned in [23,24].

For 5G communications and high accuracy MIMO radar, m-MIMO antenna systems are required to achieve high data rates and high angular resolution. Taking MIMO radar as an example, the higher the number of equivalent channels, the higher the angular resolution of the system, and the better the performance of the corresponding DOA estimation method. Therefore, theoretically, the proposed method can be applied to MIMO systems with infinite array elements. However, limited by the complexity of hardware implementation, the number of array elements in the actual MIMO radar is often limited. For the measured data of this paper, we use a $12T \times 16R$ MIMO antenna system. We also made some simulations for m-MIMO antennas system. Due to the memory limitation of the simulation platform, the maximum number of MIMO elements in our simulation is $64T \times 64R$. The results show that the proposed method can still achieve good performance.

5. Conclusions

In this paper, a super-resolution DOA estimation method for MIMO radar based on ADMM-qSPICE was proposed. This approach has two significant advantages. First, compared with the ADMM-LASSO, the proposed method employs the weight matrix of the qSPICE to eliminate a user parameter and yield a higher resolution and stronger sidelobe suppression ability. Second, compared with qSPICE, the computational complexity of the proposed method is reduced from $O\left(J_3(MN)^3\right)$ to $O\left(K^3\right)$ without performance loss. Furthermore, the proposed method can also be applied to MIMO radar imaging, SAR imaging and massive MIMO systems in 5G communication. In future work, we will further improve the proposed method to achieve completely parameter adaptation.

Author Contributions: Conceptualization, Y.Z. (Yongwei Zhang) and Y.Z. (Yongchao Zhang); methodology, Y.Z. (Yongwei Zhang); software, J.L.; validation, Y.Z. (Yongchao Zhang) and Y.Z. (Yongwei Zhang); formal analysis, Y.Z. (Yongwei Zhang); investigation, Y.Z. (Yongwei Zhang) and J.Y. (Jianan Yan); resources, Y.Z. (Yongwei Zhang); data curation, Y.Z. (Yongwei Zhang); writing—original draft preparation, Y.Z. (Yongwei Zhang); writing—review and editing, J.Y. (Jianan Yan); visualization, Y.Z. (Yongchao Zhang); supervision, Y.Z. (Yongchao Zhang); project administration, Y.Z. (Yongchao Zhang); funding acquisition, Y.Z. (Yongchao Zhang), Y.Z. (Yin Zhang), Y.H. and J.Y. (Jianyu Yang). All authors have read and agreed to the published version of the manuscript.

Funding: This work was supported in part by the National Natural Science Foundation of China under grants 61901092, 61901090, and was supported by the Municipal Government of Quzhou under Grant Number 2022D011 and 2022D036.

Data Availability Statement: This study did not report any data.

Conflicts of Interest: The authors declare no conflict of interest.

Abbreviations

The following abbreviations are used in this manuscript:

DOA	directions of arrival
MIMO	multiple-input multiple-output
LASSO	least absolute shrinkage and selection operator
ADMM	alternating-direction method of multipliers
qSPICE	generalized SParse Iterative Covariance-based Estimation
DAS	delay-and-sum
FFT	fast Fourier transform
SIMO	single-input multiple-output
ESPRIT	estimation of signal parameters via rotational invariance techniques
MUSIC	multiple signal classification
PARAFAC	parallel factor analysis
MVDR	Minimum variance distortionless response
IAA	iterative adaptive method
SLIM	sparse learning via iterative minimization
SPICE	sparse Iterative Covariance-based Estimation
NS	Neumann series
MIA	matrix inversion approximation
CS	compressed sensing
IPM	interior point method
ALM	augmented Lagrange multiplier
ULA	uniform linear array
RMSE	root mean square error
CRB	Cramer-Rao bound
UAV	unmanned aerial vehicle
SNR	signal-to-noise ratio
IE	Image entropy

References

1. Sharma, P.; Tiwari, R.N.; Singh, P.; Kumar, P.; Kanaujia, B.K. MIMO Antennas: Design Approaches, Techniques and Applications. *Sensors* **2022**, *22*, 7813. [CrossRef] [PubMed]
2. Hu, X.; Tong, N.; Wang, J.; Ding, S.; Zhao, X. Matrix completion-based MIMO radar imaging with sparse planar array. *Signal Process.* **2017**, *131*, 49–57. [CrossRef]
3. Fishler, E.; Haimovich, A.; Blum, R.; Chizhik, D.; Cimini, L.; Valenzuela, R. MIMO radar: An idea whose time has come. In Proceedings of the 2004 IEEE Radar Conference (IEEE Cat. No. 04CH37509), Philadelphia, PA, USA, 29 April 2004; pp. 71–78.
4. Dai, J.; So, H.C. Real-valued sparse Bayesian learning for DOA estimation with arbitrary linear arrays. *IEEE Trans. Signal Process.* **2021**, *69*, 4977–4990. [CrossRef]
5. Liang, Y.; Liu, W.; Shen, Q.; Cui, W.; Wu, S. A review of closed-form Cramér-Rao bounds for DOA estimation in the presence of Gaussian noise under a unified framework. *IEEE Access* **2020**, *8*, 175101–175124. [CrossRef]
6. Zhang, Y.; Luo, J.; Zhang, Y.; Huang, Y.; Cai, X.; Yang, J.; Mao, D.; Li, J.; Tuo, X.; Zhang, Y. Resolution Enhancement for Large-Scale Real Beam Mapping Based on Adaptive Low-Rank Approximation. *IEEE Trans. Geosci. Remote Sens.* **2022**, *60*, 5116921. [CrossRef]
7. Zhang, Y.; Luo, J.; Li, J.; Mao, D.; Zhang, Y.; Huang, Y.; Yang, J. Fast inverse-scattering reconstruction for airborne high-squint radar imagery based on Doppler centroid compensation. *IEEE Trans. Geosci. Remote Sens.* **2021**, *60*, 5205517. [CrossRef]
8. Qiu, S.; Ma, X.; Zhang, R.; Han, Y.; Sheng, W. A dual-resolution unitary ESPRIT method for DOA estimation based on sparse co-prime MIMO radar. *Signal Process.* **2023**, *202*, 108753. [CrossRef]
9. Meng, F.X.; Li, Z.T.; Xu-Tao, Y.; Zhang, Z.C. Quantum algorithm for MUSIC-based DOA estimation in hybrid MIMO systems. *Quantum Sci. Technol.* **2022**, *7*, 025002. [CrossRef]
10. Ruan, N.J.; Wen, F.q.; Ai, L.; Xie, K. A PARAFAC decomposition algorithm for DOA estimation in colocated MIMO radar with imperfect waveforms. *IEEE Access* **2019**, *7*, 14680–14688. [CrossRef]
11. Hassanien, A.; Vorobyov, S.A. Transmit energy focusing for DOA estimation in MIMO radar with colocated antennas. *IEEE Trans. Signal Process.* **2011**, *59*, 2669–2682. [CrossRef]
12. Huang, H.; Yang, J.; Huang, H.; Song, Y.; Gui, G. Deep learning for super-resolution channel estimation and DOA estimation based massive MIMO system. *IEEE Trans. Veh. Technol.* **2018**, *67*, 8549–8560. [CrossRef]
13. Roberts, W.; Stoica, P.; Li, J.; Yardibi, T.; Sadjadi, F.A. Iterative adaptive approaches to MIMO radar imaging. *IEEE J. Sel. Top. Signal Process.* **2010**, *4*, 5–20. [CrossRef]
14. Shang, X.; Liu, J.; Li, J. Multiple object localization and vital sign monitoring using IR-UWB MIMO radar. *IEEE Trans. Aerosp. Electron. Syst.* **2020**, *56*, 4437–4450. [CrossRef]
15. Tan, X.; Roberts, W.; Li, J.; Stoica, P. Sparse learning via iterative minimization with application to MIMO radar imaging. *IEEE Trans. Signal Process.* **2010**, *59*, 1088–1101. [CrossRef]
16. Stoica, P.; Babu, P.; Li, J. New method of sparse parameter estimation in separable models and its use for spectral analysis of irregularly sampled data. *IEEE Trans. Signal Process.* **2010**, *59*, 35–47. [CrossRef]
17. Luo, J.; Zhang, Y.; Yang, J.; Zhang, D.; Zhang, Y.; Zhang, Y.; Huang, Y.; Jakobsson, A. Online Sparse DOA Estimation Based on Sub–Aperture Recursive LASSO for TDM–MIMO Radar. *Remote Sens.* **2022**, *14*, 2133. [CrossRef]
18. Swärd, J.; Adalbjörnsson, S.I.; Jakobsson, A. Generalized sparse covariance-based estimation. *Signal Process.* **2018**, *143*, 311–319. [CrossRef]
19. Khan, I.; Al-Wesabi, F.N.; Obayya, M.; Hilal, A.M.; Hamza, M.A.; Rizwanullah, M.; Al-Zahrani, F.A.; Amano, H.; Mostafa, S.M. An Optimized Algorithm for CR-MIMO Wireless Networks. *CMC Comput. Mater. Contin.* **2022**, *71*, 697–715.
20. Khan, I.; Rodrigues, J.J.; Al-Muhtadi, J.; Khattak, M.I.; Khan, Y.; Altaf, F.; Mirjavadi, S.S.; Choi, B.J. A robust channel estimation scheme for 5G massive MIMO systems. *Wirel. Commun. Mob. Comput.* **2019**, *2019*, 3469413. [CrossRef]
21. Karad, K.V.; Hendre, V.S. Review of Antenna Array for 5G Technology Using mmWave Massive MIMO. In *Recent Trends in Electronics and Communication*; Springer: Singapore, 2022; pp. 775–787.
22. Al-Tarifi, M.A.; Sharawi, M.S.; Shamim, A. Massive MIMO antenna system for 5G base stations with directive ports and switched beamsteering capabilities. *IET Microwaves Antennas Propag.* **2018**, *12*, 1709–1718. [CrossRef]
23. Alja'Afreh, S.S.; Altarawneh, B.; Alshamaileh, M.H.; E'qab, R.A.; Hussain, R.; Sharawi, M.S.; Xing, L.; Xu, Q. Ten antenna array using a small footprint capacitive-coupled-shorted loop antenna for 3.5 GHz 5G smartphone applications. *IEEE Access* **2021**, *9*, 33796–33810. [CrossRef]
24. Jaglan, N.; Gupta, S.D.; Kanaujia, B.K.; Sharawi, M.S. 10 element sub-6-GHz multi-band double-T based MIMO antenna system for 5G smartphones. *IEEE Access* **2021**, *9*, 118662–118672. [CrossRef]
25. Elbir, A.M.; Mishra, K.V.; Chatzinotas, S. Terahertz-band joint ultra-massive MIMO radar-communications: Model-based and model-free hybrid beamforming. *IEEE J. Sel. Top. Signal Process.* **2021**, *15*, 1468–1483. [CrossRef]
26. Li, X.; Wang, B. Coherence optimised subarray partition for hybrid MIMO phased array radar. *IET Microwaves Antennas Propag.* **2021**, *15*, 1441–1457. [CrossRef]
27. Zhu, D.; Li, B.; Liang, P. On the matrix inversion approximation based on Neumann series in massive MIMO systems. In Proceedings of the 2015 IEEE International Conference on Communications (ICC), London, UK, 8–12 June 2015; pp. 1763–1769.
28. Albreem, M.A. Approximate matrix inversion methods for massive mimo detectors. In Proceedings of the 2019 IEEE 23rd International Symposium on Consumer Technologies (ISCT), Ancona, Italy, 19–21 June 2019; pp. 87–92.

29. Zhang, X.; Zeng, H.; Ji, B.; Zhang, G. Low Complexity Implicit Detection for Massive MIMO Using Neumann Series. *IEEE Trans. Veh. Technol.* **2022**, *71*, 9044–9049. [CrossRef]
30. Prabhu, H.; Rodrigues, J.; Edfors, O.; Rusek, F. Approximative matrix inverse computations for very-large MIMO and applications to linear pre-coding systems. In Proceedings of the 2013 IEEE Wireless Communications and Networking Conference (WCNC), Shanghai, China, 7–10 April 2013; pp. 2710–2715.
31. Song, W.; Chen, X.; Wang, L.; Lu, X. Joint conjugate gradient and Jacobi iteration based low complexity precoding for massive MIMO systems. In Proceedings of the 2016 IEEE/CIC International Conference on Communications in China (ICCC), Chengdu, China, 27–29 July 2016; pp. 1–5.
32. Kouni, V.; Paraskevopoulos, G.; Rauhut, H.; Alexandropoulos, G.C. ADMM-DAD net: A deep unfolding network for analysis compressed sensing. In Proceedings of the ICASSP 2022—2022 IEEE International Conference on Acoustics, Speech and Signal Processing (ICASSP), Singapore, 23–27 May 2022; pp. 1506–1510.
33. Qu, W.; Xiu, X.; Zhang, H.; Fan, J. An efficient semi-proximal ADMM algorithm for low-rank and sparse regularized matrix minimization problems with real-world applications. *J. Comput. Appl. Math.* **2022**, *424*, 115007. [CrossRef]
34. Li, X.; Bai, X.; Zhou, F. High-resolution ISAR imaging and autofocusing via 2d-ADMM-net. *Remote Sens.* **2021**, *13*, 2326. [CrossRef]
35. Shang, F.; Xu, T.; Liu, Y.; Liu, H.; Shen, L.; Gong, M. Differentially private ADMM algorithms for machine learning. *IEEE Trans. Inf. Forensics Secur.* **2021**, *16*, 4733–4745. [CrossRef]
36. Elgabli, A.; Elghariani, A.; Al-Abbasi, A.O.; Bell, M. Two-stage LASSO ADMM signal detection algorithm for large scale MIMO. In Proceedings of the 2017 51st Asilomar Conference on Signals, Systems, and Computers, Pacific Grove, CA, USA, 29 October–1 November 2017; pp. 1660–1664.
37. Wang, Q.; Zhao, Z.; Chen, Z. Fast compressive sensing DOA estimation via ADMM solver. In Proceedings of the 2017 IEEE International Conference on Information and Automation (ICIA), Macao, China, 18–20 July 2017; pp. 53–57.
38. Chen, J.; Zhang, C.; Fu, S.; Li, J. Robust Reweighted $\ell_{2,1}$-Norm Based Approach for DOA Estimation in MIMO Radar Under Array Sensor Failures. *IEEE Sens. J.* **2021**, *21*, 27858–27867. [CrossRef]
39. Hashempour, H.R. Sparsity-driven ISAR imaging based on two-dimensional ADMM. *IEEE Sens. J.* **2020**, *20*, 13349–13356. [CrossRef]
40. Li, J.; Stoica, P. *MIMO Radar Signal Processing*; John Wiley & Sons: Hoboken, NJ, USA, 2008.
41. Wang, X.; Wang, L.; Li, X.; Bi, G. Nuclear norm minimization framework for DOA estimation in MIMO radar. *Signal Process.* **2017**, *135*, 147–152. [CrossRef]
42. Tibshirani, R. Regression shrinkage and selection via the lasso. *J. R. Stat. Soc. Ser. B* **1996**, *58*, 267–288. [CrossRef]
43. Boyd, S.; Parikh, N.; Chu, E.; Peleato, B.; Eckstein, J. Distributed optimization and statistical learning via the alternating direction method of multipliers. *Found. Trends® Mach. Learn.* **2011**, *3*, 1–122.
44. Rojas, C.R.; Katselis, D.; Hjalmarsson, H. A note on the SPICE method. *IEEE Trans. Signal Process.* **2013**, *61*, 4545–4551. [CrossRef]
45. Stoica, P.; Zachariah, D.; Li, J. Weighted SPICE: A unifying approach for hyperparameter-free sparse estimation. *Digit. Signal Process.* **2014**, *33*, 1–12. [CrossRef]
46. Wright, S.J.; Nowak, R.D.; Figueiredo, M.A. Sparse reconstruction by separable approximation. *IEEE Trans. Signal Process.* **2009**, *57*, 2479–2493. [CrossRef]
47. Yang, X.; Zhang, Y.; Mao, D.; Bu, Y.; Yang, H.; Shi, J. Sparse Reconstruction for Synthetic Aperture Radar VIA Generalized Sparse Covariance Fitting. In Proceedings of the IGARSS 2019—2019 IEEE International Geoscience and Remote Sensing Symposium, Yokohama, Japan, 28 July–2 August 2019; pp. 767–770.
48. Ali, E.; Ismail, M.; Nordin, R.; Abdulah, N.F. Beamforming techniques for massive MIMO systems in 5G: Overview, classification, and trends for future research. *Front. Inf. Technol. Electron. Eng.* **2017**, *18*, 753–772. [CrossRef]

remote sensing

Article

Ground Clutter Mitigation for Slow-Time MIMO Radar Using Independent Component Analysis

Fawei Yang [1], Jinpeng Guo [1], Rui Zhu [1,*], Julien Le Kernec [2], Quanhua Liu [1,3] and Tao Zeng [1]

[1] Radar Research Lab, School of Information and Electronics, Beijing Institute of Technology, Beijing 100081, China
[2] James Watt School of Engineering, University of Glasgow, University Avenue, Glasgow G12 8QQ, UK
[3] Beijing Institute of Technology Chongqing Innovation Center, Chongqing 401120, China
* Correspondence: zhurui_bit2020@bit.edu.cn

Abstract: The detection of low, slow and small (LSS) targets, such as small drones, is a developing area of research in radar, wherein the presence of ground clutter can be quite challenging. LSS targets, because of their unusual flying mode, can be easily shadowed by ground clutter, leading to poor radar detection performance. In this study, we investigated the feasibility and performance of a ground clutter mitigation method combining slow-time multiple-input multiple-output (st-MIMO) waveforms and independent component analysis (ICA) in a ground-based MIMO radar focusing on LSS target detection. The modeling of ground clutter under the framework of st-MIMO was first defined. Combining the spatial and temporal steering vector of st-MIMO, a universal signal model including the target, ground clutter, and noise was established. The compliance of the signal model for conducting ICA to separate the target was analyzed. Based on this, a st-MIMO-ICA processing scheme was proposed to mitigate ground clutter. The effectiveness of the proposed method was verified with simulation and experimental data collected from an S-band st-MIMO radar system with a desirable target output signal-to-clutter-plus-noise ratio (SCNR). This work can shed light on the use of ground clutter mitigation techniques for MIMO radar to tackle LSS targets.

Keywords: ground clutter mitigation; independent component analysis; slow-time MIMO radar

1. Introduction

Small drone detection using radar has attracted enormous attention in recent years [1–3]. With the rapid growth of the consumer drone market, unmanned aerial vehicles (UAVs) have become a significant threat to civil aviation, anti-terrorism, and private security. As a powerful sensor that can operate regardless the time and weather, radar plays an important role in tackling these low, slow and small (LSS) targets. Many systems [4–13] and techniques [14–19] focusing on LSS target detection in the field of radar have been researched and developed.

The early LSS target surveillance radar systems were modified from the maritime radar systems such as the Merlin[TM] Radar System from DeTect Inc. [20] and the initial product of Robin Radar Systems Inc. [21]. They consist of two maritime radar antennas that rotate along the azimuth and elevation to achieve quasi-3D detection. With the development of phased array radar, more LSS target detection systems have turned to utilize antenna array and digital beamforming technology to obtain better target detection performance [4,5]. In recent years, some relatively new radar concepts, such as multiple-input multiple-output (MIMO) radar [6–8], multistatic radar [9,10], and ubiquitous radar (which is also named holographic or staring radar) [11–13], have also been introduced to the field of LSS target detection.

There are two main challenges in detecting LSS targets. The first challenge is the poor target signal-to-clutter-plus-noise ratio (SCNR). Due to the low flying altitude of the small drones, the radar beam must have a rather small grazing angle. This significantly

Citation: Yang, F.; Guo, J.; Zhu, R.; Le Kernec, J.; Liu, Q.; Zeng, T. Ground Clutter Mitigation for Slow-Time MIMO Radar Using Independent Component Analysis. *Remote Sens.* **2022**, *14*, 6098. https://doi.org/10.3390/rs14236098

Academic Editor: Guolong Cui

Received: 31 October 2022
Accepted: 29 November 2022
Published: 1 December 2022

Publisher's Note: MDPI stays neutral with regard to jurisdictional claims in published maps and institutional affiliations.

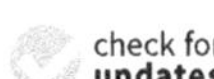

raises the ground clutter energy in the received signal. Considering the small radar cross-section (RCS) of LSS targets, the target SCNR is significantly reduced. Secondly, the slow flying, or hovering, velocity of the target renders the detection of LSS in slow-moving ground clutter via conventional techniques such as moving target indication (MTI) or moving target detector (MTD) ineffective. This is because ground clutter mainly consists of buildings, trees, cars, etc. The velocity spectrum spread of this clutter ranges from zero to tens of meters per second, which overshadows the speed of LSS targets. Thus, it is worth investigating clutter mitigation methods for LSS target detection.

Independent component analysis (ICA) [22–24] has aroused worldwide research interest in the field of signal processing since the 1990s. It has extensive applications in many fields such as communication, radar, image processing, acoustic processing, biomedical signal processing, and even financial data analysis [25]. Based on a MIMO system, the purpose of ICA is to simultaneously separate independent non-Gaussian components from observed multi-channel signals. This process can be used to help solve the ground clutter mitigation problem.

MIMO radar, as a typical multi-channel system, utilizes omnidirectional antennas and an orthogonal waveform in the transmitting stage, and then it forms a synthesized MIMO beam in the receiving stage. MIMO radar leverages waveform diversity to further increase the scale of the virtual array and spatial diversity [25–27]. A slow-time MIMO (st-MIMO) waveform [28–30] expands a conventional radar waveform by phase-coding the pulses of different channels to achieve orthogonal transmission and MIMO demodulation after pulse-Doppler (PD) processing. Although the unambiguous Doppler speed is divided by the number of Doppler sub-bands [31], st-MIMO is still feasible for slow-moving target detection with acceptable orthogonality and bandwidth efficiency.

Over the last few years, various techniques have been developed to tackle LSS targets from the radar detection stage to the parameter estimation stage and the target classification stage. Regarding methods used to enhance LSS target detection performance at the range-Doppler (r-D) level, one study [14] used the stationary point concentration technique to reduce the noise floor caused by transmitter leakage and increase the signal-to-noise ratio (SNR) of a UAV. Another study [15] adopted an iterative adaptive approach to enhance the Doppler resolution in an r-D map, which led to improved target detectability. For the parameter estimation stage, the authors of [16] proposed a long-time coherent integration method for maneuvering LSS targets to improve target estimation accuracy. Regarding the target classification stage, many works [2,17,18] focused on the micro-Doppler signature (m-DS) of UAVs. The numbers of m-DSs have been thoroughly analyzed among different types of UAVs and utilized to conduct target classification via multiple classifiers or neural networks.

There have been many studies in the field of clutter mitigation involving the aforementioned st-MIMO or ICA methods separately. Nevertheless, the combination of these two techniques while focusing on ground clutter mitigation in detecting LSS targets has been limited. For clutter mitigation involving st-MIMO, the authors of [28] developed spatial-time adaptive processing (STAP) in st-MIMO radar to mitigate multipath clutter. The spatial-time structure of st-MIMO was established, and a data covariance model, including target, direct path clutter, multipath clutter, jamming, and white noise, was defined. In [32], the beamspace st-MIMO, which can form virtual transmit nulls in directions that would result in multipath clutter returns in the main lobe of a radar system, was investigated. In [33], ground clutter was removed via principal components analysis (PCA) to enhance micro-Doppler feature extraction. For clutter mitigation involving the ICA or BSS methods, many works have focused on the multi-mode clutter suppression of over-the-horizon (OTH) radar. In [34,35], the spread-Doppler clutter caused by multi-mode propagation in OTH radar was suppressed via the second-order blind identification (SOBI) method. In addition, some works applied the BSS method in the field of main lobe jamming suppression [36].

In this study, we investigated the feasibility and performance of a ground clutter mitigation method combining the st-MIMO and ICA techniques in a ground-based MIMO radar focusing on LSS target detection. The main contributions of this work are as follows. Firstly, we propose a way to modulate ground clutter under the framework of st-MIMO. The ground clutter covariance was derived based on a Gaussian-shaped power spectrum. Combining the spatial and temporal steering vector of st-MIMO, a universal signal modelling including the target, ground clutter, and noise is provided. Secondly, we propose a st-MIMO-ICA processing scheme to separate the target signal from the received data including ground clutter plus noise. Mathematical proof that the signal model fits the framework of the ICA problem is provided. The multi-targets can be simultaneously separated via the proposed method. Finally, we validated our proposed st-MIMO-ICA method and evaluated its performance with both simulations and experiments using an S-band st-MIMO radar system developed in our previous work [6]. Comparisons with the PCA and adaptive techniques were performed, and the st-MIMO-ICA method showed the highest target output SCNR of the tested approaches.

The remainder of this paper is organized as follows. Section 2 describes the signal modeling of the st-MIMO radar and the ground clutter modulation in st-MIMO. Section 3 describes the signal modeling for conducting ICA and proposes the st-MIMO-ICA method. Section 4 presents the simulation and experimental results for the performance of the st-MIMO-ICA method. Finally, Section 5 concludes the paper and outlines possible future work.

2. Signal Modeling of St-MIMO Radar

In this section, the signal modeling of st-MIMO is described. Furthermore, the modulation of ground clutter under the framework of st-MIMO is derived.

2.1. St-MIMO Waveform Modeling and Processing

Consider a co-located 1D linear antenna array with M transmitting elements and N receiving elements that are omnidirectional. The element distance to the reference antenna of the $m^{th}, m = 0, \cdots, M - 1$ element in the transmitting array is d_m. Likewise, the element distance to the reference antenna of the $n^{th}, n = 0, \cdots, N - 1$ element in the receiving array is d_n. Note that this definition of the array is general regardless of whether the array is uniform or not. Figure 1 shows the geometry of the signal modeling setup. The operating frequency is f_0 and the operating wavelength is λ_0. There are K pulses in one coherent processing interval (CPI) and the pulse repetition frequency (PRF) $f_r = 1/T_r$, where T_r refers to pulse repetition interval (PRI).

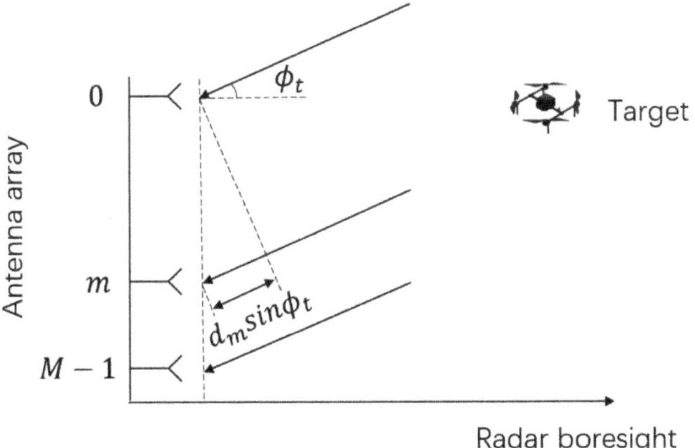

Figure 1. The geometry of signal modeling.

The slow-time MIMO approach split the whole Doppler PRF into M orthogonal Doppler sub-bands with a bandwidth of $\Delta f_{sub} = f_r/M$ via slow-time phase coding. Here, we set the number of the Doppler sub-bands as the same as the number of transmitting antennas to simplify the modeling. Furthermore, the Doppler sub-bands can be redundant, leading to some empty Doppler sub-bands that can be utilized to enlarge the velocity measurement range. The baseband pulse waveform $u_p(t)$ of each transmitting element is an identical linear frequency modulated (LFM) signal but with varying starting phases $\varphi(m,k)$, which is a function of the transmitting element index m and the pulse index (slow-time) k. Let ρ_t represent the constant transmit amplitude in each antenna without beamforming transmission. Then, the transmitting waveform of the mth element is:

$$s_m(t) = \rho_t \sum_{k=0}^{K-1} u_p(t - kT_r)e^{j2\pi(f_0 t + \varphi(m,k))} \tag{1}$$

Herein, let $\varphi(m,k)$ have a linear form:

$$\varphi(m,k) = \alpha_m kT_r \tag{2}$$

In this way, the Doppler domain is divided into M identical sub-bands. The instantaneous Doppler frequency f_d^m of the mth transmitting antenna is the derivative of the slow time variable kT_r:

$$f_d^m = \frac{\partial \varphi(m,k)}{\partial kT_r} = \alpha_m \tag{3}$$

Let the Doppler frequency of each transmitting antenna be evenly distributed in the Doppler domain and consider that the sign of the Doppler, α_m, has the form of:

$$\alpha_m = \frac{m}{M} \cdot f_r \tag{4}$$

One of the advantages of the slow-time MIMO approach is the good hardware compatibility to traditional phased array radar systems. Orthogonal transmission is realized in the Doppler domain via time-varying starting phases, while the carrier frequency remains f_0 for each channel. This makes it easy to implement to existing array radar systems. Note that the time-varying starting phases between antennas also presumably steer the beam as a function of the slow-time pulse k, which means that the main lobe direction of the transmitted beampattern sweeps the angle domain with a period of M pulses [30]. The ratio $f_r/\Delta f_{sub}$, i.e., the number of the Doppler sub-bands, can be modified to adjust the sweep rate of the main lobe.

Next, consider a far-field moving target with target speed v_t and corresponding Doppler shift $f_t = 2v_t/\lambda_0$. The target is located at an angle ϕ_t related to the array boresight. The amplitude of the echo is ρ_r. Thus, the backscattered signal of the nth receiving element via the mth transmitting element is shown in (5).

$$s_{mn}(t) = \rho_r \sum_{k=0}^{K-1} u_p(t - \tau_{mn} - kT_r)e^{j2\pi(f_0 + f_t)(t - \tau_{mn})}e^{j2\pi\alpha_m kT_r} \tag{5}$$

The round-trip delay τ_{mn} in (5) takes the form of (6), which consists of the time delay from the mth transmitting element TX_m to the target and the time delay from the target to the nth receiving element TX_n [37]:

$$\begin{aligned}
\tau_{mn} &= \tau_{tm} + \tau_{rn} \\
&= \left(\frac{R_t}{c} - \frac{d_m \sin\phi_t}{c}\right) + \left(\frac{R_t}{c} - \frac{d_n \sin\phi_t}{c}\right) \\
&= \frac{2R_t}{c} - \frac{d_m \sin\phi_t}{c} - \frac{d_n \sin\phi_t}{c}
\end{aligned} \tag{6}$$

where R_t is the target range based on the first element of the array. Under the conditions of a narrow band and slow target velocity, the following assumptions can be made:

$$\tau_{mn} \approx \tau_c = \frac{2R_t}{c} \tag{7}$$

and $e^{j2\pi f_t \tau_{mn}} \approx 1$.

Then, $s_{mn}(t)$ can be approximated as:

$$s_{mn}(t) \approx \rho_r \sum_{k=0}^{K-1} u_p(t - \tau_c - kT_r) e^{j2\pi(f_0+f_t)t} e^{-j2\pi f_0 \tau_{mn}} e^{j2\pi \alpha_m kT_r} \tag{8}$$

The output of the nth receiving element is the sum of the M transmit waveforms and can be expressed as:

$$\begin{aligned}
s_n(t) &= \sum_{m=0}^{M-1} s_{mn}(t) \\
&= \rho_r \sum_{m=0}^{M-1} \sum_{k=0}^{K-1} u_p(t - \tau_c - kT_r) e^{j2\pi f_0 t} e^{j2\pi f_t t} e^{-j2\pi f_0 \tau_{mn}} e^{j2\pi \alpha_m kT_r}
\end{aligned} \tag{9}$$

After down-converting by multiplying the echo with $e^{-j2\pi f_0 t}$, the baseband receiving signal can be expressed as:

$$\begin{aligned}
s\prime_n(t) &= s_n(t) \cdot e^{-j2\pi f_0 t} \\
&= \rho_r \sum_{m=0}^{M-1} \sum_{k=0}^{K-1} u_p(t - \tau_c - kT_r)\, e^{j2\pi f_t t} e^{-j2\pi f_0 \frac{2R_t}{c}} e^{j2\pi \frac{d_m}{\lambda_0} \sin \phi_t} e^{j2\pi \frac{d_n}{\lambda_0} \sin \phi_t} e^{j2\pi \alpha_m kT_r}
\end{aligned} \tag{10}$$

Next, conduct matched filtering on $s\prime_n(t)$ using a baseband matched filter $h(t) = u_p^*(-t)$, where the superscript * represents the conjugate operation. The output signal after matched filtering can be derived as:

$$\begin{aligned}
X_n(t) &= \int_\infty^\infty s\prime_n(x) h(t - x)\,dx \\
&= \rho_r e^{-j2\pi f_0 \frac{2R_t}{c}} e^{j2\pi \frac{d_n}{\lambda_0} \sin \phi_t} \sum_{m=0}^{M-1} \sum_{k=0}^{K-1} e^{j2\pi \frac{d_m}{\lambda_0} \sin \phi_t} e^{j2\pi \alpha_m kT_r} \ldots \\
&\quad \int_\infty^\infty u_p(x - \tau_c - kT_r) e^{j2\pi f_t x} u_p^*(-(t - x))\,dx
\end{aligned} \tag{11}$$

After conducting variable substitution and assuming $e^{j2\pi f_t \tau_c} \approx 1$, (11) can be derived as:

$$\begin{aligned}
X_n(t) &\approx \rho_r e^{-j2\pi f_0 \frac{2R_t}{c}} e^{j2\pi \frac{d_n}{\lambda_0} \sin \phi_t} \sum_{m=0}^{M-1} \sum_{k=0}^{K-1} e^{j2\pi \frac{d_m}{\lambda_0} \sin \phi_t} e^{j2\pi \alpha_m kT_r} \ldots \\
&\quad \int_\infty^\infty u_p(\beta) e^{j2\pi f_t \beta} e^{j2\pi f_t kT_r} u_p^*(\beta + \tau_c + kT_r - t)\,d\beta \\
&= \rho_r e^{-j2\pi f_0 \frac{2R_t}{c}} e^{j2\pi \frac{d_n}{\lambda_0} \sin \phi_t} \sum_{m=0}^{M-1} \sum_{k=0}^{K-1} e^{j2\pi \frac{d_m}{\lambda_0} \sin \phi_t} \ldots \\
&\quad e^{j2\pi(\alpha_m + f_t)kT_r} \chi(t - \tau_c - kT_r, f_t)
\end{aligned} \tag{12}$$

where

$$\chi(\tau, f_t) = \int_\infty^\infty u_p(\beta) u_p^*(\beta - \tau) e^{j2\pi f_t \beta}\,d\beta \tag{13}$$

is the ambiguity function of $u_p(t)$ with the time lag being τ and the Doppler frequency being f_t.

Next, we focus on the fast time of $t_k = \tau_c + kT_r$ in each pulse, which corresponds to the time lag of the target. Then, the ambiguity function in (12) is $\chi(0, f_t)$. Assume $\chi(0, f_t) \approx 1$;

because of the high Doppler tolerance of the LFM signal, the response of each t_k in (12) can be expressed as:

$$
\begin{aligned}
X_{nk} &= X_n(t_k) \\
&= \xi_t e^{j2\pi \frac{d_n}{\lambda_0} \sin \phi_t} \sum_{m\prime=0}^{M-1} e^{j2\pi \frac{d_{m\prime}}{\lambda_0} \sin \phi_t} e^{j2\pi(\alpha_{m\prime}+f_t)kT_r}
\end{aligned}
\tag{14}
$$

where $\xi_t = \rho_r e^{-j2\pi f_0 \frac{2R_t}{c}}$.

In order to separate the response of the mth transmitting element and achieve MIMO demodulation, we first shift the central Doppler frequency of the mth Doppler sub-band to zero-Doppler by multiplying $e^{-j2\pi\alpha_m kT_r}$ to (14) which yields:

$$
\begin{aligned}
X_{nk,m} &= X_{nk} \cdot e^{-j2\pi\alpha_m kT_r} \\
&= \xi_t e^{j2\pi \frac{d_n}{\lambda_0} \sin \phi_t} e^{j2\pi f_t kT_r} e^{j2\pi \frac{d_m}{\lambda_0} \sin \phi_t} \\
&+ \xi_t e^{j2\pi \frac{d_n}{\lambda_0} \sin \phi_t} e^{j2\pi f_t kT_r} \sum_{m=0,m\prime\neq m}^{M-1} e^{j2\pi \frac{d_{m\prime}}{\lambda_0} \sin \phi_t} e^{j2\pi(\alpha_{m\prime}-\alpha_m)kT_r}
\end{aligned}
\tag{15}
$$

For certain n and m, $X_{nk,m}$ is only related to the slow-time index k. Next, conduct a discrete Fourier transform (DFT) of $X_{nk,m}$ to obtain the Doppler spectrum, which yields:

$$
X_{n,k\prime,m} = \sum_{k=0}^{K-1} X_{nk,m} e^{-j\frac{2\pi k}{K}k\prime}
\tag{16}
$$

where $k\prime$ represents the index in the Doppler domain. Then, apply a low-pass Doppler filter to remove all the other $M-1$ Doppler sub-bands in $X_{nk,m}$, i.e., the second term in (15) [38]. Thus, the low-pass Doppler filter $\mathbf{H}_{LP}$ has a pass-band from $-f_r/2M$ to $f_r/2M$. Conduct an inverse discrete Fourier transform (IDFT) to obtain the temporal output after Doppler filtering, which yields:

$$
\begin{aligned}
X\prime_{n,k,m} &= \sum_{k\prime=0}^{K-1} X_{n,k\prime,m} H_{LP}(k\prime) e^{j\frac{2\pi k\prime}{K}k} \\
&\approx \xi_t e^{j2\pi \frac{d_n}{\lambda_0} \sin \phi_t} e^{j2\pi f_t kT_r} e^{j2\pi \frac{d_m}{\lambda_0} \sin \phi_t}
\end{aligned}
\tag{17}
$$

Note that $X\prime_{n,k,m}$ can be expressed as the Kronecker product of separable vectors:

$$
\mathbf{X}_t(\phi_t, f_t) = \mathbf{a}_r(\phi_t) \otimes \mathbf{b}(f_t) \otimes \mathbf{a}_t(\phi_t)
\tag{18}
$$

where $\mathbf{a}_r(\phi_t)$ and $\mathbf{a}_t(\phi_t)$ are the receiving and transmitting spatial steering vectors, respectively, of the target as the function of ϕ_t and $\mathbf{b}(f_t)$ is the temporal steering vector of the target as the function of f_t:

$$
\mathbf{a}_r(\phi_t) = \left[1, e^{j\frac{2\pi}{\lambda_0}d_1 \sin \phi_t}, \cdots, e^{j\frac{2\pi}{\lambda_0}d_n \sin \phi_t}, \cdots, e^{j\frac{2\pi}{\lambda_0}d_{N-1} \sin \phi_t}\right]^T
\tag{19}
$$

$$
\mathbf{b}(f_t) = \left[1, e^{j2\pi f_t T_r}, \cdots, e^{j2\pi f_t kT_r}, \cdots, e^{j2\pi f_t(K-1)T_r}\right]^T
\tag{20}
$$

$$
\mathbf{a}_t(\phi_t) = \left[1, e^{j\frac{2\pi}{\lambda_0}d_1 \sin \phi_t}, \cdots, e^{j\frac{2\pi}{\lambda_0}d_m \sin \phi_t}, \cdots, e^{j\frac{2\pi}{\lambda_0}d_{M-1} \sin \phi_t}\right]^T
\tag{21}
$$

Because the Doppler bandwidth of $\mathbf{b}(f_t)$ is narrowed to f_r/M after the low-pass filtering rather than the entire f_r, M-times decimation can be conducted on $\mathbf{b}(f_t)$ to reduce the dimension of the vector, which yields:

$$
\mathbf{b}_{deci}(f_t) = \left[1, e^{j2\pi \cdot M \cdot f_t T_r}, \cdots, e^{j2\pi \cdot (K/M-1)M \cdot f_t T_r}\right]^T
\tag{22}
$$

At this point, all the responses of the M transmitting elements are extracted from the N receiving elements. The final data cube of the target signal for one snapshot has the dimensions of $N \times (K/M) \times M$ and can be expressed as:

$$\zeta_s = \mathbf{a}_r(\phi_t) \otimes \mathbf{b}_{deci}(f_t) \otimes \mathbf{a}_t(\phi_t) \tag{23}$$

Furthermore, the radar return from a target-plus-clutter-plus-noise environment for one snapshot can then be expressed as:

$$\zeta_x = \zeta_s + \zeta_c + \zeta_n \tag{24}$$

where ζ_c and ζ_n are the clutter vector and noise vector with dimensions of $N \times (K/M) \times M$, respectively.

2.2. Ground Clutter Modeling under St-MIMO Framework

In this section, the modeling of the clutter matrix ζ_c based on the Doppler distributed clutter (DDC) model [39] is introduced. The fundamental principle of DDC modelling is to first decide the power spectrum $S_{ddc}(f)$ of the ground clutter. Then, the autocorrelation $r_{ddc}(\tau)$ and the covariance matrix $\mathbf{R}_{ddc}$ of the clutter can be derived using an inverse Fourier transform (IFT).

The power spectrum $S_{ddc}(f)$ of the ground clutter caused by the internal motion of the clutter itself is commonly assumed to be Gaussian-shaped [19] with the form of:

$$S_{ddc}(f) = \frac{P_c}{\sqrt{2\pi\sigma_c^2}} \exp\left[-\frac{(f - f_d)^2}{2\sigma_c^2}\right] \tag{25}$$

where P_c is the power of the clutter, σ_c is the standard deviation, and f_d is the central frequency of the clutter spectrum. We chose $f_d = 0$ for the ground clutter, and the 3 dB velocity spectrum width σ_v has the following relationship with the standard deviation σ_c:

$$\sigma_c = \frac{2\sigma_v}{\lambda_0} \tag{26}$$

By conducting an IFT of Equation (25), the autocorrelation function $r_{ddc}(\tau)$ of the clutter can be derived as:

$$r_{ddc}(\tau) = P_c \exp(j2\pi f_d\iota)\exp\left(-2\pi^2\sigma_c^2\tau^2\right) \tag{27}$$

Accordingly, the covariance matrix of the clutter can be represented as:

$$\mathbf{R}_{ddc}(k,l) = P_c^2 e^{j2\pi f_d T_r(k-l)} e^{-2\pi^2\sigma_c^2 T_r^2(k-l)^2}$$
$$k = 1, \cdots, K/M, \, l = 1, \cdots, K/M \tag{28}$$

The temporal sequence of the clutter $\mathbf{x}_c$ can be generated using the following equation:

$$\mathbf{x}_c = \mathbf{R}_{ddc}^{\frac{1}{2}}\mathbf{x}_n \tag{29}$$

where $\mathbf{x}_n$ is the K/M white Gaussian noise vector with a variance of 1.

For the ground clutter at the target's range bin, assume that there are N_C clutter directions of arrival (DOA) in the received data. Then, the clutter snapshot ζ_c can be expressed as:

$$\zeta_c = \sum_{i=1}^{N_C} \mathbf{a}_r(\phi_{ci}) \otimes \mathbf{x}_{ci} \otimes \mathbf{a}_t(\phi_{ci}) \tag{30}$$

where $i = 1, \cdots, N_C$ represents the ith clutter patch with the DOA of ϕ_{ci} and $\mathbf{x}_{ci}$ represents the temporal vector of the ith clutter patch. At this point, ground clutter modeling in st-MIMO is achieved.

3. St-MIMO-ICA Processing

In this section, the proposed st-MIMO-ICA processing method is illustrated. Firstly, the feasibility of using the ICA method to mitigate ground clutter with a st-MIMO system is demonstrated. Then, the proposed st-MIMO-ICA processing method is illustrated based on a complex fixed-point algorithm.

3.1. Signal Modeling under ICA Compliance

The previous sections illustrated the signal and clutter model under the st-MIMO framework. Now, rewrite the sequence of the terms in (23) as:

$$\zeta_s = \mathbf{b}_{deci}(f_t) \otimes \mathbf{a}(\phi_t) \tag{31}$$

where $\mathbf{a}(\phi_t) = \mathbf{a}_r(\phi_t) \otimes \mathbf{a}_t(\phi_t)$ is the $N \times M$ two-way spatial vector of the st-MIMO waveform. From this point, the virtual array of the MIMO radar is established to achieve narrower beamwidth and realize more degrees of freedom along the spatial dimension compared with a conventional phased array radar. Generally, in multiple target situations, Equation (31) can be expressed as:

$$\zeta_s = \sum_{j=1}^{N_T} \mathbf{b}_{deci}(f_{tj}) \otimes \mathbf{a}(\phi_{tj}) \tag{32}$$

where $j = 1, \cdots, N_T$ represents the jth target with an associated Doppler frequency of f_{tj} and target DOAs of ϕ_{tj}.

Regarding ground clutter, the same reasoning can be used and the clutter snapshot ζ_c can be expressed as:

$$\zeta_c = \sum_{i=1}^{N_C} \mathbf{x}_{ci} \otimes \mathbf{a}(\phi_{ci}) \tag{33}$$

Further exploiting the structures of Equations (32) and (33) leads to a universal signal model, including the target and clutter in a matrix form:

$$\mathbf{X} = \mathbf{A}\mathbf{S}_{Dop} \tag{34}$$

In Equation (34), the columns of matrix $\mathbf{A}$ represent the spatial vector of the target together with the clutter:

$$\mathbf{A} = \left[\mathbf{a}_{t1}, \mathbf{a}_{t2}, \cdots \mathbf{a}_{tN_T}, \mathbf{a}_{c1}, \mathbf{a}_{c2}, \cdots \mathbf{a}_{cN_C} \right] \tag{35}$$

and the rows of matrix $\mathbf{S}_{Dop}$ represent the frequency vectors in the Doppler domain of the target and the clutter:

$$\mathbf{S}_{Dop} = \left[\mathbf{d}_{t1}^H, \cdots, \mathbf{d}_{tj}^H, \cdots \mathbf{d}_{tN_T}^H, \mathbf{d}_{c1}^H, \cdots, \mathbf{d}_{ci}^H, \cdots \mathbf{d}_{cN_C}^H \right]^T \tag{36}$$

where $\mathbf{d}_{tj}$ and $\mathbf{d}_{ci}$ are the DFT of the temporal vectors $\mathbf{b}_{deci}(f_{tj})$ and $\mathbf{x}_{ci}$, respectively.

Equation (34) shows that the signal model of the st-MIMO waveform in a ground clutter environment enables the use of the ICA method to extract targets from clutter, that is: (1) multi-channel observations of the mixed signal are obtained, (2) sources are linearly mixed and statistically independent from each other, and (3) sources have non-Gaussian distributions [40]. By exploiting the structure of Equation (34), the received signal $\mathbf{X}$ can be regarded as multi-channel inputs for ICA. Meanwhile, the number of the input channels is enlarged from N to $M \times N$ via st-MIMO demodulation. The spatial matrix $\mathbf{A}$ can be regarded as the mixing matrix that linearly combines all the sources in the frequency matrix $\mathbf{S}_{Dop}$, of which the vectors are non-Gaussian in the Doppler domain. Thus, using the ICA method for ground clutter mitigation with a st-MIMO system is feasible.

3.2. St-MIMO-ICA Processing

ICA can be illustrated in a general framework that consists two major parts, the separation criteria and the optimization method [22]. In the application of array radar signal processing, a number of ICA methods have been proposed based on variable criteria such as maximum likelihood (ML), information-maximization (Infomax), and the maximization of non-Gaussianity (MN). In the branch of MN, kurtosis and negentropy are two commonly used criteria used in some popular ICA methods such as JADE [24] and FastICA [40], respectively.

The famous FastICA method utilizes negentropy as the cost function and the fixed-point algorithm for optimization. Furthermore, the FastICA method has been derived from complex domains. However, the complex FastICA (c-FastICA) method does not perform well with noncircular sources [41]. In this study, we adopted the noncircular FastICA (nc-FastICA) method to separate the target signal from clutter and noise. By adding second-order information in a fixed-point update, the nc-FastICA method can provide an improved separation performance with noncircular sources.

Based on (34), the linear signal mixture model of the MIMO array with $Q = M \times N$ sensors (channels) and $N_s = N_T + N_C$ sources can be expressed as:

$$
\begin{aligned}
\zeta_x &= \zeta_s + \zeta_c + \zeta_n \\
&= \mathbf{A}\mathbf{S}_{Dop} + \zeta_n
\end{aligned}
\tag{37}
$$

where ζ_x can be considered to be the observation matrix, which contains the aforementioned signal matrix $\mathbf{A}\mathbf{S}_{Dop}$ and the Gaussian white noise with zero-mean ζ_n. Then, the nc-FastICA method can be conducted via the following steps.

1. Whitening:
   The covariance matrix of the observed signal can be expressed as:

$$
\mathbf{R}_Y = E\left\{ \zeta_x \zeta_x{}^H \right\}
\tag{38}
$$

   By applying eigenvalue decomposition to the covariance matrix, one can obtain:

$$
\mathbf{R}_Y = \mathbf{U}\mathbf{\Lambda}\mathbf{U}^H
\tag{39}
$$

   Then, the whitening matrix can be denoted as:

$$
\mathbf{V} = \mathbf{\Lambda}^{-\frac{1}{2}}\mathbf{U}^H
\tag{40}
$$

   and the whitened matrix $\mathbf{Z} = \mathbf{V}\zeta_x$ can be achieved.

2. Formulating the optimization problem
   A cost function based on maximizing the negentropy can be express as:

$$
J(\mathbf{w}) = E\left\{ G\left(\left| \mathbf{w}^H \mathbf{Z} \right|^2 \right) \right\}
\tag{41}
$$

   where $G : \mathbb{R} \to \mathbb{R}$ is a smooth even function and $\mathbf{w} \in \mathbb{C}^Q$ is the $Q \times Q$ weight matrix used to de-mix the observed signal and obtain the estimated source matrix $\mathbf{e} = \mathbf{w}^H \mathbf{Z}$. Then, the optimization problem is formulated as:

$$
\mathbf{w}_{opt} = \arg \max_{\|\mathbf{w}\|^2 = 1} E\left\{ G\left(\left| \mathbf{w}^H \mathbf{Z} \right|^2 \right) \right\}
\tag{42}
$$

Here, we chose $G(u) = u^2/2$ as the function motivated by kurtosis.

3. Fixed-point updating process

The fixed-point algorithm is utilized to update the weight matrix $\mathbf{w}$ in each iteration and can be expressed as:

$$
\begin{aligned}
\mathbf{w}_{n+1} &= -E\Big\{g\Big(|\mathbf{e}|^2\Big)\mathbf{e}^*\mathbf{Z}\Big\} + E\Big\{g'\Big(|\mathbf{e}|^2\Big)|\mathbf{e}|^2 + g\Big(|\mathbf{e}|^2\Big)\Big\}\mathbf{w}_n \\
&\quad + E\big\{\mathbf{e}\mathbf{e}^T\big\}E\Big\{g'\Big(|\mathbf{e}|^2\Big)\mathbf{e}^{*2}\Big\}\mathbf{w}_n^*
\end{aligned}
\tag{43}
$$

where $g(u) = dG(u)/du$ and $g'(u) = dg(u)/du$. The third term of Equation (43) includes the second-order information in terms of the pseudo-covariance matrix $E\{\mathbf{e}\mathbf{e}^T\}$, which is non-zero if the sources are noncircular. This modification ensures that the nc-FastICA method has an improved separation performance with noncircular sources.

4. Obtain the weight matrix and estimated source matrix

The estimation of the observation matrix ζ_x can be expressed as:

$$
\mathbf{e} = \mathbf{w}^H\mathbf{Z} = \mathbf{w}^H\mathbf{V}\zeta_x
\tag{44}
$$

where $\mathbf{e} = [e_1, e_2, \ldots, e_Q]$ is the estimated source matrix with Q source vectors and $\mathbf{w} = [w_1, w_2, \ldots, w_Q]^T$ is the weight matrix with Q corresponding weight vectors. It is worth noting that the number of sources separated by the ICA method is no more than the channel number Q. When the number of sources in the mixed signal is less than the number of the channels, those redundant channels will contain noise signals [22]. In this case, the PCA method can be utilized to decrease the dimensions of the data for the following ICA processing. However, this part is beyond the scope of this paper.

4. Experimental Results

For this section, simulation and field experiment were conducted based on the MIMO radar system developed by the Beijing Institute of Technology (BIT) [6,7], which is shown in Figure 2. The ground-based radar operates at the S-band and has a co-located MIMO architecture. The antenna array is organized along the elevation axis with six elements transmitting st-MIMO waveforms. Each antenna has a wide beam of $90°$ in elevation, where the MIMO is formed, and a narrow beam of $3°$ in azimuth. For the azimuth, mechanical scanning is utilized to achieve full airspace coverage. The system is mainly focused on LSS target detection and tracking, such as that of small drones or birds. The st-MIMO-ICA method for ground clutter mitigation is first demonstrated with simulation results using parameters corresponding to a real radar system. Then, the field experiment results are presented.

Figure 2. The ground-based MIMO radar system for LSS target detection.

4.1. Simulation Results

The simulations were conducted in MATLAB. They were employed to verify the performance of the proposed method. To reflect the performance of a real radar system used in the field, consider a co-located MIMO antenna array with $M = N = 6$ linear displaced elements along the elevation axis. Each element transmits a st-MIMO waveform. The simulated radar parameters are shown in Table 1.

Table 1. Radar parameters.

Parameter	Value
Operating frequency	3 GHz
Bandwidth	40 MHz
Pulse repetition interval	56 us
Number of pulses in one CPI	900
Number of Doppler sub-band	6
Maximum detect range	8.4 km
Velocity measurement range	±72 m/s

Furthermore, the antenna array was chosen to be a non-uniform sparse array to make it cost-effective with a larger aperture and fewer elements. The element location was optimized via a genetic algorithm (GA) to suppress the grating lobes of the antenna beampattern caused by the sparse arrangement. A detailed illustration and analysis of this non-uniform and sparse arrangement can be found in [6]. The array parameters are shown in Table 2, and the beam pattern of the array after GA optimization is shown in Figure 3.

Table 2. Array parameters.

Parameter	Value
Number of antenna	6
Number of channels	36
Array aperture	0.5 m
Virtual array aperture	1 m
Element position	[0 0.1821 0.2681 0.3555 0.4206 0.5] m
Peak sidelobe level	-20 dB
3 dB beamwidth	$6.2°$

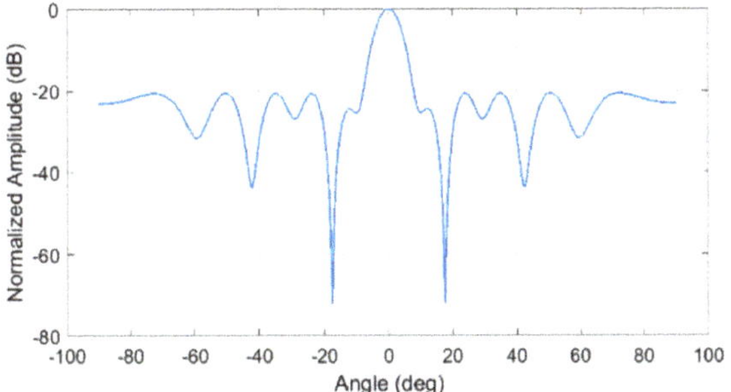

Figure 3. Array beam pattern after GA optimization. The 3 dB beamwidth is 6.2° and the side lobes are all below -20 dB without a gating lobe.

The simulation experiment setup is shown in Figure 4. The radar was ground-based, and the MIMO was achieved in elevation. The ground clutter came from the horizontal direction and Target 1 to Target N_T had the target DOA 1 to DOA N_T. A typical LSS target could be a small drone such as a DJI Phantom series drone. This type of target has a relatively slow speed and low flying altitude, which means the radar detection performance is affected by strong ground clutter. To simplify the simulation without losing generality, assume that there are three targets in the same range bin with different SNRs, speed, and DOAs. Regarding the ground clutter, we used the clutter model in Section 1 with four closely distributed DOAs to simulate a real ground environment containing trees and buildings of different heights. We set proper target SNRs to keep the target SCNRs of all the targets at -20 dB. Detailed parameters of the target and clutter setup are given in Table 3.

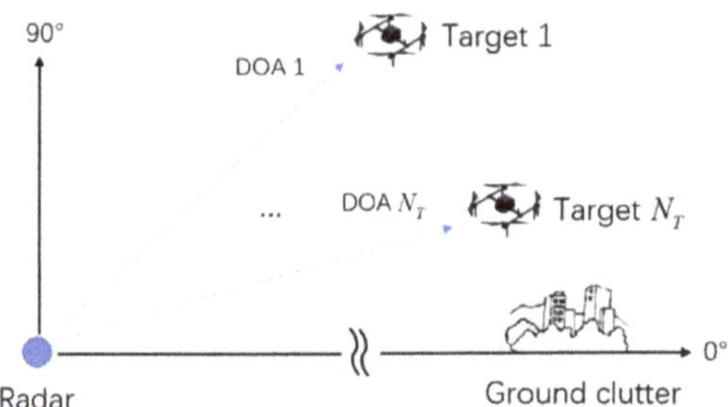

Figure 4. The experimental setup for the simulation.

Table 3. Target and clutter parameters.

Type	Parameter	Value
Target 1	Range	500 m
	Speed	2 m/s
	Elevation	4°
	SNR	40 dB
	SCNR	−20 dB
Target 2	Range	500 m
	Speed	−4 m/s
	Elevation	20°
	SNR	30 dB
	SCNR	−20 dB
Target 3	Range	500 m
	Speed	6 m/s
	Elevation	10°
	SNR	20 dB
	SCNR	−20 dB
Ground clutter	Spectrum center	0 m/s
	3 dB Spectrum width	±2 m/s
	DOAs	[0° 0.1° 0.2° 0.3°]
	CNR	60 dB

The results of signal modeling are shown in Figure 5. The range-velocity map of the received signal in antenna 1 after pulse-compression and PD processing is displayed in Figure 5a, from where the waveform diversity of st-MIMO has formed and the echoes from the six transmitting antenna, TX1 to TX6, have been shifted into different Doppler sub-bands. Moreover, the velocity spectrum at targets' range bin is displayed in Figure 5b. where the black boxes indicate six Doppler sub-bands, DS1 to DS6. Together with the received signals in the other five antennas, all the $Q = M \times N = 36$ transmitting-receiving paths, named as Channel 1 to Channel 36, can be established. The velocity spectrum of Channel 1 after MIMO demodulation is displayed in blue line in Figure 5c, and the red line represent the targets plus noise in order to indicate the targets information clearly. The CNR is 60 dB after pulse-compression and PD processing, and different target SNRs are set to keep the target SCNR as -20 dB.

The st-MIMO-ICA output of the mixed signal is shown in Figure 6. Some representative results among all the 36 output channels are displayed. Excluding the uncertainty of the sequence of the output signals [22], the three targets are separated from the mixed targets plus clutter and noise signal and located in Channel 32, 33, and 35 respectively. Channel 34 contains the combination of the ground clutter. Because the clutter DOAs are closely distributed, the method fails to separate those clutter components. Lastly, the remaining channels are noise signals. Figure 7 shows the detailed information of the three separated targets after st-MIMO-ICA processing. All the three targets are precisely separated with correct velocity and desirable output SCNR of 29.8 dB, 23.4 dB, and 33.2 dB, respectively. We compare the results of st-MIMO-ICA with two other approaches which are also frequently utilized to mitigate the ground clutter, that is PCA and adaptive digital beamforming (ADBF). Among the ADBF approaches, the sample matrix inversion (SMI) technique is chosen. The results of these two approaches are also displayed in Figure 7, and it can be seen that the ICA method enjoys the highest target output SCNR. The exact target SCNR values of the three approaches are listed in Table 4.

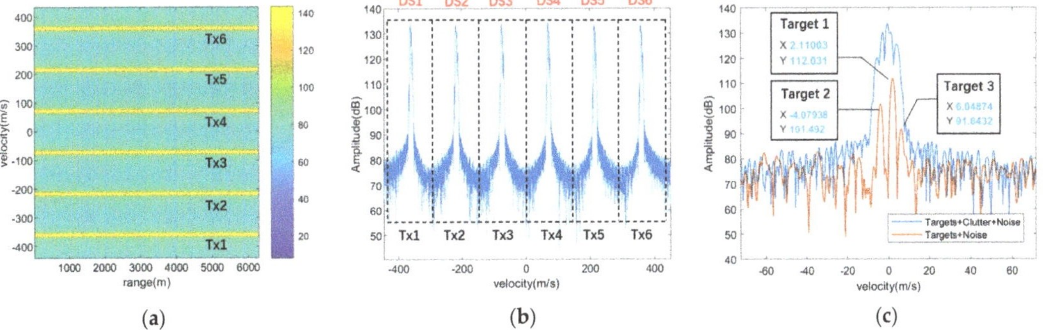

(a)   (b)   (c)

Figure 5. Results of signal modeling: (**a**) Range–velocity map of the received signal in antenna 1 after PD processing, where waveform diversity was formed via st-MIMO; (**b**) velocity spectrum at targets' range bin in antenna 1, where six Doppler sub-bands (DS1–DS6) were established and echoes from the six transmitting antenna (Tx1–Tx6) can be seen; (**c**) velocity spectrum of Channel 1 after MIMO demodulation, where three targets can be seen in the red line of targets plus noise.

Blind beamforming could be achieved using the weight matrix generated by the st-MIMO-ICA method. Figure 8 displays the beampatterns corresponding to the three targets, with w_{32}, w_{33}, and w_{35} as the weight vectors. All the beampatterns could form a deep null in the clutter DOA, which effectively mitigated the strong ground clutter. Meanwhile, when the target DOA was close to the clutter DOA, the blind beamforming also suffered from distortion in the main beam and the loss of the target SCNR, which is shown in Figure 8a. Figure 8 also displays the equivalent beampatterns of PCA that were generated from the eigenvectors corresponding to the three targets and the adaptive beampatterns of SMI with the pre-known steering vectors pointing at the three targets. The st-MIMO-ICA method showed the deepest null level at the clutter's DOA compared with the other two approaches, which illustrates the outplayed performance of the proposed method in ground clutter mitigation. The detailed null levels of the three approaches are also listed in Table 4.

To investigate the performance boundary of the proposed st-MIMO-ICA method, a simulation of the output target SCNR under each target DOA was conducted, which is shown in Figure 9. We still focused on the three aforementioned targets: Target 1, Target 2, and Target 3. We let the DOA of the targets change from $-10°$ to $10°$ while other parameters remained the same. To simplify the simulation, we also let the clutter DOA be $0°$. The number of Monte Carlo simulations was chosen to be 100 for each target DOA. The output target SCNR well-corresponded to the input target SNR plus the coherent processing gain of the MIMO array, which was 15.5 dB ($Q = 36$). However, when the target's DOA was close to the clutter's DOA, the output target SCNR was decreased due to the increase in the cross-correlation of the two components in the spatial domain. The black dashed line in the zoomed-out figure of Figure 9 indicates the sufficient detection threshold of 13 dB, and the corresponding target DOA was $0.4°$, which was 6.5% of the MIMO beamwidth.

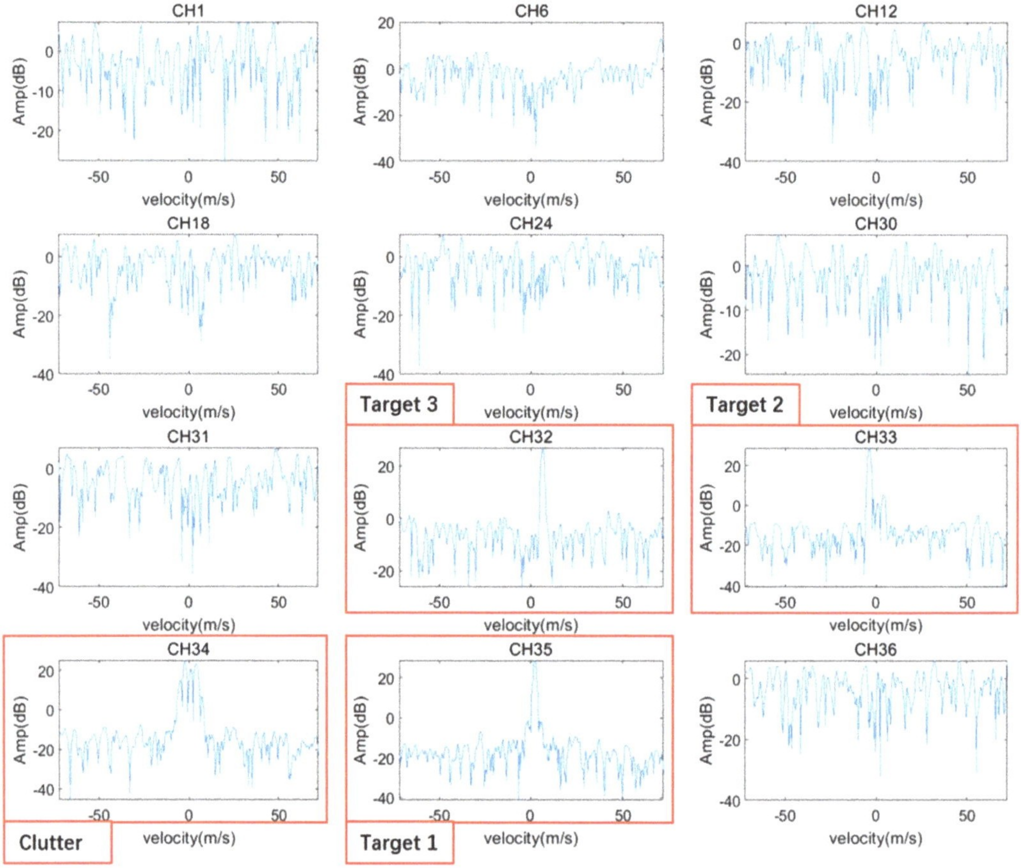

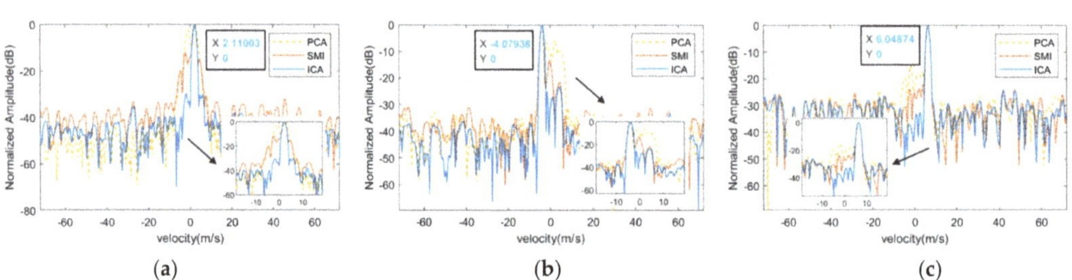

Figure 6. The st−MIMO−ICA output of the mixed signal, where the three targets were separated and were located in Channels 32, 33, and 35; Channel 34 was the ground clutter combination, and the other channels contained noise signals.

Figure 7. The three separated targets after st−MIMO−ICA processing together with the results of the other two approaches: (**a**) Target 1's velocity spectrum; (**b**) Target 2's velocity spectrum; (**c**) Target 3's velocity spectrum.

Table 4. Comparison of simulation results.

Results (dB)		Target 1	Target 2	Target 3
PCA	SCNR	2.78	5.68	15.17
	Null level	−28.10	−34.82	−53.20
SMI	SCNR	11.79	13.24	23.82
	Null level	−34.27	−44.87	−59.90
ICA	SCNR	29.77	23.42	33.20
	Null level	−52.97	−49.97	−75.46

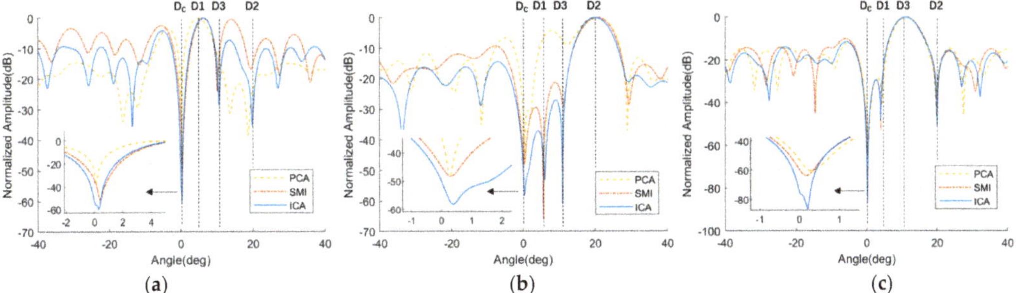

Figure 8. The ICA beampatterns corresponding to the three targets together with the equivalent beampatterns of PCA and the adaptive beampatterns of SMI; the DOAs of the targets and clutter are displayed with black dashed lines, Dc, D1, D2, and D3 represent the DOAs of the clutter, Target 1, Target 2, and Target 3, respectively: (**a**) Target 1′s beampattern; (**b**) Target 2′s beampattern; (**c**) Target 3′s beampattern.

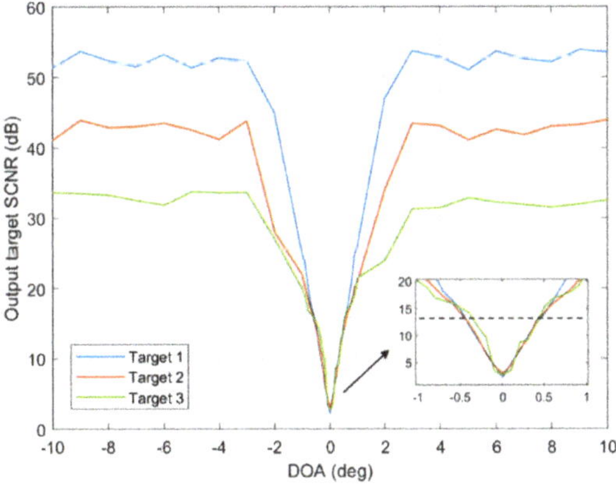

Figure 9. The output target SCNR as a function of DOA with Target 1, Target 2, and Target 3; the black dashed line indicates the normal detection threshold of 13 dB.

4.2. Field Experimental Results

Field experiments were conducted in a complex urban environment to verify the performance of the proposed method. The radar system is deployed in an industrial city in southeast China where factory buildings, trees, moving cars constitute the sources of the ground clutter. The satellite image of the radar location and surroundings courtesy of Google Maps is shown in Figure 10a, and the radar field of view is shown in Figure 10b. The parameters of the radar system were given in Tables 1 and 2. The target information is listed in Table 5. A small drone of DJI Phantom IV is used as a real flying target in this section. The diagonal size of the small drone is 350mm and can be treated as a point target compared to the range resolution of the radar system. There are two range bins chosen as the field experiment location, which are 3041 m and 5585 m. The small drone has a flying altitude of 200 m at both locations, resulting in different target elevations of 3.8° and 2°, respectively.

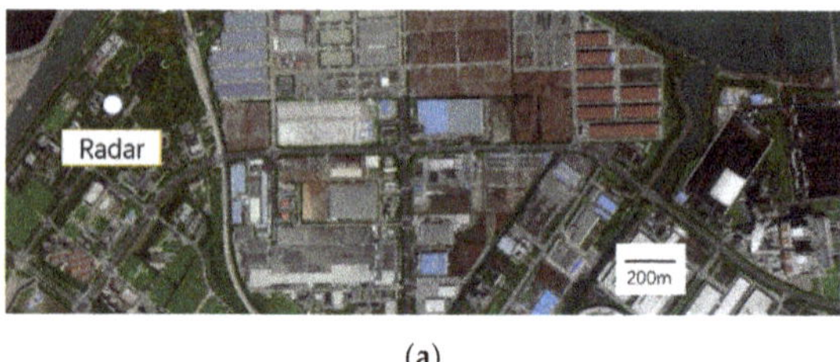

(a)

(b)

Figure 10. Field experiment scene: (**a**) the satellite image of the radar location and surroundings (courtesy of Google Maps); (**b**) the actual view from the radar.

Table 5. Target parameters in field test.

Type	Parameter	Value
Target 4	Range	3041 m
	Speed	2 m/s
	Height	200 m
	Elevation	3.8°
Target 5	Range	5585 m
	Speed	4 m/s
	Height	200 m
	Elevation	2°

In the first experiment of 3041 m, the small drone has the speed of 2 m/s. The range-velocity map in Channel 1 after pulse-compression and PD processing is shown in Figure 11, where the data label indicates the target location, speed, and amplitude. The target is submerged in the spectrum of strong ground clutter and cannot be detected. The result of the st-MIMO-ICA processing of Target 4 is shown in Figure 12. The velocity spectrum at Target 4's range bin before st-MIMO-ICA processing is shown in Figure 12a. The black dashed line indicates the target location in the mixed signal. The target is shadowed by strong ground clutter and cannot be detected via conventional method such as CFAR. The velocity spectrum after st-MIMO-ICA processing is shown in Figure 12b, where the target can be effectively extracted from the strong ground clutter with the highest output target SCNR of 33.9 dB. The other two approaches, PCA and SMI, are also conducted as a comparison with output target SCNR of 14.5 dB and 17.8dB, respectively. The output of the PCA approach also has a false alarm located at −1 m/s.

In the second experiment of 5585 m, the small drone has the speed of 4 m/s and can be seen in the range-velocity map together with some competing ground clutters which belong to the buildings and moving cars. The range-velocity map in Channel 1 after pulse-compression and PD processing is shown in Figure 13, where the data label indicates the target location, speed, and amplitude. The result of the st-MIMO-ICA processing of Target 5 is shown in Figure 14. The velocity spectrum at Target 5's range bin before st-MIMO-ICA processing is shown in Figure 14a. The target had a SCNR of -10dB compared to the ground clutter located at 0 m/s and -6.5 m/s. The velocity spectrum after st-MIMO-ICA processing is shown in Figure 14b, where the target can be effectively extracted from the competing clutters with output target SCNR of 37.8 dB, and the other two remarkable ground clutters have been successfully suppressed. The results of PCA and SMI are also displayed with output target SCNR of 26.6 dB and 15.1dB, respectively. The output of the PCA approach also has a false alarm in the clutter location of −6.5 m/s.

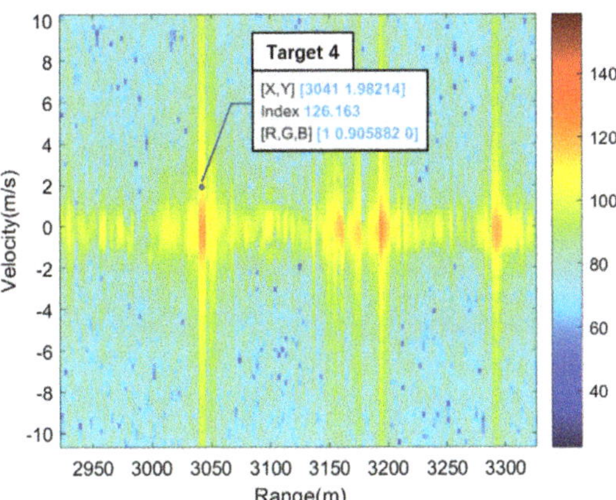

Figure 11. The range−velocity map in Channel 1 for the first field experiment. Target 4′s information is indicated via a data label; the target was submerged in strong ground clutter.

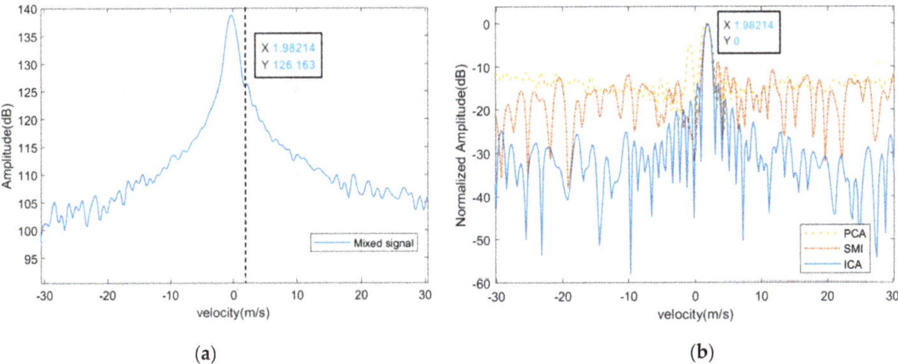

(**a**) (**b**)

Figure 12. The st−MIMO−ICA result of Target 4: (**a**) velocity spectrum at Target 4′s range bin before st−MIMO−ICA processing, where the target was submerged in strong ground clutter; (**b**) velocity spectrum comparison.

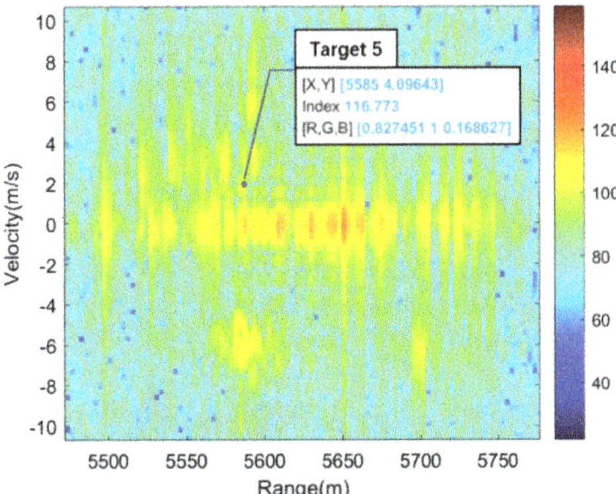

Figure 13. The range−velocity map in Channel 1 for the second field experiment. Target 5′s information is indicated via a data label, and the target can be seen together with some competing ground clutter.

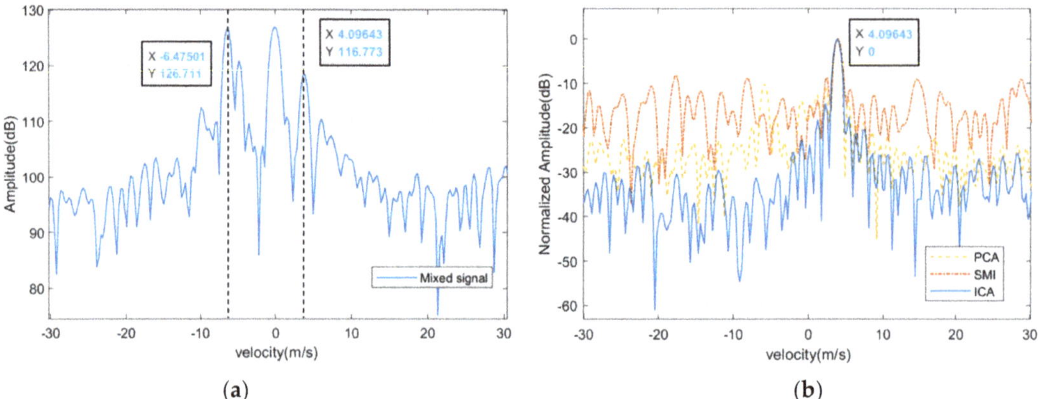

Figure 14. The st−MIMO−ICA result of Target 5: (**a**) velocity spectrum at Target 5′s range bin before st−MIMO−ICA processing, where some competing ground clutter can be seen; (**b**) velocity spectrum comparison.

5. Discussion

A ground clutter mitigation method for slow-time MIMO radar using independent component analysis was researched in this study. Firstly, a ground clutter model under a st-MIMO scheme was provided. The clutter covariance was derived based on a Gaussian-shaped power spectrum, and a universal signal modelling including the target, ground clutter, and noise was provided. Secondly, the compliance for conducting ICA was discussed, and the st-MIMO-ICA processing scheme was proposed. Lastly, the performance of the proposed method was verified with simulation and field experiments via an S-band MIMO radar system.

The simulation results indicated the feasibility of the proposed st-MIMO-ICA method, as the st-MIMO-ICA output showed the highest target SCNR compared with two other conventional clutter mitigation approaches of PCA and SMI. The three test targets with an input SCNR of −20 dB were precisely separated with the correct velocity and desirable output SCNRs of 29.8 dB, 23.4 dB, and 33.2 dB. The st-MIMO-ICA method also showed the deepest null level at the clutter's DOA compared with the other two approaches. The performance boundary of the method as validated with a separable DOA difference of 6.5% of the MIMO beamwidth. The field experimental results further proved the effectiveness of the proposed method in LSS target detection with strong ground clutter compared with conventional methods.

It is worth noting that the PCA approach only utilizes the eigenvectors of the signal covariance matrix after conducting eigenvalue decomposition. As is noted in Section 1, PCA is often utilized as a preliminary step of ICA to decrease the dimensions of the data. Regarding the SMI technique, we chose to use target-plus-clutter-plus-noise signals in this study as the training data to calculate the adaptive weight vector, which limited the performance of the SMI technique. Generally, it is not feasible to obtain independent and identical distributed clutter samples in real scenes. The ICA method, however, utilizes negentropy as the cost function and the fixed-point algorithm for optimization to achieve better source separation results, which corresponds to desirable ground clutter mitigation performance.

There is still work to be conducted. It is worth further investigating the character of ground clutter, especially regarding the number of the sources of clutter. In this paper, we assumed that the DOAs of ground clutter for a certain snapshot were limited even though actual clutter characteristics are more complex. Another work will focus on the adaption of the detection threshold to further improve the detection performance for LSS targets [42,43].

Remote Sens. **2022**, *14*, 6098

Author Contributions: Conceptualization, F.Y. and Q.L.; methodology, F.Y.; software, F.Y.; validation, F.Y. and J.G.; formal analysis, F.Y. and R.Z.; investigation, F.Y. and J.G.; writing—original draft preparation, F.Y.; writing—review and editing, R.Z. and J.L.K.; visualization, F.Y. and J.L.K.; supervision, T.Z.; project administration, R.Z.; funding acquisition, Q.L. and T.Z. All authors have read and agreed to the published version of the manuscript.

Funding: This research was funded the National Key R&D Program of China (Grant No. 2018YFE0202101 and Grant No. 2018YFE0202103), the Natural Science Foundation of Chongqing, China (Grant No. 2020ZX3100039), and the National Natural Science Foundation of China (Grant No. 62201048).

Institutional Review Board Statement: Not applicable.

Informed Consent Statement: Not applicable.

Data Availability Statement: The data that support the findings of this study are available from the corresponding author, R.Z., upon reasonable request.

Acknowledgments: The authors would like to thank the editor and anonymous reviewers for their helpful comments and suggestions.

Conflicts of Interest: The authors declare no conflict of interest.

References

1. Patel, J.; Fioranelli, F.; Anderson, D. Review of radar classification and RCS characterisation techniques for small UAVs or drones. *IET Radar Sonar Navigat.* **2018**, *12*, 911–919. [CrossRef]
2. Zhao, C.; Luo, G.; Wang, Y.; Chen, C.; Wu, Z. UAV Recognition Based on Micro-Doppler Dynamic Attribute-Guided Augmentation Algorithm. *Remote Sens.* **2021**, *13*, 1205. [CrossRef]
3. Raja Abdullah, R.S.A.; Alhaji Musa, S.; Abdul Rashid, N.E.; Sali, A.; Salah, A.A.; Ismail, A. Passive Forward-Scattering Radar Using Digital Video Broadcasting Satellite Signal for Drone Detection. *Remote Sens.* **2020**, *12*, 3075. [CrossRef]
4. Aldowesh, A.; BinKhamis, T.; Alnuaim, T.; Alzogaiby, A. Low power digital array radar for drone detection and micro-Doppler classification. In Proceedings of the 2019 Signal Processing Symposium (SPSympo), Krakow, Poland, 17–19 September 2019; pp. 203–206.
5. Aldowesh, A.; Alnuaim, T.; Alzogaiby, A. Slow-moving micro-UAV detection with a small scale digital array radar. In Proceedings of the 2019 IEEE Radar Conference (RadarConf), Boston, MA, USA, 22–26 April 2019; pp. 1–5.
6. Yang, F.; Qu, K.; Hao, M.; Liu, Q.; Chen, X.; Xu, F. Practical investigation of a MIMO radar system for small drones detection. In Proceedings of the 2019 International Radar Conference (RADAR), Toulon, France, 23–27 September 2019; pp. 1–5.
7. Yang, F.; Xu, F.; Fioranelli, F.; Le Kernec, J.; Chang, S.; Long, T. Practical investigation of a MIMO radar system capabilities for small drones detection. *IET Radar Sonar Navigat.* **2021**, *15*, 760–774. [CrossRef]
8. Hao, M.; Yang, F.; Liu, Q. Slow-time MIMO radar waveform generator with experimental results. In Proceedings of the 2019 IEEE International Conference on Signal, Information and Data Processing (ICSIDP), Chongqing, China, 11–13 December 2019; pp. 1–5.
9. Ritchie, M.; Fioranelli, F.; Griffiths, H.; Torvik, B. Monostatic and bistatic radar measurements of birds and micro-drone. In Proceedings of the 2016 IEEE Radar Conference (RadarConf), Philadelphia, PA, USA, 2–6 May 2016; pp. 1–5.
10. Palamà, R.; Fioranelli, F.; Ritchie, M.; Inggs, M.; Lewis, S.; Griffiths, H. Measurements and discrimination of drones and birds with a multi-frequency multistatic radar system. *IET Radar Sonar Navigat.* **2021**, *15*, 841–852. [CrossRef]
11. de Quevedo, Á.D.; Urzaiz, F.I.; Menoyo, J.G.; López, A.A. Drone detection and radar-cross-section measurements by RAD-DAR. *IET Radar Sonar Navigat.* **2019**, *13*, 1437–1447. [CrossRef]
12. Jahangir, M.; Baker, C. Persistence surveillance of difficult to detect micro-drones with L-band 3-D holographic radar™. In Proceedings of the 2016 CIE International Conference on Radar (RADAR), Guangzhou, China, 10–13 October 2016; pp. 1–5.
13. Jahangir, M.; Baker, C.J.; Oswald, G.A. Doppler characteristics of micro-drones with L-Band multibeam staring radar. In Proceedings of the 2017 IEEE Radar Conference (RadarConf), Seattle, WA, USA, 8–12 May 2017; pp. 1052–1057.
14. Park, J.; Jung, D.-H.; Bae, K.-B.; Park, S.-O. Range-Doppler map improvement in FMCW radar for small moving drone detection using the stationary point concentration technique. *IEEE Trans. Microw. Theory Technol.* **2020**, *68*, 1858–1871. [CrossRef]
15. Sun, H.; Oh, B.-S.; Guo, X.; Lin, Z. Improving the Doppler resolution of ground-based surveillance radar for drone detection. *IEEE Trans. Aerosp. Electron. Syst.* **2019**, *55*, 3667–3673. [CrossRef]
16. Chen, X.; Guan, J.; Chen, W.; Zhang, L.; Yu, X. Sparse long-time coherent integration-based detection method for radar low-observable manoeuvring target. *IET Radar Sonar Navigat.* **2020**, *14*, 538–546. [CrossRef]
17. Sun, Y.; Abeywickrama, S.; Jayasinghe, L.; Yuen, C.; Chen, J.; Zhang, M. Micro-Doppler signature-based detection, classification, and localization of small UAV with long short-term memory neural network. *IEEE Trans. Geosci. Remote Sens.* **2021**, *59*, 6285–6300. [CrossRef]
18. Oh, B.-S.; Lin, Z. Extraction of global and local micro-Doppler signature features from FMCW radar returns for UAV detection. *IEEE Trans. Aerosp. Electron. Syst.* **2021**, *57*, 1351–1360. [CrossRef]

19. Guo, J.; Chang, S.; Yang, F.; Cai, J.; Liu, Q.; Long, T. Low-slow-small target detection using stepped-frequency signals in a strong folded clutter environment. *IET Radar Sonar Navigat.* **2021**, *15*, 1030–1044. [CrossRef]

20. Aircraft Birdstrike Avoidance Radar. Available online: https://detect-inc.com/aircraft-birdstrike-avoidance-radar (accessed on 1 September 2021).

21. Smart, Affordable Drone Detection Radar. Available online: https://www.robinradar.com/iris-counter-drone-radar (accessed on 1 March 2020).

22. Hyvärinen, A.; Karhunen, J.; Oja, E. *Independent Component Analysis*; John Wiley & Sons, Inc.: Hoboken, NJ, USA, 2001; pp. 151–154.

23. Cardoso, J.-F. Blind signal separation: Statistical principles. *Proc. IEEE Proc.* **1998**, *86*, 2009–2025. [CrossRef]

24. Cardoso, J.-F. Higher order contrast for independent component analysis. *Neural Comput.* **1999**, *11*, 157–193. [CrossRef]

25. Cui, G.; Yu, X.; Carotenuto, V.; Kong, L. Space-Time Transmit Code and Receive Filter Design for Colocated MIMO Radar. *IEEE Signal Process.* **2017**, *65*, 1116–1129. [CrossRef]

26. Yu, X.; Cui, G.; Yang, J.; Kong, L. MIMO Radar Transmit–Receive Design for Moving Target Detection in Signal-Dependent Clutter. *IEEE Trans. Veh. Technol.* **2020**, *69*, 522–536. [CrossRef]

27. Cheng, Z.; He, Z.; Liao, B.; Fang, M. MIMO Radar Waveform Design With PAPR and Similarity Constraints. *IEEE Signal Process.* **2018**, *66*, 968–981. [CrossRef]

28. Mecca, V.F.; Ramakrishnan, D.; Krolik, J.L. MIMO radar space-time adaptive processing for multipath clutter mitigation. In Proceedings of the Fourth IEEE Workshop on Sensor Array and Multichannel Processing, Waltham, MA, USA, 12–14 July 2006; pp. 249–253.

29. Bliss, D.W.; Forsythe, K.W.; Davis, S.K.; Fawcett, G.S.; Rabideau, D.J.; Horowitz, L.L.; Kraut, S. GMTI MIMO radar. In Proceedings of the 2009 International Waveform Diversity and Design Conference, Kissimmee, FL, USA, 8–13 February 2009; pp. 118–122.

30. Li, J.; Stoica, P. *MIMO Radar Signal Processing*; John Wiley & Sons, Inc.: Hoboken, NJ, USA, 2009; pp. 293–297.

31. Mecca, V.F.; Krolik, J.L. Slow-time MIMO STAP with improved power efficiency. In Proceedings of the 2007 Conference Record of the Forty-First Asilomar Conference on Signals, Systems and Computers, Pacific Grove, CA, USA, 4–7 November 2007; pp. 202–206.

32. Mecca, V.F.; Krolik, J.L.; Robey, F.C. Beamspace slow-time MIMO radar for multipath clutter mitigation. In Proceedings of the 2008 IEEE International Conference on Acoustics, Speech and Signal Processing, Las Vegas, NV, USA, 31 March–4 April 2008; pp. 2313–2316.

33. Zhu, L.; Zhang, S.; Ma, Q.; Zhao, H.; Chen, S.; Wei, D. Classification of UAV-to-ground targets based on enhanced micro-Doppler features extracted via PCA and compressed sensing. *IEEE Sens. J.* **2020**, *20*, 14360–14368. [CrossRef]

34. Dou, D.; Li, M.; He, Z. Multi-mode clutter suppression of multiple-input–multiple-output over-the-horizon radar based on blind source separation. *IET Radar Sonar Navigat.* **2015**, *9*, 956–966. [CrossRef]

35. Yu, W.; Chen, J.; Bao, Z. Multi-mode propagation mode localisation and spread-Doppler clutter suppression method for multiple-input multiple-output over-the-horizon radar. *IET Radar Sonar Navigat.* **2019**, *13*, 1214–1224. [CrossRef]

36. Ge, M.; Cui, G.; Yu, X.; Kong, L. Main lobe jamming suppression via blind source separation sparse signal recovery with subarray configuration. *IET Radar Sonar Navig.* **2020**, *14*, 431–438. [CrossRef]

37. Robey, F.C.; Coutts, S.; Weikle, D.; McHarg, J.C.; Cuomo, K. MIMO radar theory and experimental results. In Proceedings of the Thirty-Eighth Asilomar Conference on Signals, Systems and Computers, Pacific Grove, CA, USA, 7–10 November 2004; pp. 300–304.

38. Wen, C.; Huang, Y.; Peng, J.; Wu, J.; Zheng, G.; Zhang, Y. Slow-time FDA-MIMO technique with application to STAP radar. *IEEE Trans. Aerosp. Electron. Syst.* **2022**, *58*, 74–95. [CrossRef]

39. Xu, J.; Peng, Y.; Wan, Q.; Wang, X.; Xia, X. Doppler distributed clutter model of airborne radar and its parameters estimation. *Sci. China Inf. Sci.* **2004**, *47*, 577–586. [CrossRef]

40. Hyvärinen, A.; Oja, E. Independent component analysis: Algorithms and applications. *Neural Netw.* **2000**, *13*, 411–430. [CrossRef] [PubMed]

41. Novey, M.; Adali, T. On extending the complex FastICA algorithm to noncircular sources. *IEEE Trans. Signal Process.* **2008**, *56*, 2148–2154. [CrossRef]

42. Zhang, H.; Liu, W.; Shi, J.; Fei, J.; Zong, B. Joint detection threshold optimization and illumination time allocation strategy for cognitive tracking in a networked radar system. *IEEE Trans. Signal Process.* 2008; *early access*.

43. Yan, J.; Pu, W.; Zhou, S.; Liu, H.; Bao, Z. Collaborative detection and power allocation framework for target tracking in multiple radar system. *Inf. Fusion* **2020**, *55*, 173–183. [CrossRef]

Article

A Robust Sparse Imaging Algorithm Using Joint MIMO Array Manifold and Array Channel Outliers

Jieru Ding [1], Zhiyi Wang [1,2], Xinghui Wu [1] and Min Wang [1,*]

[1] National Laboratory of Radar Signal Processing, Xidian University, Xi'an 710071, China
[2] Center for Information and Educational Technology, Xi'an University of Finance and Economics, Xi'an 710064, China
* Correspondence: wangmin@xidian.edu.cn

Abstract: The multiple-input multiple-output (MIMO) radar imaging technology has attracted many scholars due to its many inherent advantages, such as avoiding complex motion compensation and imaging a quickly maneuvering target, compared to inverse synthetic aperture radar (ISAR) imaging. Although some imaging algorithms, such as the 2D fast iterative shrinkage thresholding algorithm (2D-FISTA), can meet the demand for super-resolution, they are not directly suited to MIMO radar imaging, for which the MIMO manifold needs to be considered. In this paper, based on the above questions, we propose the MIMO radar imaging algorithm, utilizing the sparsity of the scattering map in space and the MIMO array manifold, even achieving a good performance in the presence of MIMO channel error. The sparse reconstruction algorithm is developed with the alternative direction method of multipliers (ADMM) with the help of 2D-FISTA and the l_p-norm. Then, two algorithms are derived: one is the exact sparse recovery algorithm, and the other is the inexact sparse recovery algorithm. Although the exact sparse recovery algorithm can converge to a more accurate precision than the inexact algorithm, the latter can converge at a faster speed. Finally, the results on simulation data validated the effectiveness of the algorithm.

Keywords: alternative direction method of multipliers (ADMM); multiple input multiple-output (MIMO) radar; l_p-norm; low-rank matrix completion; Schatten p-norm

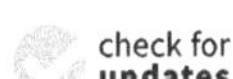

check for updates

Citation: Ding, J.; Wang, Z.; Wu, X.; Wang, M. A Robust Sparse Imaging Algorithm Using Joint MIMO Array Manifold and Array Channel Outliers. *Remote Sens.* **2022**, *14*, 4120. https://doi.org/10.3390/rs14164120

Academic Editors: Guolong Cui, Bin Liao, Yong Yang and Xianxiang Yu

Received: 18 July 2022
Accepted: 18 August 2022
Published: 22 August 2022

Publisher's Note: MDPI stays neutral with regard to jurisdictional claims in published maps and institutional affiliations.

1. Introduction

The inverse synthetic aperture radar (ISAR) technology is an important method to estimate the distribution of moving targets in space, which utilizes the signal bandwidth and the coherent accumulation time to improve the range resolution and the cross-range resolution [1–3]. Interferometric ISAR (InISAR) can achieve three-dimensional (3D) imaging now, but complex image registration, motion compensation, and other algorithms need to be considered seriously in practice [4]. Compared to the imaging technology based on relative motion, real aperture radar imaging can achieve fast imaging, avoid complex motion compensation, and image a quickly maneuvering target, but real radar imaging achieves a high resolution by increasing the array aperture, which increases the hardware complexity of the system [5].

Multiple-input multiple-output (MIMO) radar imaging is one of the real aperture imaging methods. MIMO radar can achieve a higher degree of freedom by transmitting several orthogonal waveforms, these being the time-division multiplexing (TDM) signal, frequency-division multiplexing (FDM) signal, and code-division multiplexing (CDM) signal [6–9]. It can also achieve a larger array aperture compared to the traditional phased array radar (PAR). Especially, MIMO radar imaging can directly be used to image with one snapshot to avoid the complex motion compensation compared to ISAR technology, and it has attracted many scholars because of its inherent advantages with respect to ISAR imaging technology. In this paper, we researched collocated MIMO radar imaging, improving the performance by the virtual array aperture technology according to the equivalent phase center principle (PCA). Two-dimensional (2D) imaging with MIMO radar has been studied by many researchers [10–12]. Recently, 3D MIMO imaging methods have drawn the attention of many scholars [13–15]. One of the core questions is how to achieve a better resolution along with the finite array elements in MIMO radar. As the number of array elements increases, on the one hand, the resolution along with the cross-range dimension can be improved, but on the other hand, the hardware cost and the calculation amount of signal processing will simultaneously increase rapidly. The authors especially point out that compressive sensing (CS) is utilized to achieve the super-resolution of both cross-range dimensions with a finite number of array elements [15,16].

The 2D fast Fourier transform (2D-FFT) is commonly used to finish focusing along both cross-range dimensions, but the imaging resolution is poor, especially under the limited virtual aperture in MIMO radar. CS can recover a sparse signal from far fewer observed samples, which has drawn many researchers' attention during the last decade [17]. Many sparse recovery algorithms have been derived, such as greedy iterative algorithms [18], sparse Bayesian learning (SBL) [19], convex optimization algorithms, iterative thresholding algorithms [20,21], etc. The threshold iteration algorithm is widely used in the field of signal processing because of its fast convergence and sufficient theoretical guarantee. The radar super-resolution imaging algorithm based on CS has been widely studied in synthetic aperture radar (SAR) imaging [22], ISAR imaging [23,24], and MIMO radar imaging [12,25,26] due to the sparsity of the imaging scene. However, it is essential for these imaging algorithms to make the observed matrix and grid matrix vectorized [27], which leads to the consequence that the dimension of the measurement matrix of CS is tremendously large and the computational complexity increases sharply. The 2D fast iterative shrinkage thresholding algorithm (2D-FISTA) and sequential smoothed L0 (SL0) can quickly converge, and they can achieve super-resolution withoutthe computational complexity of converting the 2D observed matrix and grid map into vectors [28–30]. We utilized some prior conditions, the sparsity of scattering points in space and the MIMO array manifold, to establish a new sparse imaging method in this paper. Besides, there is limited research on MIMO radar imaging in the presence of outliers, and in practice, this can be caused by radio interference, miscalibrated sensors, and other aspects of the MIMO radar system [31–34].

In this paper, we research a MIMO radar super-resolution imaging method, considering the array manifold and the outliers. We firstly reformulated the question as minimizing the l_1-norm and l_p-norm and subjecting themto the signal model. On the one hand, there is a guarantee that the targets' location in space can be obtained by the l_1-norm; on the other hand, the l_p-norm is robust to outliers with $0 < p < 1$. However, the l_p-norm is a non-convex question, and to solve this, a complex generalized iterated shrinkage algorithm (CGISA) is developed to resist outliers in the snapshot matrix. To the best of our knowledge, we are the first to consider sparse imaging exploiting the array manifold and outlier noise in transreceivers in MIMO imaging. Comparisons with state-of-the-art algorithms show that our methods are superior in terms of robustness and resolution, and the proposed algorithm is validated on a public dataset.

This paper is organized as follows. Section 2 gives a brief introduction to the MIMO imaging model, and a detailed description of the proposed sparse recovery algorithm is given in Section 3. Section 4 gives the simulation experiments and the imaging result of the ISAR simulated data. Finally, Section 5 presents the conclusion and introduces the future work.

In this paper, we use the following notation. We use $\|\cdot\|_F$, $\|\cdot\|_1$ and $\|\cdot\|_p$ to denote the Frobenius norm, the l_1-norm, and the l_p-norm of a matrix, respectively. The notations $(\cdot)^T$, $(\cdot)^H$, and $(\cdot)^{-1}$ represent the transpose, the Hermitian transpose, and the inverse operation, respectively. The symbols ∇ and $tr(\cdot)$ stand for the gradient and trace of a matrix, respectively. Boldface lower-case and upper-case letters represent vectors and matrices, respectively. $\Re(\cdot)$ and $\Im(\cdot)$ represent the real and imaginary parts of a complex number vector or matrix, respectively. $\max(a, b)$ indicates the maximum between a and b. Finally, $\mathbb{R}$ and $\mathbb{C}$ are used to denote the set of real and complex numbers, respectively.

2. Methods

2.1. MIMO Imaging Model

Consider the MIMO radar with a 2D planar antenna array; an imaging model of it is given in Figure 1, where there are M_t^2 transmitted elements and N_r^2 received elements. The interval spaces between adjacent transmitters and receivers are $N_r d$ and d, respectively. The central frequency is f_c, and the bandwidth is B. Suppose that M^2 orthogonal mutually phase code modulation signals with the same center frequency and bandwidth are transmitted, and the m-th transmitted signal can be expressed by

$$S_m(t) = A\varphi_m(t)\exp(j2\pi f_c t) \tag{1}$$

where $\varphi_m(t)$ denotes the phase code function, A is the amplitude of the transmit signal, and $m \in \{1, 2, \cdots, M^2\}$. Assume that there are K scatterers in the imaging scene and $\mathbf{O}$ is the imaging center. The distance between the m-th transmitted element and the k-th scattering points is R_m^k; the distance between the n-th received element and the k-th scattering points is R_n^k; then, the radar echo can be shown as

$$S_n(t) = \sum_{k=1}^{K} \sum_{m=1}^{M^2} \sigma_k \varphi_m\left(t - \tau_{mn}^k\right) \exp\left(-j2\pi f_c \tau_{mn}^k\right) \tag{2}$$

$$(n \in \{1, 2, \cdots, N^2\})$$

where σ_k is the backscattering coefficient and time delay $\tau_{mn}^k = \left(R_m^k + R_n^k\right)/c$.

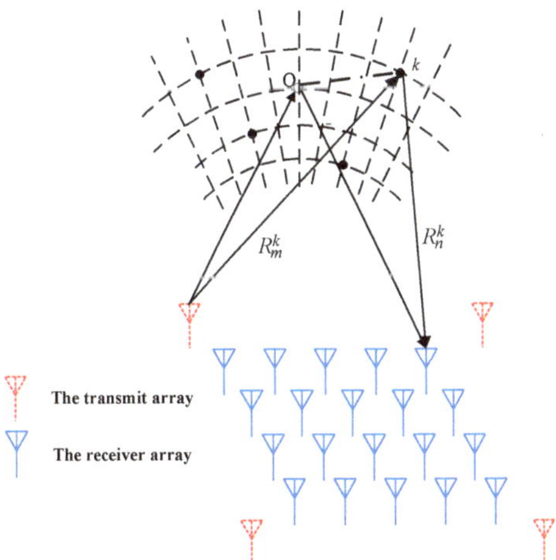

Figure 1. The MIMO radar 3D imaging model.

The echo signal can be separated into M^2 by a group of matched filter banks for signal $S_n(t)$, and then, the signal that the m-th transmitted element transmits and the n-th received element can be denoted as

$$S_{mn}(t) = \sum_{k=1}^{K} \sigma_k p_m\left(t - \left(R_m^k + R_n^k\right)/c\right)$$
$$\times \exp\left(-j2\pi\left(R_m^k + R_n^k\right)/\lambda\right) \tag{3}$$

where $p(\cdot)$ is the autocorrelation function of the m-th transmitted signal, c is the propagation velocity of the electromagnetic wave in space, and $\lambda = c/f_c$ is the wavelength of the electromagnetic wave in space.

Suppose that target scattering points in the far field are considered. The approximation conditions meet that $R_m^k + R_n^k - R_m^o - R_n^o = 2\vec{OK}^T \mathbf{n}_0$, where R_m^o and R_n^o are the distance between the m-th transmitted element and imaging center $\mathbf{O}$ and the distance between the n-th received element and imaging center $\mathbf{O}$, respectively. $\mathbf{n}_0$ is a unit vector between the imaging geometry center and the array center, and it can be understood as the coordinate originof the virtual array. Then, the radar echo can be updated as

$$S_n(t) \approx \sum_{k=1}^{K} \sum_{m=1}^{M^2} \sigma_k \varphi_m\left(t - 2\mathbf{q}^T\mathbf{n}_0/c\right)$$
$$\times \exp\left(-j2\pi\left(R_m^k + R_n^k\right)/\lambda\right) \tag{4}$$

Based on the description in the paper [13,15], the radar echo can be rewritten as

$$y_{mn}(t) = \sum_{k=1}^{K} \sigma_k p\left(t - \frac{2\mathbf{q}^T\mathbf{n}_0}{c}\right) \exp\left(\frac{j2\pi\Delta R}{\lambda}\right)$$
$$\times \exp\left(\frac{j4\pi d}{\lambda r}(ax_k + by_k)\right) \tag{5}$$

where $\Delta R = T_0O + R_0O - T_0Q - R_0Q$ is a constant and $\mathbf{q} = \vec{OK}$. r is the distance from the imaging center O to the reference virtual element. Assume that the signal of the k-th scattering point location (x_k, y_k, z_k) in space can be shown as

$$y_{mn}(t) = \sum_{k=1}^{K} \delta(t - z_k) \exp\left(\frac{j4\pi dax_k}{\lambda r}\right) \exp\left(\frac{j4\pi dby_k}{\lambda r}\right) \tag{6}$$

where $\delta(t - z_k) = \sigma_k p\left(t - \frac{2z_k}{c}\right)$, $a, b \in \{1, 2, \cdots, MN\}$ is the row and column of the virtual array, and r is the distance from the imaging center $\mathbf{O}$ to the reference virtual element.

The 2D-FFT is employed along the cross-range of the MIMO virtual array, and the MIMO imaging model about two cross-range dimensions can be rewritten as

$$\mathbf{Y} = \mathbf{A}\boldsymbol{\Sigma}\mathbf{B}^T \tag{7}$$

where $\mathbf{A}$ and $\mathbf{B}$ are the overcomplete Fourier matrix, which is related to the MIMO virtual array manifold, $\boldsymbol{\Sigma} \in \mathbb{C}^{P \times Q}$ is the 2D scattering coefficients map in space, and $\mathbf{Y}$ is the snapshot matrix corresponding to the virtual array. $\mathbf{A} = [\mathbf{a}_1, \mathbf{a}_2, \cdots, \mathbf{a}_P] \in \mathbb{C}^{M_t N_r \times P}$, and $\mathbf{B} = [\mathbf{b}_1, \mathbf{b}_2, \cdots, \mathbf{b}_Q] \in \mathbb{C}^{M_t N_r \times Q}$, which are given by

$$\mathbf{a}_p = \left[0, \exp\left(j\frac{4\pi dx_p}{\lambda r}\right), \cdots, \exp\left(j\frac{4\pi(M_t N_r - 1)dx_p}{\lambda r}\right)\right]^T$$
$$\mathbf{b}_q = \left[0, \exp\left(j\frac{4\pi dy_q}{\lambda r}\right), \cdots, \exp\left(j\frac{4\pi(M_t N_r - 1)dy_q}{\lambda r}\right)\right]^T . \tag{8}$$

2.2. The Composite Optimization

Many signal processing approaches can be established as a composite optimization model, which is given by

$$\min_{x} f(x) := g(x) + h(x) \tag{9}$$

where $g(x)$ is a convex, continuously differentiable function and $h(x)$ is a convex, continuous, but not differentiable penalty function [20,35]. The FISTA algorithm is one of the important methods to solve optimization, and the most notable feature of this algorithm is the fast convergence speed. The FISAT algorithm is also used in low-rank matrix completion (MC), sparse recovery, and other signal processing approaches [36].

2.3. The Proposed Imaging Method

In this subsection, we clarify the signal model in the presence of outliers and provide the detailed pseudo-code of the two proposed algorithms.

Based on the above description, we exploited some prior conditions, such as the sparsity of scattering points in space, to establish the 2D sparse recovery model, in order to reduce the computational complexity and improve the effectiveness of sparse recovery. We used the fast composite optimization algorithm and threshold iteration algorithm to obtain the sparse location of scattering points. The signal model can be seen as

$$\min_{\boldsymbol{\Sigma}} \|\boldsymbol{\Sigma}\|_0$$
$$\text{s.t.} \ \mathbf{Y} = \mathbf{A}\boldsymbol{\Sigma}\mathbf{B}^T \tag{10}$$

(10) is a non-convex optimization problem and an NP-hard problem. In [29], the authors proposed that the above (10) can be relaxed to a convex optimization problem by the l_1-norm. When the impulsive signal is considered in MIMO virtual channels, the observed snapshot matrix contains some outliers. In [37], the authors pointed out that the l_p-norm is robust to outliers, and thereby, the l_p regularization terms can be added in (10) to minimize the outliers' error. (10) can be updated as

$$\min_{\boldsymbol{\Sigma}} \lambda\|\boldsymbol{\Sigma}\|_1 + \|\mathbf{E}\|_p^p$$
$$\text{s.t.} \ \mathbf{E} = \mathbf{Y} - \mathbf{A}\boldsymbol{\Sigma}\mathbf{B}^T \tag{11}$$

where λ is the regularization parameter and $p \in (0, 1)$. The entries of $\mathbf{E}$ are expressed as

$$\mathbf{E}_{i,j} = \begin{cases} \rho_{i,j}e^{j\phi_{i,j}}, & (i,j) \in \Omega \\ 0, & (i,j) \notin \Omega \end{cases} \tag{12}$$

where Ω denotes the entries of the outliers in the snapshot matrix and ρ and ϕ represent the amplitude and phase of the outliers, respectively. $\|\mathbf{E}\|_p^p$ can be defined as

$$\|\mathbf{E}\|_p^p = \left(\sum_{i,j} |E_{i,j}|^p\right)^{1/p} \tag{13}$$

The optimization problem in (11) can be solved with the help of the alternative direction method of multipliers (ADMM), and the augmented Lagrangian function associated with Problem (11) is given by

$$\mathcal{L}(\boldsymbol{\Sigma}, \mathbf{E}, \mathbf{R}, \mu) = \lambda\|\boldsymbol{\Sigma}\|_1 + \|\mathbf{E}\|_p^p + \frac{\mu}{2}\|\mathbf{Y} - \mathbf{A}\boldsymbol{\Sigma}\mathbf{B}^T - \mathbf{E}\|_F^2$$
$$+ \left\langle \mathbf{R}, \mathbf{Y} - \mathbf{A}\boldsymbol{\Sigma}\mathbf{B}^T - \mathbf{E} \right\rangle \tag{14}$$

where $\mathbf{R}$ is the Lagrange constant and μ is a penalty coefficient. Then, the ADMM algorithm is employed to estimate the optimal variable $\boldsymbol{\Sigma}$, $\mathbf{E}$, $\mathbf{R}$, and μ alternately, until the convergence criterion is satisfied. Every sub-optimization problem can be formulated as

$$
\begin{cases}
\boldsymbol{\Sigma}^{k+1} = \arg\min_{\boldsymbol{\Sigma}}\left\{\mathcal{L}\left(\boldsymbol{\Sigma},\mathbf{E}^k,\mathbf{R}^k,\mu^k\right)\right\} \\
\mathbf{E}^{k+1} = \arg\min_{\mathbf{E}}\left\{\mathcal{L}\left(\boldsymbol{\Sigma}^{k+1},\mathbf{E},\mathbf{R}^k,\mu^k\right)\right\} \\
\mathbf{R}^{k+1} = \arg\min_{\mathbf{R}}\left\{\mathcal{L}\left(\boldsymbol{\Sigma}^{k+1},\mathbf{E}^{k+1},\mathbf{R},\mu^k\right)\right\} \\
\mu^{k+1} = \rho\mu^k
\end{cases}
\tag{15}
$$

where ρ is a constant to ensure the penalty coefficient μ is increasing gradually. In the next algorithmic step, we give a detailed introduction to updating every optimal variable according to the ADMM algorithm.

Updating $\boldsymbol{\Sigma}$

The updating of $\boldsymbol{\Sigma}$ can be written as

$$
\begin{aligned}
\boldsymbol{\Sigma}^{k+1} = \arg\min_{\boldsymbol{\Sigma}}\lambda\|\boldsymbol{\Sigma}\|_1 &+ \frac{\mu}{2}\|\mathbf{Y} - \mathbf{A}\boldsymbol{\Sigma}\mathbf{B}^T - \mathbf{E}^k\|_F^2 \\
&+ \left\langle\mathbf{R}^k,\mathbf{Y} - \mathbf{A}\boldsymbol{\Sigma}\mathbf{B}^T - \mathbf{E}^k\right\rangle
\end{aligned}
\tag{16}
$$

With some constant items omitted, the optimal variable $\boldsymbol{\Sigma}^{k+1}$ can be updated as

$$
\boldsymbol{\Sigma}^{k+1} = \arg\min_{\boldsymbol{\Sigma}}\lambda\|\boldsymbol{\Sigma}\|_1 + \frac{\mu}{2}\|\mathbf{Y} - \mathbf{A}\boldsymbol{\Sigma}\mathbf{B}^T - \mathbf{E}^k + \frac{1}{\mu}\mathbf{R}^k\|_F^2
\tag{17}
$$

Let $\mathbf{D}^k = \mathbf{Y} - \mathbf{E}^k + \frac{1}{\mu_1}\mathbf{R}^k$, and then, (17) can be shown in a more concise form (18).

$$
\begin{aligned}
\boldsymbol{\Sigma}^{k+1} &= \arg\min_{\boldsymbol{\Sigma}}\lambda\|\boldsymbol{\Sigma}\|_1 + \frac{\mu}{2}\|\mathbf{D}^k - \mathbf{A}\boldsymbol{\Sigma}\mathbf{B}^T\|_F^2 \\
&= \arg\min_{\boldsymbol{\Sigma}}\lambda\|\boldsymbol{\Sigma}\|_1 + \frac{\mu}{2}tr\left\{\left(\mathbf{A}\boldsymbol{\Sigma}\mathbf{B}^T - \mathbf{D}^k\right)^H\left(\mathbf{A}\boldsymbol{\Sigma}\mathbf{B}^T - \mathbf{D}^k\right)\right\}
\end{aligned}
\tag{18}
$$

The classical problem of the nonsmooth convex optimization model can be solved by FISTA [20]. According to the description of the composite optimization, let nonsmooth function $h(\boldsymbol{\Sigma}) = \lambda\|\boldsymbol{\Sigma}\|_1$ and smooth function $g(\boldsymbol{\Sigma}) = \frac{\mu}{2}\|\mathbf{D} - \mathbf{A}\boldsymbol{\Sigma}\mathbf{B}^T\|_F^2$. The gradient of $g(\boldsymbol{\Sigma})$ can be written as (19).

$$
\begin{aligned}
\nabla g(\boldsymbol{\Sigma}) &= \nabla_{\boldsymbol{\Sigma}}\frac{\mu_1}{2}\left\{tr\left(\mathbf{A}\boldsymbol{\Sigma}\mathbf{B}^T\mathbf{B}^*\boldsymbol{\Sigma}^H\mathbf{A}^H\right) - 2tr\left(\mathbf{A}\boldsymbol{\Sigma}\mathbf{B}^H((\mathbf{D}^k)^H)\right)\right\} \\
&= \frac{\mu}{2}\left(2\mathbf{A}^H\mathbf{A}\boldsymbol{\Sigma}\mathbf{B}^T\mathbf{B}^* - 2\mathbf{A}\mathbf{D}^k\mathbf{B}^*\right) \\
&= \mu\mathbf{A}^H\left(\mathbf{A}\boldsymbol{\Sigma}\mathbf{B}^T - \mathbf{D}^k\right)\mathbf{B}^*
\end{aligned}
\tag{19}
$$

To obtain the accurate target location, the temporary variable $\mathbf{X}$ is brought in, and the relationship among $\boldsymbol{\Sigma}$, $\mathbf{X}$, and $\nabla f(\mathbf{X})$ can be shown as

$$
\boldsymbol{\Sigma} = \arg\min_{\boldsymbol{\Sigma}}\left\{h(\boldsymbol{\Sigma}) + \frac{L}{2}\left\|\boldsymbol{\Sigma} - \left(\mathbf{X} - \frac{1}{L}\nabla g(\mathbf{X})\right)\right\|_F^2\right\}
\tag{20}
$$

where L is the Lipschitz constant, enabling $\nabla f(X)$ to meet the Lipschitz continuity. The pseudo-code of this algorithm to obtain the sparse solution at the j-th iteration is shown in Algorithm 1, The algorithm can converge quickly with $\mathcal{O}(\frac{1}{j^2})$, and this was demonstrated in [20]. The soft in Algorithm 1 is the soft thresholding function, which is defined as

$$\text{soft}(\mathbf{X}, \eta) = \max\left(1 - \frac{\eta}{|\mathbf{X}|}, 0\right)\mathbf{X}.\tag{21}$$

Algorithm 1 Two-dimensional sparse solution of imaging results.

Input: Convergence accuracy ϵ; the maximum number of iterations K;
$\mathbf{D}^k = \mathbf{Y}^k - \mathbf{E}^k + \frac{1}{\mu^k}\mathbf{R}^k$

Output: $\mathbf{\Sigma}^{k+1}$.

1: **Initialization:**
2: Initiate algorithm parameters $\mathbf{X}_0 = \mathbf{0}$; $t_0 = 0$;
3: **while** (1) **do**
4:    $\mathbf{F}_{j+1} = \mathbf{X}_j - \frac{1}{L}\nabla g(\mathbf{X}_j)$
5:    $\mathbf{\Sigma}^k_{j+1} = \text{soft}(\mathbf{F}_{j+1}, \frac{\lambda}{L})$
6:    $t_{j+1} = \frac{\sqrt{1+4t_j^2}+1}{2}$
7:    $\mathbf{X}_{j+1} = \mathbf{\Sigma}^k_{j+1} + \frac{t_j-1}{t_{j+1}}\left(\mathbf{\Sigma}^k_{j+1} - \mathbf{\Sigma}^k_j\right)$
8:    **if** $\frac{\left\|\mathbf{\Sigma}^k_{j+1} - \mathbf{\Sigma}^k_j\right\|_F^2}{\left\|\mathbf{\Sigma}^k_{j+1}\right\|_F^2} < \epsilon$ or $j \geq K$ **then**
9:      $\mathbf{\Sigma}^{k+1} = \mathbf{\Sigma}^k_{j+1}$;
10:     break;
11:   **end if**
12:   $j = j + 1$
13: **end while**

Updating $\mathbf{E}$

The updating of $\mathbf{E}$ can be shown as

$$\begin{aligned}
\mathbf{E} &= \arg\min_{\mathbf{E}} \|\mathbf{E}\|_p^p + \frac{\mu}{2}\|\mathbf{Y} - \mathbf{A}\mathbf{\Sigma}^{k+1}\mathbf{B}^T - \mathbf{E}\|_F^2 \\
&\quad + \left\langle \mathbf{R}^k, \mathbf{Y} - \mathbf{A}\mathbf{\Sigma}^{k+1}\mathbf{B}^T - \mathbf{E}\right\rangle \\
&= \arg\min_{\mathbf{E}} \|\mathbf{E}\|_p^p + \frac{\mu}{2}\|\mathbf{E} - \mathbf{Y} + \mathbf{A}\mathbf{\Sigma}^{k+1}\mathbf{B}^T - \frac{1}{\mu}\mathbf{R}^k\|_F^2
\end{aligned}\tag{22}$$

Like (18), let $\mathbf{H}^k = \mathbf{Y} - \mathbf{A}\mathbf{\Sigma}^{k+1}\mathbf{B}^T + \frac{1}{\mu}\mathbf{R}^k$, and then, (22) can be defined as

$$\mathbf{E} = \arg\min_{\mathbf{E}} \|\mathbf{E}\|_p^p + \frac{\mu}{2}\|\mathbf{E} - \mathbf{H}^k\|_F^2\tag{23}$$

This is a non-convex optimization question, but fortunately, the authors proposed a generalized iterated shrinkage algorithm (GISA) for non-convex sparse coding algorithm in the paper [38], which is used to solve the l_p-norm optimization question. The method is easily implemented and has a good performance. However, the GISA algorithm is often used for real signals, and the radar data used for imaging are the complex signal. In this article, taking into account the characteristics of radar data, the GISA algorithm is separately performed on the real part and the imaginary part of the radar echo signal, respectively. The concrete algorithmic processing of GISTA is shown in Algorithm 2.

Then, $\mathbf{E}$ is written as

$$\begin{cases}
\mathbf{H}^k = \mathbf{Y} - \mathbf{A}\mathbf{\Sigma}^{k+1}\mathbf{B}^T + \frac{1}{\mu}^k\mathbf{R}^k \\
\mathbf{E}^{k+1} = \text{CGISA}\left(\mathbf{H}^k, \mu^k, p, J\right)
\end{cases}\tag{24}$$

Empirically, we set $J = 3$, in order to balance the gap between the computational complexity and the convergence precision.

Algorithm 2 Signal processing of CGISA.

Input signal: $\mathbf{x} \in \mathbb{C}^{M \times N}, \lambda, p, J$
Output signal: $\mathbf{y} \in \mathbb{C}^{M \times N}$.

1: **Initialization:**
2: $\mathbf{x}_1 = \Re(\mathbf{x}); \mathbf{x}_2 = \Im(\mathbf{x}); \mathbf{y}_1 = \text{zeros}(M, N); \mathbf{y}_2 = \text{zeros}(M, N)$
3: $\tau_p = (2\lambda(1-p))^{\frac{1}{2-p}} + \lambda p(2\lambda(1-p))^{\frac{p-1}{2-p}}$
4: **for** $i = 1$ to J **do**
5:    **if** $|\mathbf{x}_i| < \tau_p$ **then**
6:       $\Omega_i = \text{find}(|\mathbf{x}_i| \leq \tau_p); \mathbf{y}_i(\Omega_i) = 0;$
7:    **else**
8:       $k = 0, \mathbf{x}_i^k = |\mathbf{y}_i|$
9:       **while** (1) **do**
10:          $\mathbf{x}_i^{k+1} = |\mathbf{y}_i| - \lambda p(\mathbf{x}_i^k)^{p-1};$
11:          $k = k + 1$
12:          **if** $k \geq J$ **then**
13:             break;
14:          **end if**
15:       **end while**
16:    **end if**
17:    $\overline{\Omega_i} = \text{find}(|\mathbf{x}_i| > \tau_p); \mathbf{y}_i(\overline{\Omega_i}) = \text{sgn}(\mathbf{x}_i^k(\overline{\Omega_i}))\mathbf{x}_i^k(\Omega_i)$
18: **end for**

***Updating Lagrange multiplier* R *and penalty coefficient* μ**

Lagrange multiplier **R** and Lagrange penalty parameter μ can be updated as follows:

$$\begin{cases} \mathbf{R}^{k+1} = \mathbf{Y} - \mathbf{A}\mathbf{\Sigma}^{k+1}\mathbf{B}^T - \mathbf{E}^{k+1} \\ \mu^{k+1} = \rho\mu^k \end{cases} \tag{25}$$

The pseudo-code of the proposed super-resolution imaging algorithm is shown in Algorithm 3, called the exact sparse recovery algorithm. It will consume much times, and we extended the algorithm in Algorithm 3 to the algorithm in Algorithm 4, which is dubbed the inexact sparse recovery algorithm. Compared to the exact sparse recovery algorithm, the inexact algorithm reduces the iterative process of $\mathbf{\Sigma}$. The algorithm was omitted in the iterative process of $\mathbf{\Sigma}$ inside the algorithm, and thereby, it can reduce the algorithm's runtime.

Algorithm 3 The exact sparse algorithm.

Input: $\mathbf{Y}$, $\mathbf{A}$, and $\mathbf{B}$
Output: Σ.

1: **Initialization:**
2: Initiate some parameters, $\Sigma_0 = \mathbf{X}_0 = \mathbf{0}^{P \times Q}$, $\mathbf{R}_0 = \mathbf{0}^{P \times Q}$, $L > 0$,
$\quad \epsilon = 1 \times 10^{-6}$, $\lambda > 0$, $\mu_0 > 0$ $\rho > 0$, and the maximum iteration K.
3: **while** (1) **do**
4: $\quad$ Update Σ^{k+1} using Algorithm 1 until it converges
5: $\quad$ Update $\mathbf{E}^{k+1}$ with (24)
6: $\quad$ Update $\mathbf{R}^{k+1}$ and μ with (25), respectively
7: $\quad$ **if** $\dfrac{\left\| \Sigma^k_{j+1} - \Sigma^k_j \right\|^2_F}{\left\| \Sigma^k_{j+1} \right\|^2_F} < \epsilon$ or $j \geq K$ **then**
8: $\quad\quad$ $\Sigma^{k+1} = \Sigma^k_{j+1}$;
9: $\quad\quad$ break;
10: $\quad$ **end if**
11: $\quad$ $k = k + 1$
12: **end while**
13: $\Sigma = \Sigma_k$

Algorithm 4 The inexact sparse algorithm.

Input: $\mathbf{Y}$, $\mathbf{A}$, and $\mathbf{B}$
Output: Σ.

1: **Initialization:**
2: Initiate some parameters, $\Sigma_0 = \mathbf{X}_0 = \mathbf{0}^{P \times Q}$, $\mathbf{R}_0 = \mathbf{0}^{P \times Q}$, $L > 0$,
$\quad \epsilon = 1 \times 10^{-6}$, $\lambda > 0$, $\mu_0 > 0$ $\rho > 0$, and the maximum iteration K.
3: **while** (1) **do**
4: $\quad$ Update $\mathbf{D}^k = \mathbf{Y}^k - \mathbf{E}^k + \frac{1}{\mu^k}\mathbf{R}^k$
5: $\quad$ Update $\mathbf{F}^k = \mathbf{X}^k - \frac{1}{L}\nabla f\left(\mathbf{X}^k\right)$
6: $\quad$ Update $\Sigma^{k+1} = \mathrm{soft}\left(\mathbf{F}^k, \frac{\lambda}{L}\right)$
7: $\quad$ Update $t^{k+1} = \frac{\sqrt{1+4t^{k^2}}+1}{2}$
8: $\quad$ Update $\mathbf{X}^{k+1} = \mathbf{X}^k + \frac{t^k-1}{t^{k+1}}\left(\Sigma^{k+1} - \mathbf{X}^k\right)$
9: $\quad$ Update $\mathbf{E}^{k+1}$ with (24)
10: $\quad$ Update $\mathbf{R}^{k+1}$ and μ with (25), respectively
11: $\quad$ **if** $\dfrac{\left\| \Sigma^k_{j+1} - \Sigma^k_j \right\|^2_F}{\left\| \Sigma^k_{j+1} \right\|^2_F} < \epsilon$ or $j \geq K$ **then**
12: $\quad\quad$ $\Sigma^{k+1} = \Sigma^k_{j+1}$; break;
13: $\quad$ **end if**
14: $\quad$ $k = k + 1$
15: **end while**
16: $\Sigma = \Sigma_k$

---

3. Results

In this section, we perform a detailed simulation of MIMO radar imaging, present the imaging result, and analyze the difference between the algorithm we propose and the other existing algorithms.

4. Simulation Experiments

We used the phase-coded signal as the transmitted signal, and the number of transmitted arrays was 2×2, while the number of the received arrays was 20×20. The carrier frequency of the radar was 12 GHz; the bandwidth was 500 MHz; the sampling frequency was 1 GHz. The simulation parameters of the MIMO radar can be seen in Table 1.

Table 1. MIMO radar simulation parameters.

Parameters	Symbol	Value
Center frequency	f_c	12 GHz
Bandwidth	B	500 MHz
Sampling frequency	f_s	1 GHz
The number of transmitters	$M_t \times M_t$	2×2
The number of receivers	$N_r \times N_r$	20×20

Suppose that the location coordinate of the center of the scatterers is $O(4000, 3000, 5000)$. The targets' locations in the Cartesian coordinate system with respect to $\mathbf{O}$ are $(6, 0, 5)$, $(4, 0, 4)$, $(0, 0, 0)$, $(-3, -3, 1)$, $(-3, 3, 1)$, $(-6, -7, -1)$, $(-6, 7, -1)$, $(-4, 0, -3)$, $(-8, 3, -6)$, $(-6, 0, -6)$, and $(-8, -3, -6)$. The target distribution in space and the top view of the target distribution are shown as Figure 2a,b, respectively.

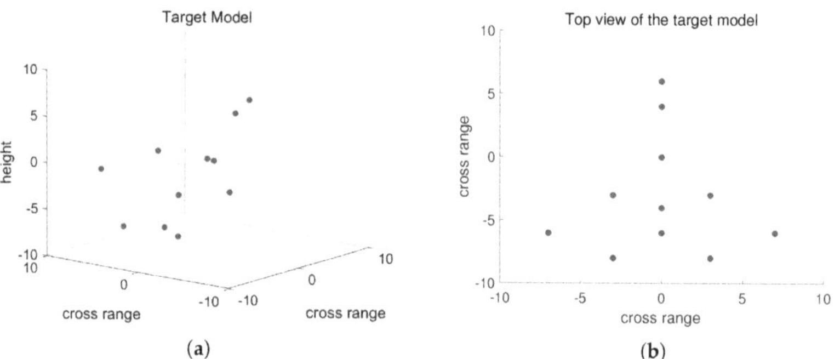

Figure 2. (a) is the space distribution of the target scattering points; (b) is the distribution along the cross-range.

Here, we made the assumption that the signal can be separated by the matched filter for convenience. We extracted the observed snapshot matrix from the virtual array in the MIMO radar, which contains impulsive signals caused by the array channels. Assume that the target location in space $\mathbf{\Sigma}$ recovered by the algorithms can be denoted by $\hat{\mathbf{\Sigma}}$. Two parameters were defined to evaluate the performance of the proposed algorithm. One is the normal mean-squared error (NMSE) to measure the performance of the algorithms with its definition as follows:

$$\text{NMSE} = \frac{\left\| \mathbf{\Sigma} - \hat{\mathbf{\Sigma}} \right\|_F^2}{\left\| \mathbf{\Sigma} \right\|_F^2} \tag{26}$$

where $\mathbf{\Sigma}$ and $\hat{\mathbf{\Sigma}}$ are the original sparse signal and the reconstructed sparse signal, respectively. The other is the correlation (Corr) coefficient used to evaluate the imaging performance and to measure the influence of false targets on the imaging quality. The mathematical expression of the Corr can be defined as follows:

$$\text{Corr}\left(\mathbf{\Sigma}, \hat{\mathbf{\Sigma}} \right) = \frac{\left\langle \text{vec}(\mathbf{\Sigma}), \text{vec}(\hat{\mathbf{\Sigma}}) \right\rangle}{\left\| \mathbf{\Sigma} \right\|_F \left\| \hat{\mathbf{\Sigma}} \right\|_F} \tag{27}$$

where $\text{vec}(\mathbf{\Sigma})$ and $\text{vec}(\hat{\mathbf{\Sigma}})$ denote the vector form of $\mathbf{\Sigma}$ and $\hat{\mathbf{\Sigma}}$, respectively.

All experiments were performed with MATLAB and run on a computer with an Inter(R) Core(TM) i7-8565U at @1.80 GHz, 1.99 GHz, and RAM 16 GB.

In this section, the inexact recovery algorithm was chosen to validate the algorithms' performance, and the difference between the exact recovery algorithm and the inexact recovery algorithm is discussed in the next section. The impulsive signal is generated based on (12), for which $\rho > 0$ and ϕ are randomly produced. MIMO imaging results with these recovery algorithms along the cross-range are shown in Figure 3, in which the signal-to-noise ratio (SNR) is 10 dB and the percentage of outliers in the array is about 33%. Moreover, in order to ensure the algorithms' convergence, we set $L = \left\|\mathbf{A}\mathbf{A}^H\right\|_F^2 \left\|\mathbf{B}^T\mathbf{B}^*\right\|_F^2$, the algorithm parameters $P = Q = 201$, the regularization constant $\lambda = 0.5$, and $p = 0.1, 0.3, 0.5, 0.7, 0.9$. The snapshot matrix $\mathbf{Y}$ contains some impulsive signals, which makes the the imaging results very poor when the traditional imaging algorithms are used, such as 2D-FFT, 2D-FISTA, and 2D-SL0. What we can see in Figure 3 is that there is dense background noise for 2D-FFT, 2D-FISTA, and 2D-SL0; however, the algorithm we propose can well inhibit the background noise and outliers. The concrete imaging indicator of Figure 3 can be seen in Table 2, which shows that the proposed algorithm is obviously superior to the other algorithms.

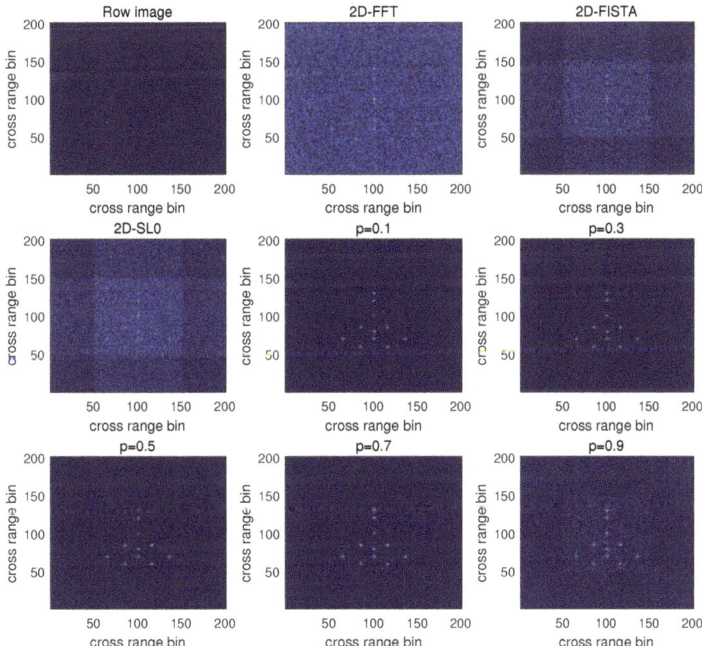

Figure 3. MIMO imaging results with 2D-FFT, 2D-FISTA, 2D-SL0, and the proposed algorithm with $p = 0.1$, $p = 0.3$, $p = 0.5$, $p = 0.7$, and $p = 0.9$.

Table 2. Imaging performance.

Algorithm	2D-FFT	2D-FISTA	2D-SL0	$p = 0.1$	$p = 0.3$	$p = 0.5$	$p = 0.7$	$p = 0.9$
NMSE	4.048	2.0923	2.4894	0.4836	0.5100	0.5443	0.5629	0.8544
Corr	0.0450	0.1868	0.1703	0.8177	0.8025	0.7827	0.7788	0.6771

Both Figures 4 and 5 illustrate the variation curve of the SNR with the NMSE and Corr, respectively, where the performance of the proposed algorithm is obviously better than the 2D-FFT, 2D-FISTA, and 2D-SL0 algorithms, especially under low SNRs. Meanwhile, the NMSE gradually decreases and the Corr ascends for all algorithms contained in Figure 4 and Figure 5, respectively, with the increase of the SNR. The result implies that the algorithm is more robust compared to the other algorithms. In our algorithm, we added the augmented Lagrange multiplier and l_p penalty function to inhibit the outliers and improve the robustness of the algorithm.

Next, the influence of the percentage of outliers in the array on imaging performance was studied, and we set the percentage from 10% to 60%. Their relationship can be seen in Figures 6 and 7, which show that our algorithm is more robust than 2D-FISTA, 2D-SL0, and 2D-FISTA. The results also imply that the proposed algorithm is more robust to outliers.

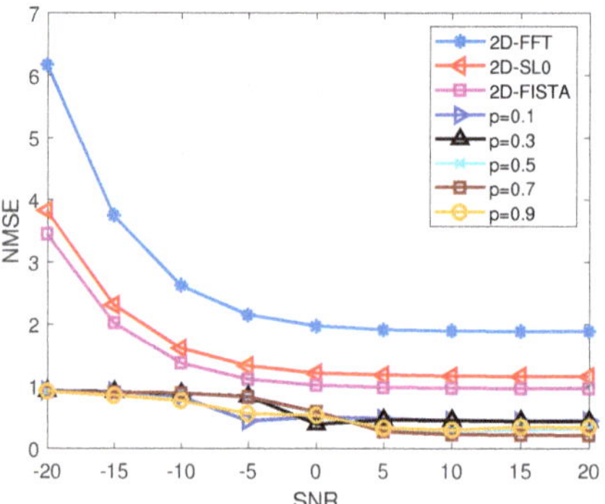

Figure 4. The relationship between the SNR and NMSE.

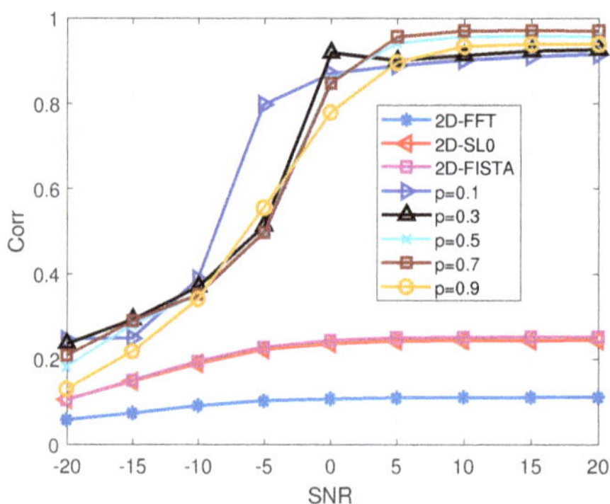

Figure 5. The relationship between the SNR and Corr.

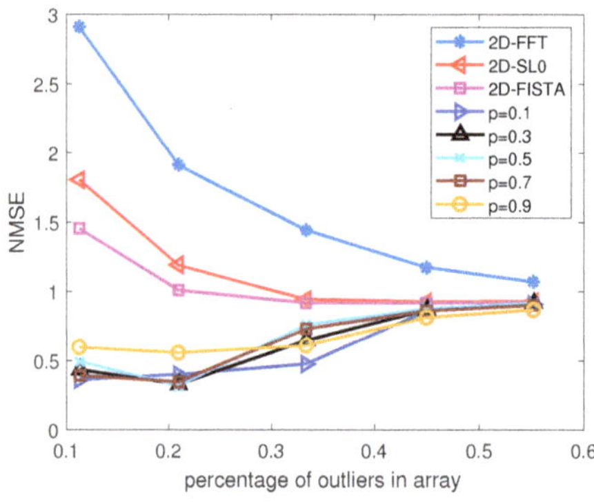

Figure 6. The relationship between the percentage of outliers and the NMSE.

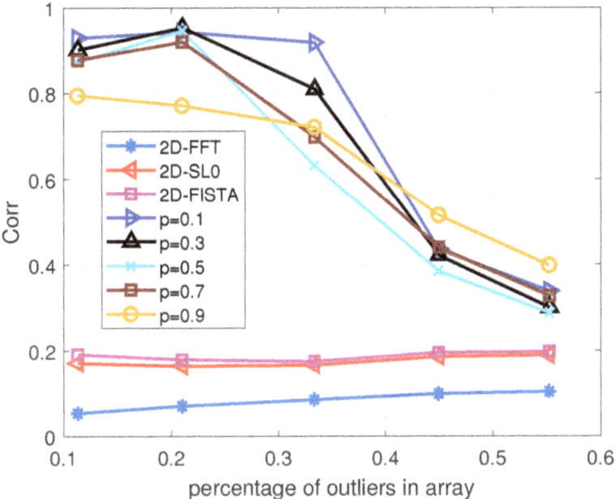

Figure 7. The relationship between the percentage of outliers and the Corr.

Public Dataset Experiment

Finally, we used a public dataset, the Boeing-727 dataset, to validate the performance of the proposed algorithm, which demonstrated that the dataset is consistent with the signal model for MIMO imaging. Suppose that every point in the radar echo matrix can be seen as the snapshot of the MIMO virtual array. The dataset is generated by an X-band (9.0 GHz) stepped frequency radar. The radar parameters are shown in the Table 3.

In this experiment, we chose the number of transmitted frequencies and the number of weeps as 64 and 256, respectively, and the algorithm parameters were $p = 0.1$, $\lambda = 0.4$, SNR $= 10\,$dB, and $P = 1024, Q = 256$. The radar imaging result can be seen in Figures 8 and 9, whose percentage of outliers were set as 12% and 39%, respectively. Our algorithm can well inhibit background noise and outliers compared to the existing algorithms.

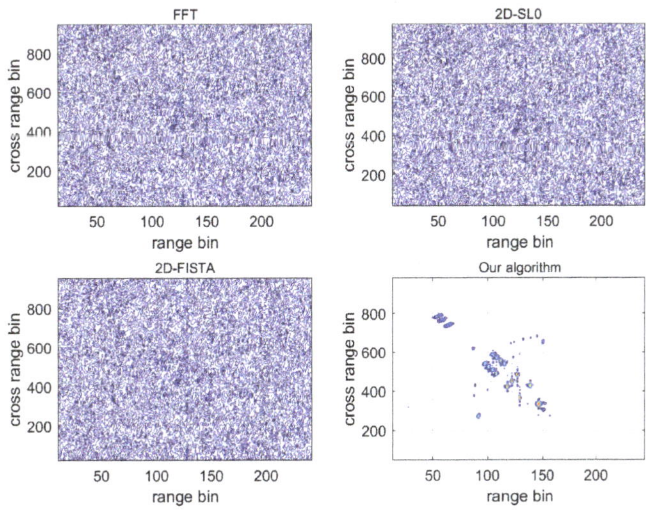

Figure 8. Imaging result with the percentage of outliers being 12%.

Table 3. The parameters of the step radar.

Parameters	Value
Center frequency	9 GHz
Bandwidth	150 MHz
Sampling points	64
Pulse number	256
SNR	10 dB

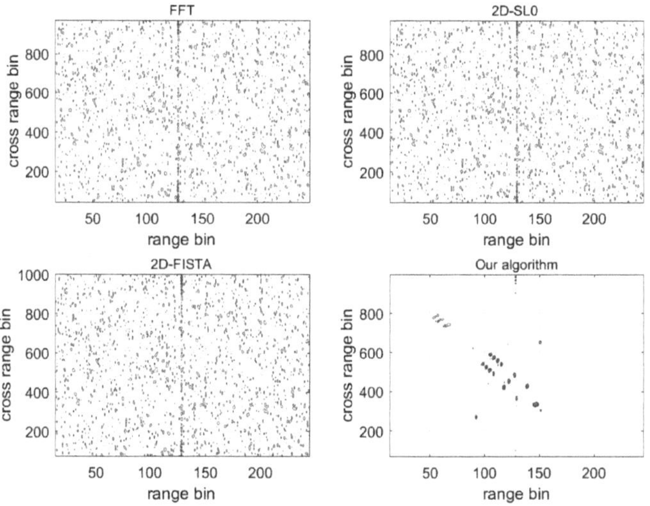

Figure 9. Imaging result with the the percentage of outliers being 39%.

5. Discussion

5.1. Algorithm Convergence Analysis

To evaluate the converge of both algorithms, the experiment conditions were $p = 0.5$, $L = 1 \times 10^6$, $\lambda = 0.5$, and SNR $= 10$ dB. In Figure 10, although the exact algorithm can converge to a more concise result compared to the inexact algorithm, it can converge at a higher speed. The runtime and convergence error of both algorithms are presented in Table 4.

In Table 4, the result shows that the exact algorithm will speed up the time until the convergence condition is met, and the inexact recovery algorithm reduces the time complexity at the expense of the convergence accuracy.

Table 4. Runtime and convergence error of both algorithms.

Algorithm	Items	$p = 0.1$	$p = 0.3$	$p = 0.5$	$p = 0.7$
Exact recovery	Time (s)	179.1385	179.1395	179.1375	179.1363
	Error	0.0032	0.0041	0.0042	0.0042
Inexact recovery	Time (s)	3.3559	3.342	3.341	3.358
	Error	0.0088	0.0088	0.0086	0.0088

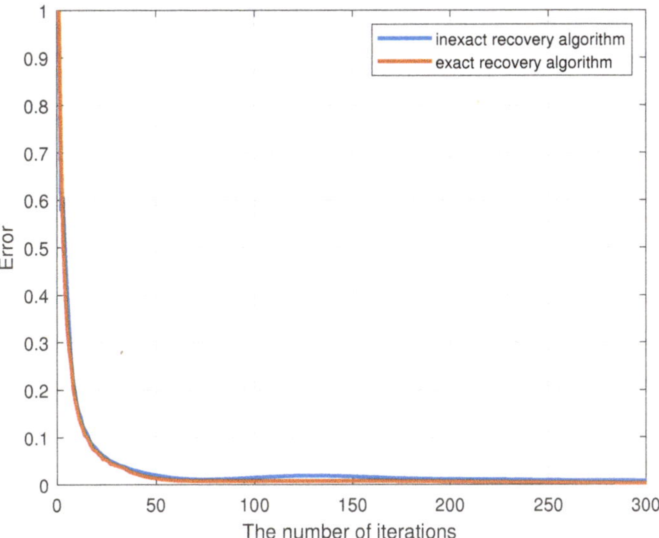

Figure 10. Convergence error curve of both algorithms.

5.2. Algorithm Complexity Analysis

On the one hand, we successfully recovered the uniform snapshot matrix from the contaminated snapshot matrix by the array outliers. On the other hand, the two algorithms, the exact recovery algorithm and the inexact recovery algorithm, were proposed to present the sparse imaging result, and Σ reflected the sparse position of the scattering points in space. In addition, the proposed algorithm was dominant in computational complexity.

Assume that the size of a contaminated snapshot matrix is $M \times N$, and the Fourier matrices are $M \times P$ and $N \times Q$. Generally, the computational complexity of SVD is around $\mathcal{O}(NM^2) + \mathcal{O}(M^3)$. The proposed algorithm in this paper was divided into three steps, and the computational complexity of every step can be seen below:

(1)   Updating Σ: $\mathcal{O}(P^2M) + \mathcal{O}(P^2Q) + \mathcal{O}(Q^2N)$;
(2)   Updating **E**: $\mathcal{O}(MPQ) + \mathcal{O}(MNQ) + \mathcal{O}(JPQ)$;
(3)   Updating **R**: $\mathcal{O}(MPQ) + \mathcal{O}(MNQ)$.

For the exact recovery algorithm, the inner algorithm in Table 1 converges when it iterates K times. Thereby, the exact recovery algorithm will speed up the time more than the inexact recovery algorithm.

6. Conclusions

For MIMO radar 3D imaging, we achieved super-resolution imaging along the cross-range. The algorithm we proposed does not convert the grid matrix into the larger matrix and the observed matrix into the 1D vector and is more robust to noise than the existing algorithms, 2D-FISTA and 2D-SL0. We proposed two sparse recovery algorithms, the exact recovery algorithm and the inexact recovery algorithm, to inhibit the impact of outliers on the imaging performance. The simulation experiments and public dataset experiment validated the effectiveness of the algorithm.

Author Contributions: Conceptualization, J.D., Z.W., X.W. and M.W.; methodology, J.D.; software, J.D. and Z.W.; validation, J.D., Z.W. and X.W.; formal analysis, J.D.; investigation, X.W.; resources, X.W. and Z.W.; data curation, J.D. and X.W.; writing—original draft preparation, J.D.; writing—review and editing, Z.W.; visualization, X.W.; supervision, M.W.; project administration, J.D.; funding acquisition, M.W. All authors have read and agreed to the published version of the manuscript.

Funding: This work was supported in part by the National Natural Science Foundation of China under Grant 61771380 U1730109 and Grant CEMEE 2017K0202B and in part by the Teaching Reform Research Project under Grant 19xcj047. The work is also supported by the Fundamental Research Funds for the Central Universities and supported by the Innovation Fund of Xidian University.

Data Availability Statement: Not applicable.

Acknowledgments: The authors sincerely thank the Editors and all Reviewers for their valuable reviews, which played an important role in improving the article's quality.

Conflicts of Interest: The authors declare no conflict of interest.

Abbreviations

The following abbreviations are used in this manuscript:

MIMO	multiple-input multiple-output
PAR	phased array radar
ADMM	alternative direction method of multipliers
ISAR	inverse synthetic aperture radar
SAR	synthetic aperture radar
MC	matrix completion
InISAR	interferometric ISAR
FDM	frequency-division multiplexing
TDM	time-division multiplexing
CDM	code-division multiplexing
PCA	phase center principle
BP	back projection
FRI	finite rate of innovation
SVT	singular-value thresholding
SBL	sparse Bayesian learning
GISA	generalized iterated shrinkage algorithm
CGISA	complex generalized iterated shrinkage algorithm
SL0	smoothed L0
ALM	augmented Lagrange multiplier method
SNR	signal-to-noise ratio
CS	compressive sensing
FFT	fast Fourier transform
FISTA	fast iterative shrinkage thresholding algorithm
SVD	singular-value decomposition
IALM	inexact augmented Lagrange multiplier
RAM	random access memory
NMSE	normal mean-squared error

References

1. Ozdemir, C. *Inverse Synthetic Aperture Radar Imaging with MATLAB Algorithms*; Wiley: New York, NY, USA, 2012.
2. Shakya, P.; Raj, A.B. Inverse Synthetic Aperture Radar Imaging Using Fourier Transform Technique. In Proceedings of the 2019 1st International Conference on Innovations in Information and Communication Technology (ICIICT), Chennai, India, 25–26 April 2019; pp. 1–4.
3. Zhang, L.; Xing, M.; Qiu, C.W.; Li, J.; Sheng, J.; Li, Y.; Bao, Z. Resolution Enhancement for Inversed Synthetic Aperture Radar Imaging Under Low SNR via Improved Compressive Sensing. *IEEE Trans. Geosci. Remote Sens.* **2010**, *48*, 3824–3838. [CrossRef]
4. Tian, B.; Lu, Z.; Liu, Y.; Li, X. Review on Interferometric ISAR 3D Imaging: Concept, Technology and Experiment. *Signal Process.* **2018**, *153*, 164–187. [CrossRef]
5. Wang, D.W.; Ma, X.Y.; Chen, A.L.; Su, Y. High-resolution imaging using a wideband MIMO radar system with two distributed arrays. *IEEE Trans. Image Process.* **2009**, *19*, 1280–1289. [CrossRef]
6. Bechter, J.; Roos, F.; Waldschmidt, C. Compensation of Motion-Induced Phase Errors in TDM MIMO Radars. *IEEE Microw. Wirel. Components Lett.* **2017**, *27*, 1164–1166. [CrossRef]
7. Jian, L.; Stoica, P. MIMO Radar with Colocated Antennas. *IEEE Signal Process. Mag.* **2007**, *24*, 106–114.
8. Miwa, T.; Ogiwara, S.; Yamakoshi, Y. MIMO Radar System for Respiratory Monitoring Using Tx and Rx Modulation with M-Sequence Codes. *IEICE Trans. Commun.* **2010**, *93*, 2416–2423. [CrossRef]

9. Deng, H. Polyphase code design for Orthogonal Netted Radar systems. *IEEE Trans. Signal Process.* **2004**, *52*, 3126–3135. [CrossRef]
10. Roberts, W.; Stoica, P.; Li, J.; Yardibi, T.; Sadjadi, F.A. Iterative Adaptive Approaches to MIMO Radar Imaging. *IEEE Jour Sel. Top. Signal Process* **2010**, *4*, 5–20. [CrossRef]
11. Tan, X.; Roberts, W.; Li, J.; Stoica, P. Sparse Learning via Iterative Minimization with Application to MIMO Radar Imaging. *IEEE Trans. Signal Process.* **2011**, *59*, 1088–1101. [CrossRef]
12. Ding, L.; Chen, W.; Zhang, W.; Poor, H.V. MIMO radar imaging with imperfect carrier synchronization: A point spread function analysis. *IEEE Trans. Aerosp. Electron. Syst.* **2015**, *51*, 2236–2247. [CrossRef]
13. Ma, C.; Yeo, T.S.; Tan, C.S.; Liu, Z. Three-Dimensional Imaging of Targets Using Colocated MIMO Radar. *IEEE Trans. Geosci. Remote Sens.* **2011**, *49*, 3009–3021. [CrossRef]
14. Zhu, Y.T.; Yi, S. A type of M 2-transmitter N 2-receiver MIMO radar array and 3D imaging theory. *Sci. China* **2011**, *54*, 2147–2157. [CrossRef]
15. Hu, X.; Tong, N.; Zhang, Y.; Huang, D. MIMO Radar Imaging With Nonorthogonal Waveforms Based on Joint-Block Sparse Recovery. *IEEE Trans. Geosci. Remote Sens.* **2018**, *56*, 5985–5996. [CrossRef]
16. Hu, X.; Tong, N.; Zhang, Y.; Wang, Y. 3D imaging using narrowband MIMO radar and ISAR technique. In Proceedings of the 2015 International Conference on Wireless Communications Signal Processing (WCSP), Nanjing, China, 15–17 October 2015; pp. 1–5. [CrossRef]
17. Candes, E.J.; Wakin, M.B. An Introduction To Compressive Sampling. *IEEE Signal Process. Mag.* **2008**, *25*, 21–30. [CrossRef]
18. Tropp, J.A.; Gilbert, A.C. Signal Recovery From Random Measurements Via Orthogonal Matching Pursuit. *IEEE Trans. Inf. Theory* **2007**, *53*, 4655–4666. [CrossRef]
19. Babacan, S.D.; Molina, R.; Katsaggelos, A.K. Bayesian Compressive Sensing Using Laplace Priors. *IEEE Trans. Image Process.* **2010**, *19*, 53–63. [CrossRef]
20. Beck, A.; Teboulle, M. A fast iterative shrinkage-thresholding algorithm for linear inverse problems. *SIAM J. Imaging Sci.* **2009**, *2*, 183–202. [CrossRef]
21. Blumensath, T.; Davies, M.E. Iterative hard thresholding for compressed sensing. *Appl. Comput. Harmon. Anal.* **2009**, *27*, 265–274. [CrossRef]
22. Wei, Z.; Zhang, B.; Xu, Z.; Han, B.; Hong, W.; Wu, Y. An Improved SAR Imaging Method Based on Nonconvex Regularization and Convex Optimization. *IEEE Geosci. Remote Sens. Lett.* **2019**, *16*, 1580–1584. [CrossRef]
23. Xu, G.; Xing, M.; Zhang, L.; Liu, Y.; Li, Y. Bayesian Inverse Synthetic Aperture Radar Imaging. *IEEE Geosci. Remote Sens. Lett.* **2011**, *8*, 1150–1154. [CrossRef]
24. Kang, H.; Li, J.; Guo, Q.; Martorella, M. Pattern Coupled Sparse Bayesian Learning Based on UTAMP for Robust High Resolution ISAR Imaging. *IEEE Sens. J.* **2020**, *20*, 13734–13742. [CrossRef]
25. Ding, J.; Wang, M.; Kang, H.; Wang, Z. MIMO Radar Super-Resolution Imaging Based on Reconstruction of the Measurement Matrix of Compressed Sensing. *IEEE Geosci. Remote Sens. Lett.* **2022**, *19*, 1–5. [CrossRef]
26. Ding, L.; Chen, W. MIMO Radar Sparse Imaging With Phase Mismatch. *IEEE Geosci. Remote Sens. Lett.* **2015**, *12*, 816–820. [CrossRef]
27. Duarte, M.F.; Baraniuk, R.G. Kronecker compressive sensing. *IEEE Trans. Image Process.* **2011**, *21*, 494–504. [CrossRef]
28. Ghaffari, A.; Babaie-Zadeh, M.; Jutten, C. Sparse decomposition of two dimensional signals. In Proceedings of the 2009 IEEE International Conference on Acoustics, Speech and Signal Processing, Taipei, Taiwan, 19–24 April 2009; pp. 3157–3160. [CrossRef]
29. Wimalajeewa, T.; Eldar, Y.C.; Varshney, P.K. Recovery of sparse matrices via matrix sketching. *arXiv* **2013**, arXiv:1311.2448.
30. Liu, Z.; You, P.; Wei, X.; Li, X. Dynamic ISAR Imaging of Maneuvering Targets Based on Sequential SL0. *IEEE Geosci. Remote Sens. Lett.* **2013**, *10*, 1041–1045. [CrossRef]
31. Jiang, X.; Yasotharan, A.; Kirubarajan, T. Robust Beamforming with Sidelobe Suppression for Impulsive Signals. *IEEE Signal Process. Lett.* **2015**, *22*, 346–350. [CrossRef]
32. Wang, B.; Zhang, Y.D.; Wang, W. Robust DOA Estimation in the Presence of Miscalibrated Sensors. *IEEE Signal Process. Lett.* **2017**, *24*, 1073–1077. [CrossRef]
33. Kou, J.; Li, M.; Jiang, C. Robust Direction-of-Arrival Estimation for Coprime Array in the Presence of Miscalibrated Sensors. *IEEE Access* **2020**, *8*, 27152–27162. [CrossRef]
34. Kou, J.X.; Li, M.; Wang, L.; Yang, K.; Jiang, C.L. Generalized weight function selection criteria for the compressive sensing based robust DOA estimation methods. *Signal Process.* **2020**, *175*, 107663. [CrossRef]
35. Rakotomamonjy, A.; Flamary, R.; Gasso, G. DC Proximal Newton for Nonconvex Optimization Problems. *IEEE Trans. Neural Netw. Learn. Syst.* **2016**, *27*, 636–647. [CrossRef] [PubMed]
36. Candes, E.J.; Plan, Y. Matrix Completion With Noise. *Proc. IEEE* **2010**, *98*, 925–936. [CrossRef]
37. Zeng, W.J.; So, H.C. Outlier-Robust Matrix Completion via ℓ_p-Minimization. *IEEE Trans. Signal Process.* **2018**, *66*, 1125–1140. [CrossRef]
38. Zuo, W.; Meng, D.; Zhang, L.; Feng, X.; Zhang, D. A Generalized Iterated Shrinkage Algorithm for Non-convex Sparse Coding. In Proceedings of the 2013 IEEE International Conference on Computer Vision, Sydney, Australia, 1–8 December 2013; pp. 217–224. [CrossRef]

Technical Note

Non-Uniform MIMO Array Design for Radar Systems Using Multi-Channel Transceivers

Eunhee Kim [1,*], Ilkyu Kim [2] and Wansik Kim [3]

[1] Department of Defence System Engineering, Sejong University, 209 Neungdong-ro, Gwangjn-gu, Seoul 05006, Republic of Korea
[2] Electrical and Electronics Engineering Department, Sejong Cyber University, 121, Gunja-ro, Gwangjn-gu, Seoul 05000, Republic of Korea
[3] LIGNex1 Co., 207 Mabuk-ro, Giheung-gu, Yongin-si 16911, Republic of Korea
* Correspondence: eunheekim@sejong.ac.kr

Abstract: Multiple-input multiple-output (MIMO) technology has recently attracted attention with regard to improving the angular resolution of small antennas such as automotive radars. If appropriately placed, the co-located transmit and receive arrays can make a large virtual aperture. This paper proposes a new method for designing arrays by adopting a structure with minimum redundancy. The proposed structure can significantly increase the virtual array aperture while keeping the transmit and receive antennas at the same size. We describe the application of the proposed method to subarray-type antennas using multi-channel transceivers, which is essential for arranging RF hardware in a small antenna operating at high frequency. Further, we present an analysis of the final beam pattern and discuss its benefits and limitations.

Keywords: MIMO array; minimum redundancy array; virtual array antenna; subarray

Citation: Kim, E.; Kim, I.; Kim, W. Non-Uniform MIMO Array Design for Radar Systems Using Multi-Channel Transceivers. *Remote Sens.* **2023**, *15*, 78. https://doi.org/10.3390/rs15010078

Academic Editors: Guolong Cui, Yong Yang, Xianxiang Yu and Bin Liao

Received: 23 November 2022
Revised: 20 December 2022
Accepted: 20 December 2022
Published: 23 December 2022

1. Introduction

An essential requirement for a radar system is its angular resolution. As the angular resolution depends on the aperture size, a small antenna system, such as an automotive radar, is challenging to achieve. Subspace-based algorithms, such as multiple signal classification (MUSIC) and estimation of signal parameters via rational invariance techniques (ESPRIT), and parameter estimation algorithms based on the maximum likelihood (ML) function have been used to achieve the high-resolution angle estimation [1,2]. However, the multiple-input multiple-output (MIMO) technology has recently attracted the most attention.

A MIMO radar, which synthesizes a virtual antenna array (VAA) using co-located transmit and receive antennas, is typically used in automotive systems [3–6]. If appropriately placed, co-located transmit and receive arrays can create a large virtual aperture with a small number of arrays. The arrangement of the VAA is determined by the spatial convolution of the transmit and receive array positions, and its aperture is the sum of each antenna aperture [7]. In the case of a uniform linear array (ULA), if the receive array has M_r elements and the transmit array has M_t elements, the VAA can become a filled ULA with $M_t \times M_r$ elements when the interelement spacing is d and $M_r \times d$, respectively. Furthermore, if the total number of arrays is $2K$, the maximum aperture is obtained when $M_r = M_t = K$. However, in this case, the sizes of the receive and transmit arrays are considerably different, i.e., $M_r \times d$ and $(M_t - 1) \times M_r \times d$. For example, if $K = 8$, the maximum VAA aperture can be $16d$ when $M_r = M_t = 4$. The apertures of receive and transmit arrays are $4d$ and $12d$, respectively. Thus, the two antennas have a three-fold size difference, and the long antenna will ultimately determine the physical dimensions of the entire antenna. Therefore, this arrangement is insufficient for the miniaturization of the antenna when considering the physical dimensions and the VAA aperture. A simple design to make the size of the transmit and receive arrays the same is to use $M_t = 2$. Then, although the aperture of the VAA is reduced to $12d$ in this example, the physical size becomes half.

Most automotive radars employ this type of transmit-and-receive array spacing, and it is difficult to find other arrangements. This paper suggests a new spacing method for VAA that provides the largest VAA aperture and makes the physical size of the transmit and receive array antennas the same by employing the co-array property of the non-uniform, minimum redundancy array.

Another trend driving module size reduction is to move to higher operating frequencies. A higher operating frequency in radar systems is preferred because of its increased bandwidth, high range resolution, and accuracy. The unlicensed industrial, scientific, and medical (ISM) frequency above 100 GHz is particularly interesting for mass-volume commercial radar-sensor applications [8]. However, when the frequency increases, arranging the RF hardware becomes challenging due to the compact antenna. Research on integrating antennas into packages or chips is ongoing to find a cost-effective solution without requiring RF signals on the printed circuit board (PCB) [9]. Several studies on D band (110~170 GHz) transceivers have been published [10–13]. The multi-channel transceiver is a single-chip solution that integrates multiple transmit and receive channels. It includes amplifiers and a phase shifter in each channel supporting analog beamforming in both transmit and receive directions. Transmission beamforming can increase the power and extend the detection range. Receive beamforming decreases the number of analog-to-digital converters (ADCs) and all subsequent digital hardware, effectively reducing size and cost of the system. However, this front-end beamforming, or subarray structure, is disadvantageous for adaptive beamforming or multiple beamforming compared with full digital arrays and affects the MIMO VAA configuration as well [14].

In this paper, we propose a design approach for MIMO VAA considering the transmit and receive subarray structure using multi-channel transceivers. We herein present an analysis of the final beam pattern using subarray structure and MIMO VAA and the application of the proposed MIMO arrangement method. The benefits and limitations are discussed.

The remainder of this paper is organized as follows. The next section derives the fundamental formula of a MIMO antenna with subarrays. Section 3 describes the non-uniform arrays, including the non-redundant array and minimum redundancy array (MRA), and proposes a new MIMO configuration. Section 4 suggests a subarray-type MIMO antenna using non-uniform spacing and, finally, Section 5 presents the conclusion.

2. MIMO with Subarrays

2.1. Basic Principle

A MIMO, in which the transmit arrays are configured using subarrays, is called phased MIMO, and the beam pattern, SNR, and SINR are discussed in [15–17]. Based on this, the following formula is derived for M transmission subarrays with L arrays shown in Figure 1.

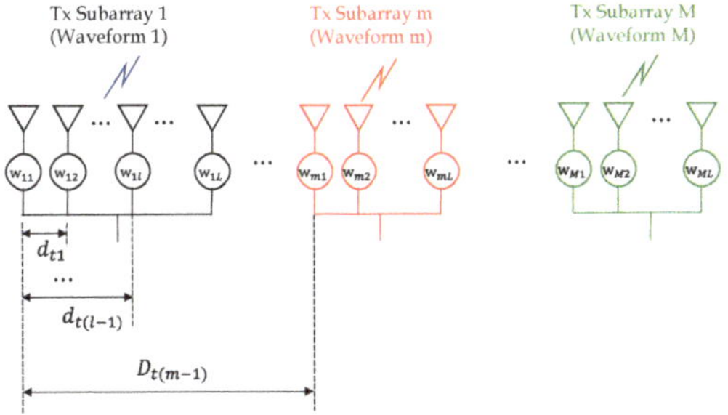

Figure 1. Configuration of the transmit antenna with subarrays.

We assume that each subarray transmits a different waveform $\phi_m(t)$, and the waveforms are orthogonal to one another.

$$\boldsymbol{\phi}(t) = \begin{bmatrix} \phi_1(t) & \phi_2(t) & \cdots & \phi_M(t) \end{bmatrix}^T \tag{1}$$

$$\int_{T_0} \boldsymbol{\phi}(t)\boldsymbol{\phi}^H(t)dt = I_M \tag{2}$$

where I_M is the $M \times M$ identity matrix. If the beamforming weight of each subarray is w_{ml}, the transmission signal in the θ direction by the m-th subarray can be expressed as follows:

$$\begin{aligned} s_m(t,\theta) &= \begin{bmatrix} w_{m1}^* & w_{m2}^*, \cdots & w_{mL}^* \end{bmatrix} \begin{bmatrix} 1 & e^{-jkd_{t1}\sin\theta} & \cdots & e^{-jkd_{t(L-1)}\sin\theta} \end{bmatrix}^T e^{-jkD_{t(m-1)}\sin\theta}\phi_m(t) \\ &= \mathbf{w}_m^H \mathbf{a}(\theta)e^{-jkD_{t(m-1)}\sin\theta}\phi_m(t) \ (m = 1, 2, \dots M) \end{aligned} \tag{3}$$

where $\mathbf{w}_m = \begin{bmatrix} w_{m1} & w_{m2}, \cdots & w_{mL} \end{bmatrix}^T \in \mathbf{C}^{L \times 1}$, $\mathbf{a}(\theta) = \begin{bmatrix} 1 & e^{-jkd_{t1}\sin\theta} & \cdots & e^{-jkd_{t(L-1)}\sin\theta} \end{bmatrix}^T$, k is the wave number, $2\pi/\lambda$ (λ : *wavelength*), and $*$ and H stand for the complex conjugate and conjugate transpose, respectively.

Then, the transmit signal becomes the sum of them.

$$\begin{aligned} s(t,\theta) &= \sum_{m=1}^M s_m(t,\theta) = \sum_{m=1}^M \mathbf{w}_m^H \mathbf{a}(\theta)e^{-jkD_{t(m-1)}\sin\theta}\phi_m(t) \\ &= \begin{bmatrix} \mathbf{w}_1^H \mathbf{a}(\theta)\mathbf{w}_2^H \mathbf{a}(\theta) \cdots \mathbf{w}_M^H \mathbf{a}(\theta) \end{bmatrix} \odot \begin{bmatrix} 1e^{(-jkD_{t1}\sin\theta)} \cdots e^{-jkD_{t(M-1)}\sin\theta} \end{bmatrix} \begin{bmatrix} \phi_1(t) \\ \phi_2(t) \\ \vdots \\ \phi_M(t) \end{bmatrix} \\ &= \bar{\mathbf{a}}^T(\theta) \odot \mathbf{d}^T(\theta)\boldsymbol{\phi}(t) \end{aligned} \tag{4}$$

where $\bar{\mathbf{a}}(\theta) = \begin{bmatrix} \mathbf{w}_1^H \mathbf{a}(\theta) \cdots \mathbf{w}_M^H \mathbf{a}(\theta) \end{bmatrix}^T \in \mathbf{C}^{M \times 1}$, $\mathbf{d}(\theta) = \begin{bmatrix} 1 & e^{-jkD_{t1}\sin\theta} & \cdots & e^{-jkD_{t(M-1)}\sin\theta} \end{bmatrix}^T$, and $\odot$ stands for the Hadamard product. Moreover, if the weights for the subarrays are identical, i.e., $\mathbf{w}_1 = \mathbf{w}_2 = \cdots = \mathbf{w}_M = \mathbf{w}$, it is simplified to

$$s(t,\theta) = \mathbf{w}^T\mathbf{a}(\theta)\mathbf{d}^T(\theta)\boldsymbol{\phi}(t) = g_t(\theta)\mathbf{d}^T(\theta)\boldsymbol{\phi}(t), \tag{5}$$

where $g_t(\theta) = \mathbf{w}^T\mathbf{a}(\theta)$.

The target reflection signal in the θ direction can be expressed by

$$r(t,\theta) = \beta\bar{\mathbf{a}}^T(\theta) \odot \mathbf{d}^T(\theta)\boldsymbol{\phi}(t-\tau) + n(t) = \beta g_t(\theta)\mathbf{d}^T(\theta)\boldsymbol{\phi}(t-\tau) + n(t) \tag{6}$$

where β is the complex reflection coefficient, τ is the delay by the target distance, and $n(t)$ is the white Gaussian noise.

Suppose the receiver, like the transmit arrays, has a structure that performs analog beamforming in subarray units and then performs digital beamforming, as shown in Figure 2. Then the output of the nth subarray can be written as follows:

$$r_n(t,\theta) = \mathbf{c}_n^H\mathbf{b}(\theta)e^{-jkD_{r(n-1)}\sin\theta} r(t,\theta) \ (n = 1, \dots N) \tag{7}$$

where $\mathbf{c}_n = \begin{bmatrix} c_{n1} & c_{n2} & \cdots & c_{nP} \end{bmatrix}^T$ are the analog beamforming weights within each subarray, and $\mathbf{b}(\theta) = \begin{bmatrix} 1 & e^{-jkd_{r1}\sin\theta} & \cdots & e^{-jkd_{r(P-1)}\sin\theta} \end{bmatrix}^T$ is the phase difference in the θ direction.

Again, if $\mathbf{c}_1 = \mathbf{c}_2 = \cdots = \mathbf{c}_N = \mathbf{c}$, then we can simplify to $\mathbf{c}_1^H \mathbf{b}(\theta) = \mathbf{c}_2^H \mathbf{b}(\theta) = \cdots \mathbf{c}_N^H \mathbf{b}(\theta) = g_r(\theta)$ and the receive signal vector from all subarray is expressed by

$$
\begin{aligned}
\mathbf{r}(r,\theta) &= [r_1(t,\theta)\, r_2(t,\theta) \ldots r_N(t,\theta)]^T \\
&= g_r(\theta)\left[1\, e^{-jkD_{r1}\sin\theta} \ldots e^{-jkD_{r(N-1)}\sin\theta}\right]^T r(t,\theta) = g_r(\theta)\mathbf{h}(\theta)r(t,\theta)
\end{aligned}
\tag{8}
$$

where $\mathbf{h}(\theta) = \left[1\, e^{-jkD_{r1}\sin\theta} \ldots e^{-jkD_{r(N-1)}\sin\theta}\right]^T$.

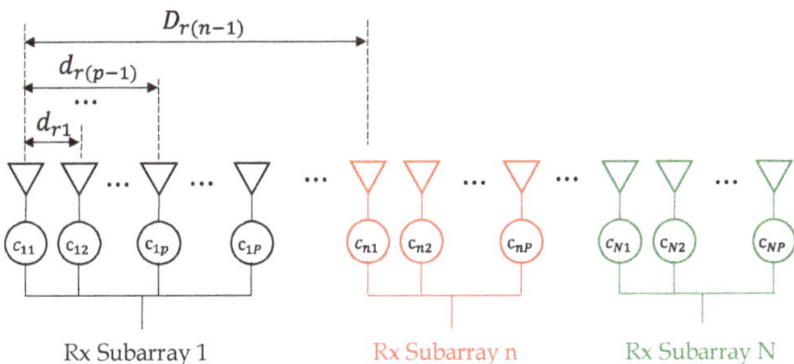

Figure 2. Configuration of the receive antenna with subarrays.

Each transmit waveform is recovered by matched filtering with $\{\phi(t)\}_{m=1}^{M}$. The m-th signal after matched filtering is

$$
\mathbf{x}_m(\theta) = \int_{T_0} \mathbf{r}(t,\theta)\phi_m^*(t)dt = \beta g_t(\theta)g_r(\theta)d_m\mathbf{h}(\theta) + \tilde{n}_m(t),\ m = 1\ldots,M
\tag{9}
$$

where d_m is the m-th element of the vector $\mathbf{d}(\theta)$ in (4). Thus, the final MIMO received signal of $NM \times 1$ is represented as follows:

$$
\mathbf{y} = \left[\mathbf{x}_1^T(\theta)\ \mathbf{x}_2^T(\theta)\ \ldots\ \mathbf{x}_M^T(\theta)\right]^T = \beta g_t(\theta)g_r(\theta)\,\mathbf{d}(\theta) \otimes \mathbf{h}(\theta) + \tilde{N}
\tag{10}
$$

The above equation is the same as the typical MIMO formula except for the subarray gains of $g_t(\theta)$ and $g_r(\theta)$. After MIMO VAA beamforming with the weight vector $\mathbf{w}_{\text{MIMO}} \in C^{NM \times 1}$, the final beam pattern is expressed by

$$
G(\theta) = \left|\mathbf{w}^H \mathbf{a}(\theta)\right|^2 \left|\mathbf{c}^H \mathbf{b}(\theta)\right|^2 \left|\mathbf{w}_{\text{MIMO}}^H [\mathbf{d}(\theta) \otimes \mathbf{h}(\theta)]\right|^2 = |g_t(\theta)|^2 |g_r(\theta)|^2 |g_{\text{MIMO}}(\theta)|^2
\tag{11}
$$

The final pattern revealed is the multiplication of the transmitter subarray pattern, the receiver subarray pattern, and the MIMO VAA beampattern.

Here, we can summarize three design factors of the subarray MIMO antenna:

- the orthogonal waveforms $\{\phi_m\}$, which are not a subject of this manuscript, but are an important issue. Conventionally, the orthogonality is obtained by time-domain multiplexing (TDM), frequency domain multiplexing (FDM), or code domain multiplexing (CDM) [18–21]. Beat-frequency multiplexing or Doppler-domain multiplexing (DDM) is also proposed in automotive radars [3,5];
- the antenna structure, including the subarrays and MIMO configuration; and
- the beamforming weights.

The antenna structure is represented by $\mathbf{a}(\theta)$, $\mathbf{b}(\theta)$, and $\mathbf{d}(\theta)$, while $\mathbf{h}(\theta)$. $\mathbf{a}(\theta)$ and $\mathbf{b}(\theta)$ are the subarray configuration, and $\mathbf{d}(\theta)$ and $\mathbf{h}(\theta)$ are the MIMO configuration which results from the structure among subarrays. The beamforming weights are $\mathbf{w}$ and $\mathbf{c}$, and $\mathbf{w}_{\text{MIMO}}$. $\mathbf{w}$ and $\mathbf{c}$ are the transmit and receive subarray weights, respectively, designed for suppressing the sidelobe level as well as steering the subarray beam direction. $\mathbf{w}_{\text{MIMO}}$ is the beamforming weight of MIMO VAA for multiple beamforming, adaptive beamforming, or any purpose.

2.2. MIMO Configuration

A MIMO antenna is typically defined as having transmit arrays with L = 1 in Figure 1 and receive arrays with $p = 1$ in Figure 2, implying that it does not have subarrays. The VAA beam pattern is determined in this case by the array spacing D_t, D_r and the MIMO beamforming weight. From Equation (10), the input is written as follows:

$$\mathbf{y} = \left[\mathbf{x}_1^T(\theta)\,\mathbf{x}_1^T(\theta)\,\cdots\,\mathbf{x}_1^T(\theta)\right]^T = \beta\mathbf{d}(\theta)\otimes\mathbf{h}(\theta) + \tilde{N} = \beta\,\mathbf{v}(\theta) + \tilde{N} \tag{12}$$

where $\mathbf{v}(\theta) = \mathbf{d}(\theta)\otimes\mathbf{h}(\theta) = [v_1(\theta), v_2(\theta),\cdots,v_{NM}(\theta)]^T$ and the element is

$$v_{[n+(m-1)N]}(\theta) = e^{-jk\sin\theta[D_{t(m-1)}+D_{r(n-1)}]},\ m = 1,\ldots M \text{ and } n = 1,\ldots,N \tag{13}$$

This equation is more commonly referred to as the spatial convolution of transmit and receive array positions. If the transmit and receive antenna has a uniform interval, then we can write that $D_{t(m-1)} = (m-1)D_t$ and $D_{r(n-1)} = (n-1)D_r$. The maximum length of the VAA that can be implemented with $(M+N)$ arrays is $MN \times D_r$, when $D_t = ND_r$ (or $D_r = MD_t$). In addition, the maximum length is achieved when the number of transmit arrays is the same as the number of receive arrays, i.e., $M = N = K$ if $M+N = 2K$. However, in this case, the physical lengths of the antennas are $(M-1) \times ND_r$, and ND_r, respectively, so the transmit antenna is $(M-1)$ times longer than the receive antenna. For example, if $M+N = 8$, the maximum length of the VAA is $16D_r$ when $M = N = 4$. The length of the transmit antenna is $12D_r$, which is three times longer than that of the receive antenna, $4D_r$. A similar physical size of the two antennas can be obtained using two transmit arrays and six receive arrays, but the length of the virtual antenna is reduced to $12D_r$. Figure 3 shows VAAs for two cases.

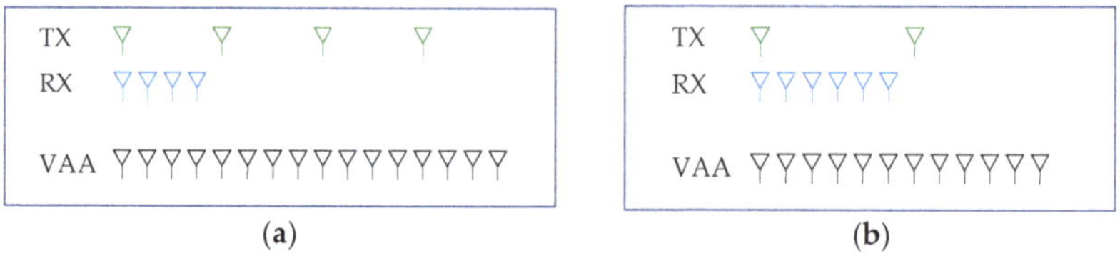

Figure 3. VAA configuration with eight arrays: (**a**) four Tx and four Rx (left) (**b**) two Tx and six Rx (right).

Although the beam pattern of the VAA varies depending on the selection of weights, if we choose $\mathbf{w}_{MIMO} = \mathbf{d}(\theta_s)\otimes\mathbf{h}(\theta_s)$, which is the conventional beamforming weight for a uniform array, the resulting beam pattern becomes the product of the transmit pattern and the receive pattern is as follows:

$$\begin{aligned}(\theta) &= \left|\mathbf{w}_{MIMO}^H[\mathbf{d}(\theta)\otimes\mathbf{h}(\theta)]\right|^2 = \left|[\mathbf{d}(\theta_s)\otimes\mathbf{h}(\theta_s)]^H[\mathbf{d}(\theta)\otimes\mathbf{h}(\theta)]\right|^2 \\ &= \left|\mathbf{d}^H(\theta_s)\mathbf{d}(\theta)\right|^2\left|\mathbf{h}^H(\theta_s)\mathbf{h}(\theta)\right|^2 = |g_{tx}(\theta)|^2|g_{rx}(\theta)|^2\end{aligned} \tag{14}$$

where θ_s is the steering direction. If the beam direction is fixed to only one angle, it is theoretically the same pattern as performing transmit beamforming by $\mathbf{d}(\theta)$ and receive beamforming by $\mathbf{h}(\theta)$ separately. However, in the MIMO approach, the beamforming is performed only in the receiver, and the number of weights increases MN. Thus, it has more degree of freedom to make multiple beams simultaneously in several directions or perform adaptive beamforming, which is usually performed digitally.

2.3. Transmit and Receive Subarray

In the case of a MIMO antenna with transmit and receive subarrays, the final pattern is the product of the subarray pattern and the MIMO pattern, as written in Equation (11). If w or c steers the subarray pattern at a specific angle, MIMO beamforming to a different angle is bound by the subarray pattern and suffers a loss. The loss increases as the beamwidth of the subarray gets smaller, i.e., as the size of the subarray becomes larger. The idea is the same as how a single array pattern constrains the array antenna pattern.

Figure 4 shows a case where each transmit array of Figure 3b is replaced by subarrays with four arrays. The space of all the arrays is set to 0.5 wavelengths. The resulting beam patterns are shown in Figure 5. The MIMO pattern is obtained by the Figure 3b configuration, and four arrays make the subarray pattern. The final pattern becomes a product of these two patterns, which has the good effect of suppressing the sidelobe but shows a loss of gain when steering at $10°$ compared with the gain at $0°$.

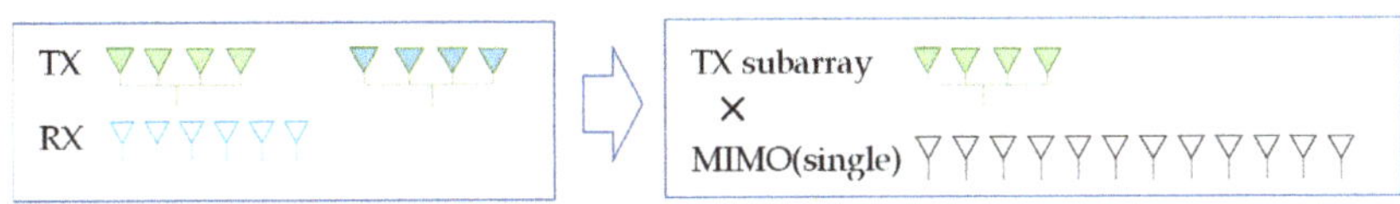

Figure 4. The final beampattern with TX subarrays and RX arrays shown on the left is the product of the beampattern of a TX subarray, and that of MIMO made by a single TX array is shown on the right. One TX subarray transmits the identical waveform.

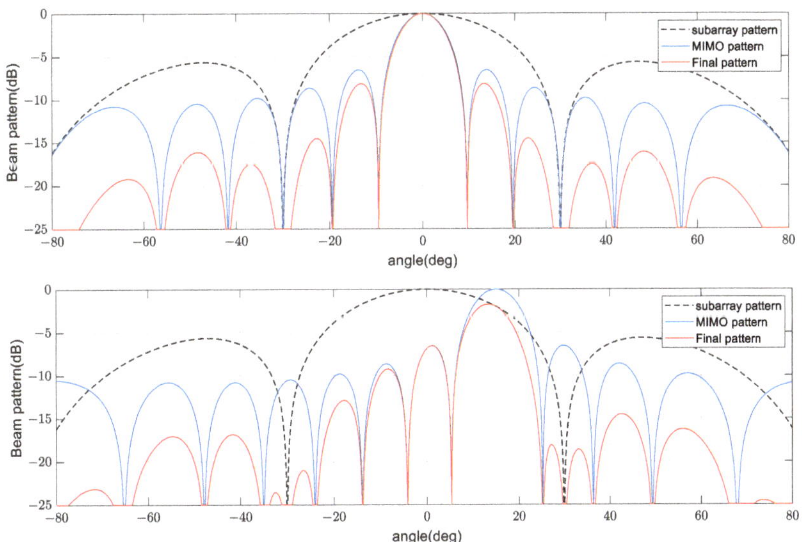

Figure 5. The final beampattern with TX subarrays in the case of steering to $0°$ (**upper**) and $15°$ (**lower**).

The adoption of the transmit subarray can improve the SNR by increasing the output power and has the effect of suppressing the sidelobe. However, because it restricts the angle of multi-beams, it is recommended to use a window to increase the beamwidth. Figure 6 shows the beampattern created by applying a Taylor window to the transmit subarray. It demonstrates that the sidelobe is still suppressed while the loss in the $10°$ beam is decreased.

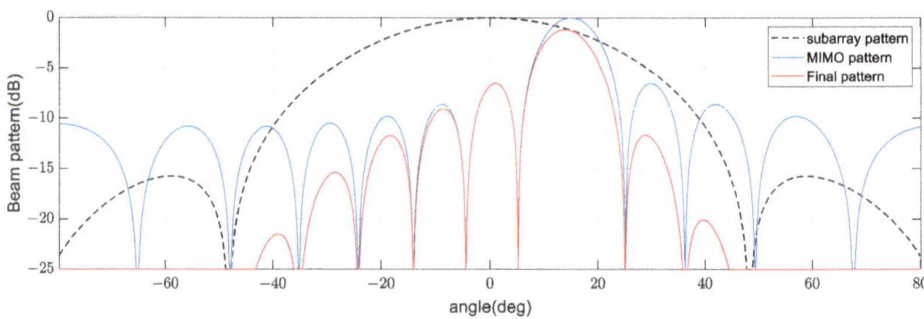

Figure 6. The final beampattern when a Taylor window is applied to subarray beamforming.

On the other hand, when the receive array is composed of an analog beamforming subarray without overlap, another consequence appears in addition to the pattern constraint. Figure 7 shows the receive antenna configured in the form of a subarray. In this case, the spacing of the MIMO receiver is no longer half a wavelength and thus results in the grating lobe of the MIMO pattern, as shown in Figure 8. The final pattern is the same as the uniform array pattern if both the subarray beamforming on an analog receiver and the MIMO beamforming are performed in the same direction. However, if the angle of the MIMO beamforming is changed in a different direction, the grating lobe and the subarray pattern modify the final pattern at the same time.

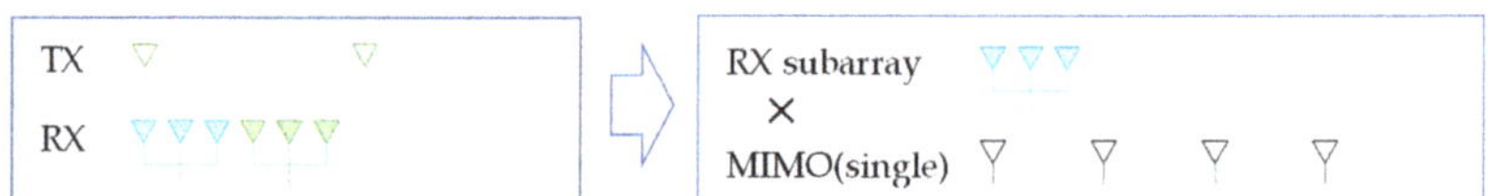

Figure 7. The final beampattern with TX arrays and RX subarrays shown on the left is the product of the beampattern of the RX subarray, and that of MIMO made by a single RX array is shown on the right. Two TX arrays transmit different waveforms.

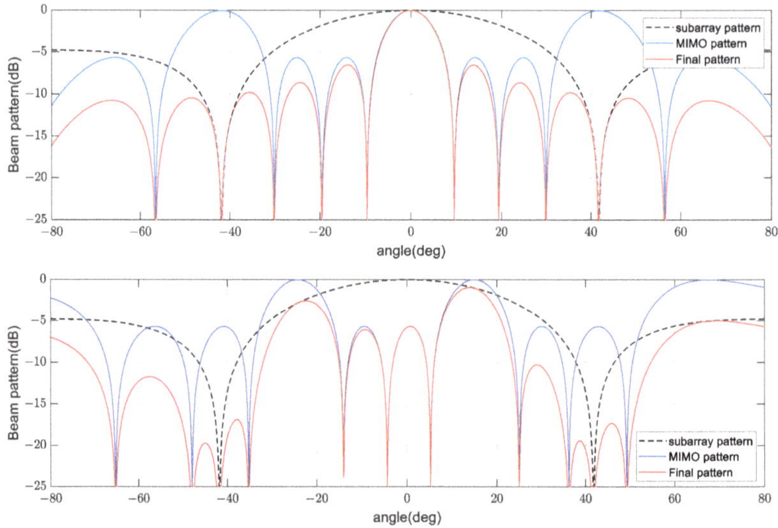

Figure 8. The final beampattern with RX subarrays in the case of steering to $0°$ (**upper**) and $15°$ (**lower**).

Although the receiver subarray is unavoidable due to the space constraint of hardware as frequency increases, the performance is somewhat limited compared with the full digital array.

3. Design of MIMO Array

3.1. Minimum Redundancy Array

The MRA is a class of non-uniform linear array designed to minimize the number of sensor pairs with the same spatial correlation lag. If this number of sensor pairs is 1, it is called a perfect array. If we assume that the arrays are located on an underlying grid with unit spacing and that w is a vector having 1 at the array position and 0 at the other location, then the number of times each spatial correlation lag is computed by the autocorrelation of the position vector with itself, which is called the co-array [22].

$$c(\gamma) = \sum_{k=1}^{N_e} w(k)w(k - \gamma) \tag{15}$$

where N_e is the size of w, $w(k)$ is k-th element of w, and $w(k) = 0$ if $k < 1$ or $k > N_e$. For example, a vector w = [1 1 0 0 1 0 1] has 4 arrays (N) and the size (N_e) is 7. As illustrated in Figure 9, in this case, all co-arrays have one except when $\gamma = 0$, indicating that the sensor pair with the same spatial correlation lag is 1. This type of co-array is called a perfect array. However, because there is no perfect array for $N > 4$, it should be chosen between non-redundant arrays that partially permit holes and MRA with no holes. A hole is a position where the co-array value is 0, and redundancy is a position where the co-array value is greater than 1, where γ is not 0. The following equation determines the final size:

$$N_e = \frac{N(N - 1)}{2} - N_R + N_H + 1 \tag{16}$$

where N_R is the number of redundancies and N_H is the number of holes.

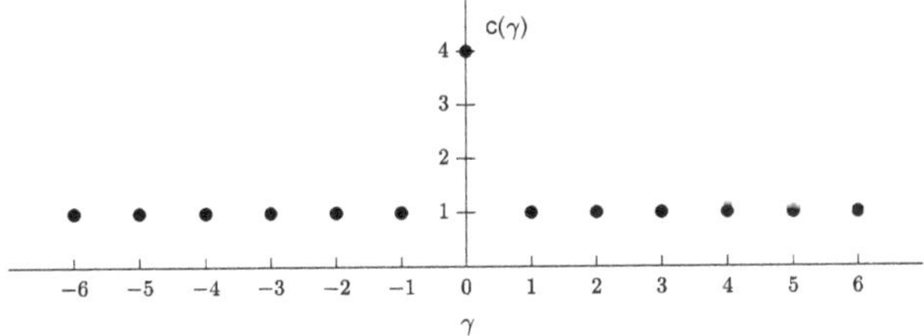

Figure 9. Co-array for the MRA when $N = 4$.

The MRA is designed to make the largest possible aperture without holes. There has been a significant amount of research on element spacing to achieve as low a redundancy as possible for arrays of up to 30 elements [23–25].

3.2. Non-Uniform MIMO Array Configuration

The co-array property is applied to a bistatic MIMO VAA. Because the location of the MIMO virtual arrays consists of the spatial convolution of the transmit and receive arrays, if we choose the receive array as the reverse of the transmit arrays, the resulting VAA is the co-array structure with a length approximately twice that of the transmit array. When the size of a single antenna is N_e, the size of the resulting VAA is as follows:

$$N_{VAA} = 2N_e - 1 \tag{17}$$

In other words, a VAA composed of an MRA generates uniform arrays, and one made up of non-redundant arrays results in sparse arrays. For example, suppose the transmit arrays and receive arrays are configured with a perfect array of $N = 4$ and $N_e = 7$ in the previous section. In that case, the VAA can have the same beam pattern as that of uniform arrays with 13 elements, as shown in Figure 10.

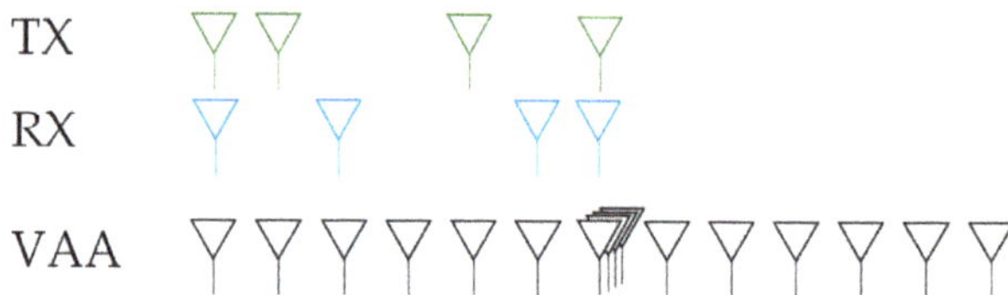

Figure 10. TX and RX antenna with an MRA structure and the resulting VAA.

For each N, the size of non-redundant arrays, $N_{e,nr}$, and the size of the MRA, $N_{e,mr}$, are summarized in Table 1. The size of the MIMO VAA for each is also listed. In addition, the VAA size for the case of using two transmit arrays and $(2N - 2)$ receive arrays is written, in which the transmit antenna and the receive antenna have the same size. Table 1 shows that when N is greater than 3, VAA apertures using non-uniform arrays are longer than the aperture using conventional two-transmit array structures. The VAA aperture ratio increases with N; when $N = 8$, the ratio is up to 2.46 and 1.68, respectively.

Table 1. List of VAA sizes for the non-uniform array configurations and number.

N	Non-Redundant Array		Minimum Redundancy Array		2-Transmit Arrays	
	$N_{e,nr}$	N_{VAA}	$N_{e,mr}$	N_{VAA}	N_{rx}	N_{VAA}
3	4	7	4	7	6	4
4	7	13	7	13	6	12
5	12	23	10	19	8	16
6	18	35	14	27	10	20
7	26	51	18	35	12	24
8	35	69	24	47	14	28

When using five transmit and receive arrays, it is possible to make a VAA with 23 sizes using non-redundant arrays with an interval of (1,3,5,2) and a VAA with 19 sizes using an MRA with an interval of (1,3,3,2). Both are larger than a VAA with 16 sizes, which is composed of 10 arrays of two transmit arrays and eight receive arrays as shown in Figure 11.

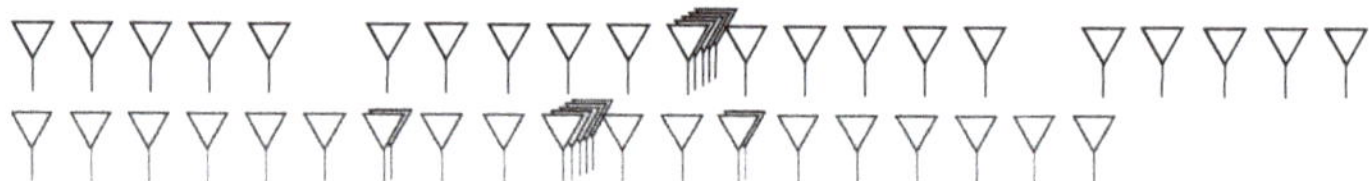

Figure 11. Non-redundant array with an interval of (1,3,5,2) can make a 23-sized VAA (**up**) and the MRA with an interval of (1,3,3,2) can make a 19-sized VAA (**down**).

Figure 12 shows the beampattern for each case. The beamforming is performed by adjusting the weight to make it the same as for uniform arrays, although the hole cannot be filled out. The non-redundant array has less beamwidth than the MRA because of the larger aperture size. However, due to the two holes, it has different positions and higher sidelobe than 23 uniform arrays. Both the MRA and non-redundant arrays have small beamwidth compared with 16 uniform arrays with two transmit arrays.

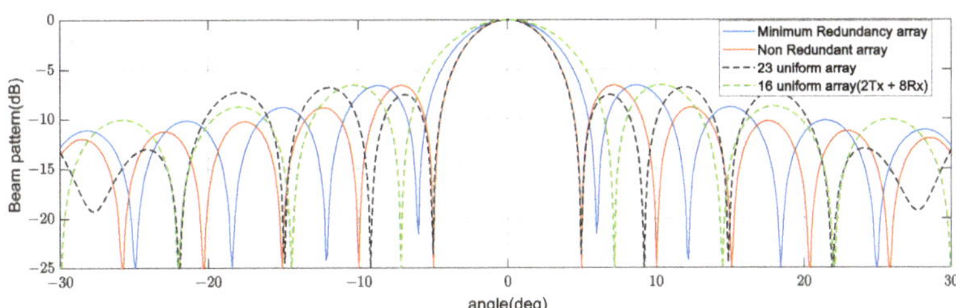

Figure 12. Comparison of beam patterns at $N = 5$. VAA's beampattern by the non-redundant array shows the smallest beamwidth, but the sidelobe is similar to that by the MRA.

4. Proposed MRA MIMO with Subarrays

The minimum redundancy configuration can be applied to D_t in Figure 1 and D_r in Figure 2 to build a subarray MIMO structure using a multi-channel transceiver. The number of arrays that comprise the subarray, L and P, depends on the hardware structure of the transceiver, which are four in this paper.

Assume five transmission waveforms and five receive antennas, as shown in Figure 13. The spacing of the arrays is 0.5 wavelength and the minimum spacing of the subarrays is 2 wavelengths. The ratio of the spacing between subarrays is 1:3:3:2. The aperture of the transmit and the receive antenna is 20 wavelengths, and the final aperture is doubled to 40 wavelengths.

Figure 13. MRA MIMO Configuration. The ratios of the spacing among subarrays are 1:3:3:2 and 2:3:3:1. Each TX and RX subarray is comprised of 4 uniform arrays.

According to Equation (11), the final beam pattern is the product of the transmit subarray beampattern, the receive subarray beampattern, and the MIMO beampattern. Again, MIMO beamforming is performed by adjusting the weight to make it the same as for uniform arrays.

When the transmit subarray beamforming, the receive subarray beamforming, and the MIMO beamforming are performed at the same boresight angle, say $0°$, we can get the same result as the uniform array, as shown in the upper graph of Figure 14. However, if we control the MIMO digital beam at $5°$, it suffers a loss by the subarray beampattern, and the grating lobe also occurs, as shown in the lower graph of Figure 14. In other words, the subarray beamwidth limits the MIMO beamforming angle. The transmit subarray pattern in the figure employs the Taylor window to broaden the beamwidth, whereas the receive subarray pattern does not.

Next, Figure 15 shows the configuration in which the non-redundant structure is applied to MIMO. Likewise, the spacing of the arrays is 0.5 wavelengths, the spacing of the subarrays has two wavelengths as the minimum spacing, and the spacing ratio is 1:3:5:2. The total aperture is 48 wavelengths longer than the minimum redundancy configuration, so the beam width is improved. However, the sidelobe characteristics deteriorate due to holes in the MIMO structure. The upper graph in Figure 16 shows the beampattern formed when the steering angle of the subarrays and the steering angle of the MIMO beam are equal to $0°$, and the lower graph shows the beampattern when only the MIMO steering angle is changed to $5°$.

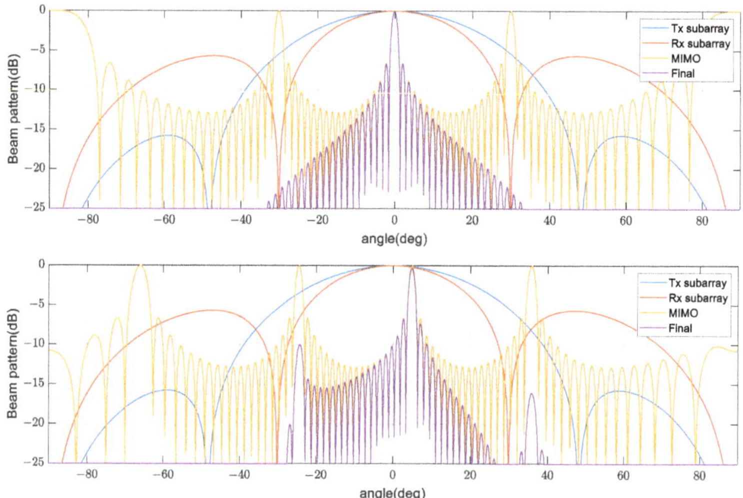

Figure 14. Final beampatterns by minimum redundant MIMO steered to $0°$ (**up**) and $5°$ (**down**).

Figure 15. Non-redundant MIMO configuration. The ratios of the spacing among subarrays are 1:3:5:2 and 2:5:3:1. Each TX and RX subarray is comprised of 4 uniform arrays.

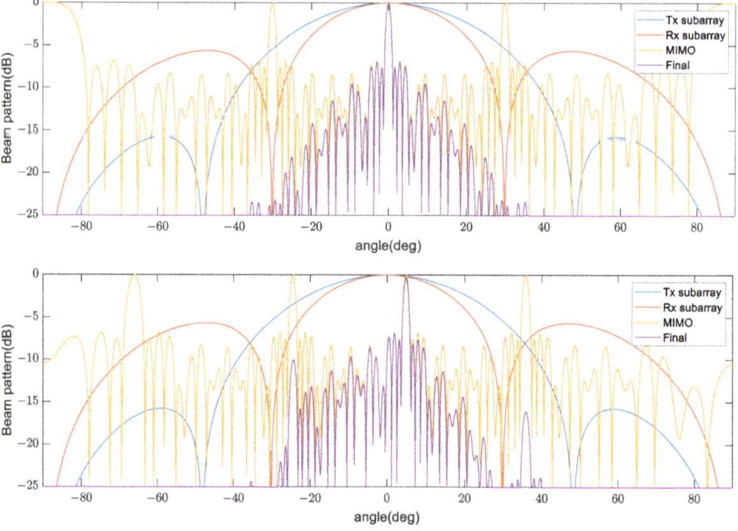

Figure 16. Final beampatterns by non-redundant MIMO steered to $0°$ (**up**) and $5°$ (**down**).

Finally, if two transmitters and eight receivers with about 18 wavelengths are configured as shown in Figure 17, the final aperture becomes a uniform array of 36 wavelengths.

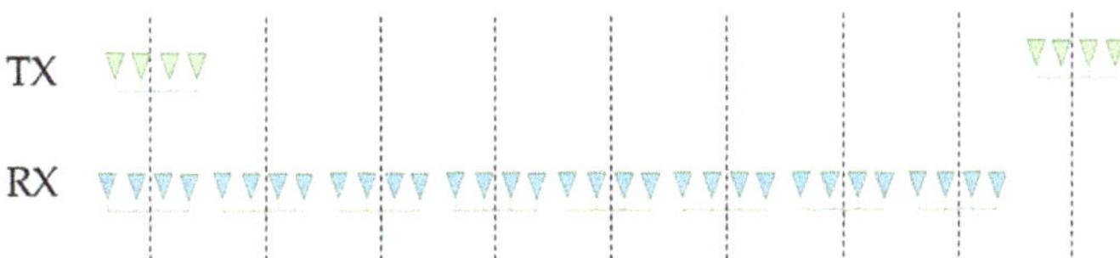

Figure 17. Two transmit waveforms and eight receive subarrays.

Figure 18 compares beampatterns for three configurations steered to 5°. As expected from the aperture length, the non-redundant array has the smallest beamwidth and the worst sidelobe characteristic. Grating lobes due to the minimum interval of the subarray are common.

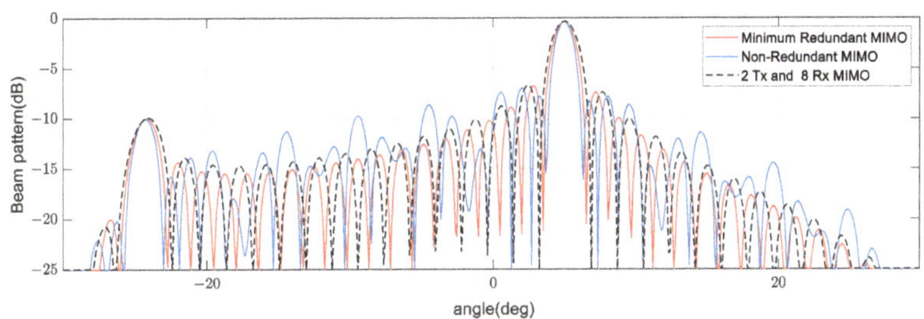

Figure 18. Comparison of beampatterns steered to 5°. The non-redundant MIMO has the smallest beamwidth but the worst sidelobe characteristic. Grating lobes are common.

5. Conclusions

We devised a new method for placing transmit and receive arrays for MIMO VAA with non-redundant or minimum redundant structures. In contrast to the conventional arrangement, wherein one of the antennas has a relatively long physical size, the proposed design can increase the VAA aperture while keeping the transmit and receive antennas at the same size.

In addition, we applied the proposed method to the MIMO antenna with subarrays and analyzed the beampatterns. Subarrays restrict the direction of multiple beamforming, and produce grating lobes if the subarrays do not overlap and the interval is substantially greater than a half waveform. However, the subarray structure is expected to be essential for small antennas with multi-channel transceivers, which are necessary for moving to a high operating frequency to improve the range resolution and miniaturize antennas. The goal of this study was to develop a D-band radar; the results will be implemented using multi-channel transceivers currently under development.

Author Contributions: Conceptualization, E.K.; methodology, E.K.; validation, I.K.; formal analysis, E.K.; investigation, I.K.; writing—original draft preparation, E.K.; writing—review and editing, I.K.; project administration, W.K.; funding acquisition, W.K. All authors have read and agreed to the published version of the manuscript.

Funding: This research was supported by the Challengeable Future Defense Technology Research and Development Program (No. 912913601) of Agency for Defense Development in 2020.

Data Availability Statement: Not applicable.

Conflicts of Interest: The authors declare no conflict of interest.

References

1. Patole, S.M.; Torlak, M.; Wang, D.; Ali, M. Automotive radars: A review of signal processing techniques. *IEEE Signal Process. Mag.* **2017**, *34*, 22–35. [CrossRef]
2. Kim, E.H.; Kim, K.H. Efficient implementation of the ML estimator for high-resolution angle estimation in an unmanned ground vehicle. *IET Radar Sonar Navig.* **2018**, *12*, 145–150. [CrossRef]
3. Waldschmidt, C.; Hasch, J.; Menzel, W. Automotive radar—From first efforts to future systems. *IEEE J. Microw.* **2021**, *1*, 135–148. [CrossRef]
4. Sun, S.; Petropulu, A.P.; Poor, H.V. MIMO radar for advanced driver-assistance systems and autonomous driving: Advantages and challenges. *IEEE Signal Process. Mag.* **2020**, *37*, 97–117. [CrossRef]
5. Hakobyan, G.; Yang, B. High-performance automotive radar: A review of signal processing algorithms and modulation schemes. *IEEE Signal Process. Mag.* **2019**, *36*, 32–44. [CrossRef]
6. Donnet, B.J.; Longstaff, I.D. MIMO radar, techniques and opportunities. In Proceedings of the Radar Conference, EuRAD 2006, 3rd European, IEEE, Manchester, UK, 13–15 September 2006; pp. 112–115.
7. Li, J.; Stoica, P. MIMO radar with colocated antennas. *IEEE Signal Process. Mag.* **2007**, *24*, 106–114. [CrossRef]
8. Issakov, V.; Bilato, A.; Kurz, V.; Englisch, D.; Geiselbrechtinger, A. A highly integrated D-band multi-channel transceiver chip for radar applications. In Proceedings of the 2019 IEEE BiCMOS and Compound semiconductor Integrated Circuits and Technology Symposium, Nashville, TN, USA, 3–6 November 2019; pp. 1–4.
9. De Kok, M.; Smolders, A.B.; Johannsen, U. A review of design and integration technologies for D-band antennas. *IEEE Open J. Antennas Propag.* **2021**, *2*, 746–758. [CrossRef]
10. Cakraborty, A.; Trotta, S.; Wuertele, J.; Weigel, R. A D-band transceiver front-end for broadband applications in a 0.35um SiGe bipolar technology. In Proceedings of the 2014 IEEE Radio Frequency Integrated Circuits Symposium, Tampa, FL, USA, 1–3 June 2014; pp. 405–408.
11. Jaeschke, T.; Bredendiek, C.; Küppers, S.; Pohl, N. High-precision D-band FMCW-radar sensor based on a wideband SiGe-transceiver MMIC. *IEEE Trans. Microw. Theory Tech.* **2014**, *62*, 3582–3597. [CrossRef]
12. Furqan, M.; Ahmed, F.; Aufinger, K.; Stelzer, A. A D-band fully-differential quadrature FMCW radar transceiver with 11 dBm output power and a 3-dB 30-GHz bandwidth in SiGe BiCMOS. In Proceedings of the 2017 IEEE MTT-S International Microwave Symposium (IMS), Honolulu, HI, USA, 4–9 June 2017; pp. 1404–1407.
13. Ng, H.J.; Kucharski, M.; Ahmad, W.; Kissinger, D. Multi-purpose fully differential 61-and 122-GHz radar transceivers for scalable MIMO sensor platforms. *IEEE J. Solid State Circuits* **2017**, *52*, 2242–2255. [CrossRef]
14. Mailloux, R.J. *Phased Array Antenna Handbook*, 3rd ed.; Artech House: London, UK, 2018.
15. Hassanien, A.; Vorobyov, S.A. Phased-MIMO radar: A tradeoff between phased-array and MIMO radars. *IEEE Trans. Signal Process.* **2010**, *58*, 3137–3151. [CrossRef]
16. Hassanien, A.; Vorobyov, S.A. Why the phased-MIMO radar outperforms the phased-array and MIMO radars. In Proceedings of the 2010 18th European Signal Processing Conference, IEEE, Aalborg, Denmark, 23–27 August 2010; pp. 1234–1238.
17. Reza, A.; Muttaqin, H.; Miftachul, U. Phased-MIMO Radar: Angular resolution. In Proceedings of the IOP Conference Series: Materials Science and Engineering 1125, Makassar, Indonesia, 16–17 October 2020; p. 012046.
18. Deng, H.; Geng, Z.; Himed, B. MIMO Radar Waveform Design for Transmit Beamforming and Orthogonality. *IEEE Trans. Aerosp. Electron. Syst.* **2016**, *52*, 1421–1433. [CrossRef]
19. Kim, E.; Kim, K. Random phase code for automotive MIMO radars using combined frequency shift keying-linear FMCW waveform. *IET Radar Sonar Navig.* **2018**, *12*, 1090–1095. [CrossRef]
20. Kim, E. MIMO FMCW Radar with Doppler-Insensitive Polyphase Codes. *Remote Sens.* **2022**, *14*, 2595. [CrossRef]
21. De Wit, J.J.M.; Van Rossum, W.L.; De Jong, A.J. Orthogonal waveforms for FMCW MIMO radar. In Proceedings of the Radar Conference (RADAR), IEEE, Kansas City, MO, USA, 23–27 May 2011; pp. 686–691.
22. Van Trees, H.L. *Optimum Array Processing: Part IV of Detection, Estimation, and Modulation Theory*; John Wiley & Sons: Hoboken, NJ, USA, 2004.
23. Pearson, D.; Pillai, S.U.; Lee, Y. An algorithm for near-optimal placement of sensor elements. *IEEE Trans. Inf. Theory* **1990**, *36*, 1280–1284. [CrossRef]
24. Linebarger, D.A.; Sudborough, I.H.; Tollis, I.G. Difference bases and sparse sensor arrays. *IEEE Trans. Inf. Theory* **1993**, *39*, 716–721. [CrossRef]
25. Abramovich, Y.I.; Spencer, N.K.; Gorokhov, A.Y. Positive-definite Toeplitz completion in DOA estimation for nonuniform linear antenna arrays. II. Partially augmentable arrays. *IEEE Trans. Signal Process.* **1999**, *47*, 1502–1521. [CrossRef]

remote sensing

Article

Phase Characteristics and Angle Deception of Frequency-Diversity-Array-Transmitted Signals Based on Time Index within Pulse

Changlin Zhou [1], Chunyang Wang [1], Jian Gong [1,*], Ming Tan [2], Lei Bao [3] and Mingjie Liu [1]

[1] Air and Missile Defense College, Air Force Engineering University, Xi'an 710051, China
[2] College of Information and Communication, National University of Defense Technology, Wuhan 430010, China
[3] Test and Training Base, National University of Defense Technology, Xi'an 710106, China
* Correspondence: drgong@aliyun.com

Abstract: The transmitted beam of frequency diversity array (FDA) has the range–angle–time coupling property, which has essential applicative potential in angle deception and active anti-jamming. In this paper, the concept of time index within pulse is introduced. Firstly, the phase characteristics of FDA-transmitted signals based on the time index within pulse concept are studied. Then, the deceptive angle performance of FDA-transmitted signals is discussed. The theoretical analysis and simulation results show that the phase characteristics of the FDA signal are not related to the range, but to the time index within pulse. With the phase center as the reference point, the phase is equal as long as the time index within the pulse is the same. Angle deception and active anti-jamming can be achieved using the optimized frequency increment of each FDA.

Keywords: frequency diversity array (FDA); time index within pulse; phase center; angle deception; active anti-jamming

check for
updates

Citation: Zhou, C.; Wang, C.; Gong, J.; Tan, M.; Bao, L.; Liu, M. Phase Characteristics and Angle Deception of Frequency-Diversity-Array-Transmitted Signals Based on Time Index within Pulse. *Remote Sens.* **2023**, *15*, 5171. https://doi.org/10.3390/rs15215171

Academic Editor: Gerardo Di Martino

Received: 6 September 2023
Revised: 16 October 2023
Accepted: 17 October 2023
Published: 30 October 2023

1. Introduction

Frequency diversity array (FDA) has attracted wide attention since it was first proposed [1]. Unlike phased array (PA), which transmits signals with the same carrier frequency, FDA introduces different frequency increments in each array element so that the transmitted beam is range–angle–time-dependent [2–4]. When the frequency increment is zero, FDA simplifies to PA so that PA is a particular case for FDA. Due to the increased freedom of range dimension and improved capacity for information processing, FDA has potential application value in many areas, such as radar detection, positioning, and anti-jamming [5,6].

When the frequency increment of FDA increases linearly, its transmitted beam is coupled in range, angle, and time [7]. Regarding de-coupling, many scholars pay attention to the transmitted beam with a non-uniform increase in the frequency increment of FDA. For example, Log frequency offset [8], cubic frequency offset [9], multi-carrier frequency [10], random frequency offset [11], time-dependent frequency offset [12,13], and genetic algorithm optimization frequency increment [14] are used to achieve beam focusing. However, these methods ignore the influence of the time factor and lead to an instantaneous beam. The literature [15] points out that these methods ignore the signal propagation process, and that the beam cannot only focus on a specific position in space and last for a particular time. The literature [16] emphasizes that the propagation process of the electromagnetic wave cannot be ignored. The literature [17–19] has obtained the corrected pulse FDA expression by analyzing the frequency–phase and time–range relationships. Furthermore, the literature [20] points out that the FDA beam is time–angle-dependent, not range-dependent, and that a reasonable signal processing scheme at the receiving end is necessary for activating

FDA distance correlation [21,22]. Ref. [23] further explains the relationship between the time index within pulse and real-time, which is similar to the relationship between radar fast time and slow time. However, ref. [20] does not give the relationship between the time index within pulse and the phase.

Radar needs not only a good detection performance, but also a good capacity for anti-jamming. With the development of digital radio frequency memory (DRFM), active jammers can produce complex and flexible jamming signals, which seriously threaten the performance of radar systems. Therefore, radar anti-jamming technology is essential. FDA radar beam can activate range correlation. Based on this, many scholars have researched the suppression of deceptive interference [24–30]. However, these methods are targeted at specific interference scenes. The literature [31] discusses the low interception performance of FDA-transmitted signals, indicating that FDA-transmitted signals are less likely to be intercepted by jammers than PA-transmitted signals, thus making it difficult for jammers to target jamming signals in arrays. The literature [32] has preliminarily discussed the possibility of FDA's resistance to interferometer-based direction of arrival (DOA) reconnaissance. On this basis, the literature [33–37] has studied the phase characteristics of FDA's transmitted signals and their cheating effect on the interferometer. However, all such studied have taken the spatial phase and range into one-to-one correspondence, have not considered the influence of wave propagation, and have ignored the time index within pulse. At the same time, it has also been ignored that interferometer direction finding is based on its measurement and the estimation of the incident signal frequency, and that it then estimates the direction of the phase center of the received signal. Inspired by the study of uniform linear PA phase centers in the literature [38], based on the time index within pulse, this paper discusses the phase characteristics of FDA-transmitted signals and the calculation method of the phase center, and uses the adjustable phase center to achieve FDA active anti-jamming.

Finally, for the jammer based on the interferometer to determine the position of the array radar, this paper proposes an active anti-jamming method so that the jammer signal cannot be aligned with the array, even by setting multiple phase center flashing so that the jammer completes the direction finding with difficulty. The main contributions of this paper are as follows:

1. The phase characteristics of the FDA-transmitted signal based on the time index within pulse are analyzed.
2. The phase center of the FDA-transmitted signal is calculated, and the theoretical basis of FDA angle deception is analyzed by adjusting the phase center.
3. An active anti-jamming algorithm based on the FDA phase center is proposed by optimizing each element's frequency increment.

The rest of this paper is arranged as follows. Sections 2 and 3 derive and analyze the phase characteristics of PA and FDA, respectively. Next, Section 4 presents the calculation method of the phase center of the array-transmitted signal and discusses how FDA implements angle deception. Finally, the theoretical analysis is verified through simulation in Section 5, and the conclusion is given in Section 6.

Notation: We use boldface for vectors **a** and matrices **A**. Scalar *a* is denoted by italics. The transpose, conjugate and conjugate transpose are denoted by the symbols $(\cdot)^T$, $(\cdot)^*$ and $(\cdot)^H$, respectively. The letter j represents the imaginary unit (i.e., $j = \sqrt{-1}$).

2. Phase Characteristics of PA

In this section, we first analyze the propagation characteristics of electromagnetic waves to fully understand the relationship between real-time and the time index within pulse, and determine the real-time–phase relationship. Then, the time–angle–phase relationship of the transmitted signal is determined by analyzing the phase of the PA signal based on the time index within pulse.

2.1. Phase Propagation Analysis

Set the time index within pulse as t' and the real-time as t. The pulse signal with the carrier frequency f_0 can be expressed as follows:

$$s(t) = rect(\frac{t}{T_p})\, e^{j2\pi f_0 t} \tag{1}$$

$$rect(\frac{t}{T_p}) = \begin{cases} 1 &, \quad 0 \le t \le T_p \\ 0 &, \qquad else \end{cases} \tag{2}$$

where T_p stands for the pulse duration when the pulse signal propagates to r_1 and r_2; the phase changes in the propagation process are shown in Figure 1.

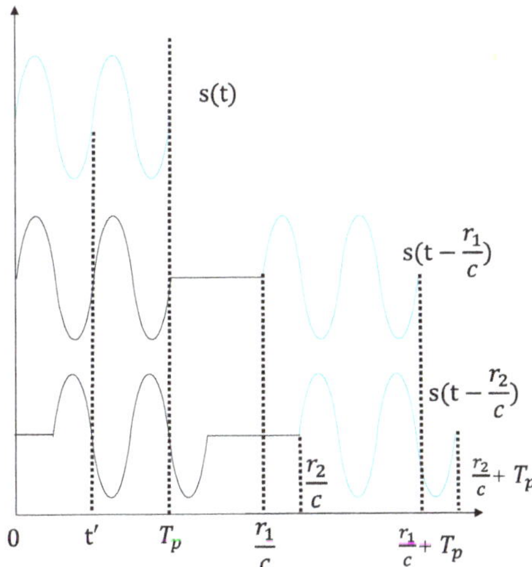

Figure 1. Phase-change process of pulse signal.

The pulse signals of the two moments can be expressed as follows:

$$s(t - r_1/c) = rect(\frac{t - r_1/c}{T_p})\, e^{j2\pi f_0(t - r_1/c)}, \frac{r_1}{c} \le t \le \frac{r_1}{c} + T_p \tag{3}$$

$$s(t - r_2/c) = rect(\frac{t - r_2/c}{T_p})\, e^{j2\pi f_0(t - r_2/c)}, \frac{r_2}{c} \le t \le \frac{r_2}{c} + T_p \tag{4}$$

Figure 1 shows that the pulse signal phase is related to its real-time spatial propagation, which can be divided into two parts: propagation delay τ and the time index within pulse t'. Further, when the time index within pulse is equal, the phases of the pulse signals are also equal. In order to highlight the vital parameter of the time index within pulse, we let $t' = t - \tau$; then, (3) and (4) can be unified into (5).

$$s(t') = rect(\frac{t'}{T_p})\, e^{j2\pi f_0 t'}, 0 \le t' \le T_p \tag{5}$$

Equation (5) can uniformly represent the pulse signals transmitted to different ranges, which means that the phase of the pulse signal is only related to the time index within pulse but has nothing to do with the propagation delay. Therefore, the time index within pulse can be used to replace the real-time when analyzing the phase of the pulse signal.

The continuous wave signal can be regarded as the pulse signal with an infinite pulse time, and its phase is only related to the time index within pulse. In a word, the phase of both the pulse signal and continuous-wave (CW) signal is only related to the time index within pulse and has nothing to do with the propagation delay. The propagation delay is related to the range, which means that the electromagnetic wave signal constantly propagates forward.

2.2. PA Signal Analysis

Through the above analysis, it is clear that the signal phase is related to the time index within pulse. In this section, we will analyze the phase characteristics of PA emission signals so that we can accurately understand the phase characteristics of FDA emission signals.

Consider a uniform linear PA with M isotropic antennas, as shown in Figure 2.

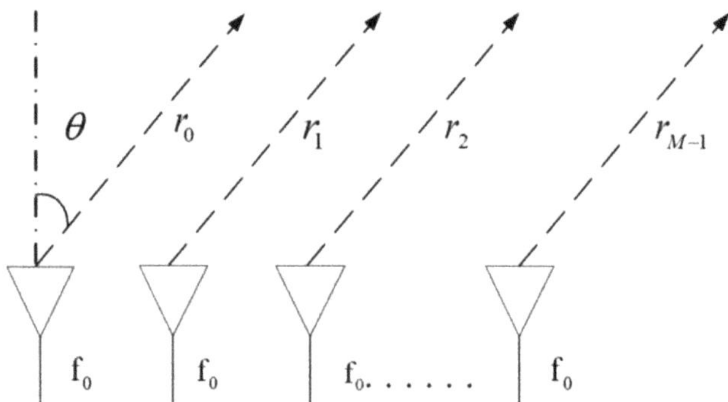

Figure 2. Uniform linear phased array.

The carrier frequency is f_0 and the array spacing is d. When it propagates to the far field with range r and angle θ, the signal can be expressed as follows:

$$y_{PA}(t) = \sum_{m=0}^{M-1} s_m(t) = \sum_{m=0}^{M-1} rect(\frac{t - r_m/c}{T_p})e^{j(2\pi f_0(t - r_m/c))} \tag{6}$$

where $r_m = r_0 - md\sin\theta$. Under the far-field narrow-band condition, the envelope is approximately invariant, as follows:

$$rect(\frac{t - r_m/c}{T_p}) \approx rect(\frac{t - r_0/c}{T_p}) \tag{7}$$

Meanwhile, we replace the real-time t with the time index within pulse t', and Equation (6) can be further expressed as follows:

$$
\begin{aligned}
y_{PA}(t) &= rect(\frac{t - r_0/c}{T_p})e^{j2\pi f_0(t - r_2/c)} \sum_{m=0}^{M-1} e^{j(2\pi f_0(m-1)d\sin\theta/c)} \\
&= rect(\frac{t'}{T_p})e^{j2\pi f_0 t'} \sum_{m=0}^{M-1} e^{j2\pi f_0 md\sin\theta/c} \\
&= rect(\frac{t'}{T_p})e^{j(2\pi f_0 t' + \pi(M-1)f_0 d\sin\theta/c)} \frac{\sin(\pi M f_0 d\sin\theta/c)}{\sin(\pi f_0 d\sin\theta/c)}
\end{aligned}
\tag{8}
$$

According to (8), the amplitude and phase of the PA-transmitted signal in the far field can be expressed as follows:

$$P_{PA}(t',\theta) = rect(\frac{t'}{T_p})\left| \frac{\sin(\pi M f_0 d\sin\theta/c)}{\sin(\pi f_0 d\sin\theta/c)} \right| \tag{9}$$

$$\varphi_{PA}(t',\theta) = rect(\frac{t'}{T_p})(2\pi f_0 t' + \pi(M-1)f_0 d\sin\theta/c) \tag{10}$$

According to (10), the far-field phase of the PA-transmitted signal is related to the time index within pulse, angle, and array spacing, but not to the range; in addition, the phase of each point in the far field changes with the change in the time index within pulse. During the reconnaissance phase, the phase measured by the jammer is as follows:

$$\varphi_{PA\_n}(t',\theta) = rect(\frac{t'}{T_p})\arctan(\frac{\sin(2\pi f_0 t' + \pi(M-1)f_0 d\sin\theta/c)}{\cos(2\pi f_0 t' + \pi(M-1)f_0 d\sin\theta/c)}) \tag{11}$$

3. FDA Phase Characteristics

Consider a uniformly linear FDA with M isotropic antennas, as shown in Figure 3.

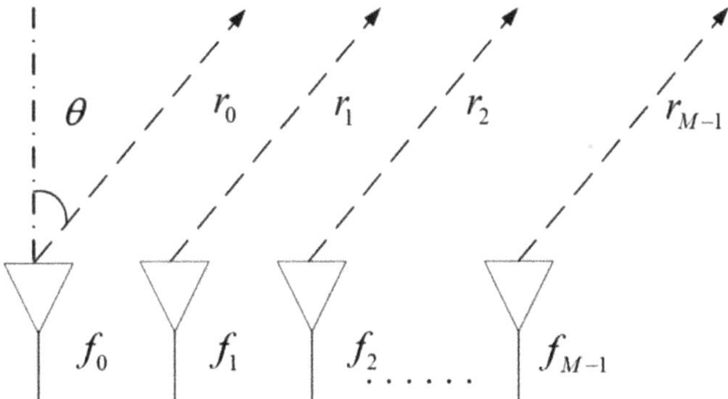

Figure 3. Uniform linear frequency diversity array.

The carrier frequency of each array is $f_M = f_0 + \Delta f(m)$, and the array spacing is d. When it propagates to the far field with range r and angle θ, the signal can be expressed as follows:

$$y_{FDA}(t) = \sum_{m=1}^{M} s_m(t) = \sum_{m=1}^{M} rect(\frac{t - r_m/c}{T_p})e^{j(2\pi f_m(t-r_m/c))} \tag{12}$$

where $r_m = r_0 - (m-1)d\sin\theta$. First, consider the FDA with linearly increasing frequency offset (LIFDA), namely $\Delta f(m) = (m-1)\Delta f$. Assuming that the envelope of the far-field condition is approximately unchanged and that the real-time t is replaced by the time index within pulse t', Equation (12) can be further expressed as follows:

$$\begin{aligned} y_{LIFDA}(t) &= rect(\frac{t-r_0/c}{T_p})e^{j2\pi f_0(t-r_0)/c}\sum_{m=1}^{M} e^{j2\pi(m-1)(\Delta f(t-r_0/c)+f_0 d\sin\theta/c)} \\ &= rect(\frac{t'}{T_p})e^{j2\pi f_0 t'}\sum_{m=1}^{M} e^{j2\pi(m-1)(\Delta f t'+f_0 d\sin\theta/c)} \\ &= rect(\frac{t'}{T_p})e^{j(2\pi f_0 t'+\pi(M-1)(\Delta f t'+f_0 d\sin\theta/c))}\frac{\sin(\pi M(\Delta f t'+f_0 d\sin\theta/c))}{\sin(\pi(\Delta f t'+f_0 d\sin\theta/c))} \end{aligned} \tag{13}$$

It can be concluded from (13) that the amplitude and phase of the LIFDA in the far field are, respectively, expressed as follows:

$$P_{LIFDA}(t',\theta) = rect(\frac{t'}{T_p})\left|\frac{\sin(\pi M(\Delta f t' + f_0 d\sin\theta/c))}{\sin(\pi(\Delta f t' + f_0 d\sin\theta/c))}\right| \tag{14}$$

$$\begin{aligned} \varphi_{LIFDA}(t',\theta) &= rect(\frac{t'}{T_p})(2\pi f_0 t' + \pi(M-1)(\Delta f t' + f_0 d\sin\theta/c)) \\ &= rect(\frac{t'}{T_p})(2\pi f_c t' + \pi(M-1)f_0 d\sin\theta/c) \end{aligned} \tag{15}$$

$$f_c = f_0 + \frac{M-1}{2}\Delta f = \frac{1}{M}\sum_{m=1}^{M} f_M \tag{16}$$

According to (15), the far-field phase characteristics of the LIFDA are similar to those of PA, which is related to the time index within pulse, angle, and array spacing, but not to the range. The phase of each point in the far field changes with the change in the time index within pulse. The different phase of PA and FDA in the same far-field location is due to their different carrier frequencies. The PA carrier frequency is f_0, while the equivalent carrier frequency of the LIFDA is f_c, which is the average of each array carrier frequency. During the reconnaissance phase, the phase measured by the jammer is as follows:

$$\varphi_{LIFDA\_n}(t',\theta) = rect(\frac{t'}{T_p})\arctan(\frac{\sin(2\pi f_c t' + \pi(M-1)f_0 d\sin\theta/c)}{\cos(2\pi f_c t' + \pi(M-1)f_0 d\sin\theta/c)}) \tag{17}$$

Further, consider an FDA with an arbitrary frequency offset (AIFDA), assuming that the frequency offset $\Delta f(m)$ can be any value. By replacing the real-time t with the time index within pulse t', (12) can be further expressed as follows:

$$\begin{aligned} y_{AIFDA}(t) &= rect(\frac{t-r_0/c}{T_p})e^{j2\pi f_0(t-r_0)/c}\sum_{m=1}^{M}e^{j2\pi(\Delta f(m)(t-r_0/c)+f_0(m-1)d\sin\theta/c)} \\ &= rect(\frac{t'}{T_p})e^{j2\pi f_0 t'}\sum_{m=1}^{M}e^{j2\pi(\Delta f(m)t'+f_0(m-1)d\sin\theta/c)} \end{aligned} \tag{18}$$

For observation (18), we found it challenging to adjust $y_{AIFDA}(t)$ to a form similar to $y_{PA}(t)$. Therefore, we used Matlab function "angle" to extract the phase of (18), which can be expressed as follows:

$$\varphi_{AIFDA}(t',\theta) = angle(y_{AIFDA}(t)) \tag{19}$$

The phase of the AFFDA in the far field can be obtained using (19). By observing (18) and (19), it can be seen that the far-field phase of the FDA-transmitted signal is related to the time index within pulse, angle, array spacing, and frequency increment, but has nothing to do with the range. The phase changes with the change in the time index within pulse. Furthermore, the phase of the far-field point can be modulated by setting an appropriate frequency increment, which is also the theoretical basis for implementing FDA angle deception and active anti-jamming.

4. Phase Center and Angle Deception

4.1. Phase Center

As shown in Figure 4, a single-baseline phase interferometer was considered to measure the phase of FDA-transmitted signals to obtain the DOA. It is worth noting that the DOA points to the phase center of the array signal.

The single-baseline phase interferometer consists of two channels. The line formed by antenna 1 and antenna 2 is called the interferometer baseline. If the array-equivalent phase center is far enough from the receiver, the electromagnetic wave received is approximately a plane wave, and the angle between the incoming wave direction and the antenna is θ. Then, the time of the plane wave arriving at antenna 1 and antenna 2 is different, and there is a phase difference φ_Δ, which is related to the carrier equivalent frequency f_e. This can be expressed as follows:

$$\varphi_\Delta = \frac{2\pi f_e l}{c}\sin\theta \tag{20}$$

If the gains of the two channels are entirely consistent, then the DOA of the array radiation signal can be written as follows after angle transformation:

$$\theta = \arcsin(\frac{c\varphi_\Delta}{2\pi f_e l}) \tag{21}$$

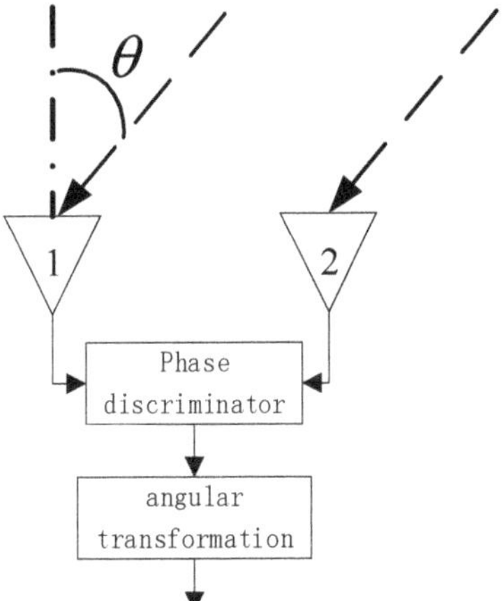

Figure 4. Principle of single-baseline phase interferometer.

Phase ambiguity is discussed in ref. [27]. This paper does not consider phase ambiguity, and the DOA points to the phase center of the array. The measurement of phase difference φ_Δ depends on the equivalent frequency f_e of the incident array signal. Therefore, the FDA has the potential to mislead the interferometer's measurement of the DOA by optimizing the phase difference in the frequency increment modulation of each array element, meaning that ultimately the jammer cannot align the array in space.

The phase center of the array plays a vital role in the direction-finding process of the jammer, so we need to analyze the phase center of the array. According to IEEE standards, the phase center is defined as the position of the point associated with the antenna. If the phase center is used as the reference point, the phase of the radiant sphere surface is constant. We can conclude that if the phase center is taken as the reference point, its phase is only related to the time index within pulse and not to the angle. If the time index within pulse is the same, the phase of the signals is equal regardless of the angle.

Let the range between the phase center and the first antenna be d_c. Firstly, PA is considered. With the phase center as the reference point, the PA-transmitted signal can be expressed as follows:

$$y_{PA}(t) = \sum_{m=0}^{M-1} s_m(t) = \sum_{m=0}^{M-1} rect(\frac{t - \frac{r_m + d_c \sin\theta}{c}}{T_p}) e^{j(2\pi f_0(t - \frac{r_m + d_c \sin\theta}{c}))} \tag{22}$$

By replacing the real-time t with the time index within pulse t', (22) can be expressed as follows:

$$\begin{aligned}
y_{PA_c}(t) &= rect(\frac{t'}{T_p}) e^{j2\pi f_0 t'} \sum_{m=0}^{M-1} e^{j\frac{2\pi f_0(md - d_c)\sin\theta}{c}} \\
&= rect(\frac{t'}{T_p}) e^{j(2\pi f_0 t' + \frac{\pi(M-1)f_0 d \sin\theta}{c} - \frac{2\pi f_0 d_c \sin\theta}{c})} \frac{\sin(\pi M f_0 d \sin\theta/c)}{\sin(\pi f_0 d \sin\theta/c)}
\end{aligned} \tag{23}$$

In this case, the far-field phase characteristics can be expressed as follows:

$$\varphi_{PA\_c}(t',\theta) = rect(\frac{t'}{T_p})(2\pi f_0 t' + \frac{\pi(M-1)f_0 d \sin\theta}{c} - \frac{2\pi f_0 d_c \sin\theta}{c}) \tag{24}$$

Because the reference point is the phase center, $\varphi_{PA\_c}(t',\theta) = \varphi_{PA\_c}(t')$ is required; the phase has nothing to do with the angle. In this case, the coefficient sum of the angle should be zero, so the phase center can be expressed as follows:

$$d_c = \frac{M-1}{2}d \tag{25}$$

Equation (25) indicates that the phase center of the PA-transmitted signal is located in the geometric center of PA, which the phase interferometer can accurately measure. Therefore, PA has no capacity for angle deception according to the phase interferometer.

Then, considering the FDA, with the phase center as the reference point, its transmitted signal can be expressed as follows:

$$y_{FDA\_c}(t) = \sum_{m=1}^{M} rect\left(\frac{t - \frac{r_m + d_c \sin\theta}{c}}{T_p}\right) e^{j(2\pi f_m(t - \frac{r_m + d_c \sin\theta}{c}))} \tag{26}$$

For the FDA, the LIFDA is first analyzed. By replacing the real-time t with the time index within pulse t', (26) can be expressed as follows:

$$y_{LIFDA\_c}(t) = rect\left(\frac{t'}{T_p}\right) e^{j2\pi f_0(t' - \frac{d_c \sin\theta}{c})} \sum_{m=1}^{M} e^{j2\pi(m-1)(\Delta f t' + \frac{f_0 d \sin\theta}{c} - \frac{\Delta f d_c \sin\theta}{c})} \tag{27}$$

According to (27), the far-field phase characteristics of the LIFDA-transmitted signals can be expressed as follows:

$$\begin{aligned}\varphi_{LIFDA\_c}(t',\theta) &= rect\left(\frac{t'}{T_p}\right)(2\pi f_0(t' - \frac{d_c \sin\theta}{c}) + \pi(M-1)(\Delta f t' + \frac{f_0 d \sin\theta}{c} - \frac{\Delta f d_c \sin\theta}{c})) \\ &= rect\left(\frac{t'}{T_p}\right)2\pi(f_c t' + (\frac{(M-1)f_0 d}{2} - \frac{(M-1)\Delta f d_c}{2} - f_0 d_c)\frac{\sin\theta}{c})\end{aligned} \tag{28}$$

where $f_c = f_0 + \frac{M-1}{2}\Delta f = \frac{1}{M}\sum_{m=1}^{M} f_M$. Because the reference point is the phase center, the phase has nothing to do with the angle, that is, $\varphi_{LIFDA\_c}(t',\theta) = \varphi_{LIFDA\_c}(t')$. Therefore, the phase center can be expressed as follows:

$$\frac{(M-1)f_0 d}{2} - \frac{(M-1)\Delta f d_c}{2} - f_0 d_c = 0 \tag{29}$$

Through further simplification, (29) can be expressed as follows:

$$d_c = \frac{(M-1)f_0 d}{2f_0 + (M-1)\Delta f} = \frac{(M-1)d}{2} \cdot \frac{f_0}{f_c} = \frac{(M-1)d}{2} \cdot \frac{f_0}{\sum_{m=1}^{M} f_M / M} \tag{30}$$

Equation (30) shows that when the frequency increment is zero, the phase center of the FDA-transmitted signal is equal to that of the PA, both located in the geometric center of the array. When $\Delta f \neq 0$, the phase center of the FDA-transmitted signal is related to the ratio of the base carrier frequency and to the center carrier frequency that treats the FDA-transmitted signal as a whole signal. The angle deception of the FDA according to the phase interferometer can be realized by taking the appropriate value of Δf.

Finally, the phase center of the AIFDA-transmitted signal is analyzed. By replacing the real-time t with the time index within pulse t', (26) can be expressed as follows:

$$y_{AIFDA\_c}(t) = rect\left(\frac{t'}{T_p}\right) e^{j2\pi f_0(t' - \frac{d_c \sin\theta}{c})} \sum_{m=1}^{M} e^{j2\pi(\Delta f(m)t' + \frac{f_0(m-1)d \sin\theta}{c} - \frac{\Delta f(m)d_c \sin\theta}{c})} \tag{31}$$

By observing (31), it can be found that it is not easy to extract the phase expression due to the arbitrariness of frequency increment, but it can be found that the phase is related to t', θ, d_c, and $\mathbf{f} = [\Delta f(1),\ldots,\Delta f(M)]$. The phase of (31) is extracted by using the function "angle", which can be expressed as follows:

$$\varphi_{AIFDA}(t',\theta,d_c,\mathbf{f}) = angle(y_{AIFDA\_c}(t)) \tag{32}$$

According to the definition of phase center, the phase of the array signal in space has nothing to do with the angle. When the time index within pulse $t' = t'_1$ and frequency increment $\mathbf{f} = \mathbf{f}_1$ are determined, the phase at each angle is equal; this can be expressed as following:

$$\varphi_{AIFDA}(t'_1, \theta_1, d_c, \mathbf{f}_1) = \varphi_{AIFDA}(t'_1, \theta_2, d_c, \mathbf{f}_1) = C \tag{33}$$

Equation (33) shows that after t', θ and $\mathbf{f}$ are determined, d_c can be calculated.

4.2. Angle Deception

Before analyzing the angle deception, we first discuss obtaining d_c through the use of (33). Then, on this basis, we set the phase center of the virtual radiation source as $d_c{}'$. Finally, by optimizing the frequency increment $\mathbf{f}$, the phase center is located at $d_c{}'$. At this time, the direction measured using the phase interferometer points to the center of the virtual radiation source, and the FDA angle deception and active anti-interference are realized.

As the parameters in (33) are coupled with each other, it is difficult to derive the phase center directly, so the particle swarm optimization (PSO) algorithm is adopted in this paper. PSO is derived from the well-developed laws of bird population activities, and it uses swarm intelligence to establish a simplified model, comparing the search space of solving problems to the flight space of birds. Each particle represents a possible solution and then solves complex optimization problems through the evolution of population particles. Its operation flow chart is shown in Figure 5.

The parameters of PSO include the particle position, particle velocity, individual optimal position, and optimal global position. Suppose that each particle has a D dimension and N particles form a population, then the position of the i-th particle can be expressed as follows:

$$\mathbf{x}_i = [x_{i1}, x_{i2}, \ldots, x_{iD}], i = 1, 2, \ldots N \tag{34}$$

The velocity of the i-th particle is expressed as follows:

$$\mathbf{v}_i = [v_{i1}, v_{i2}, \ldots, v_{iD}], i = 1, 2, \ldots N \tag{35}$$

The optimal position searched by the i-th particle so far is the extreme individual value, which can be expressed as follows:

$$\mathbf{P}_{best} = [p_{i1}, p_{i2}, \ldots, p_{iD}], i = 1, 2, \ldots N \tag{36}$$

The optimal position searched by the whole population so far is the extreme global value, expressed as follows:

$$\mathbf{g}_{best} = [g_1, g_2, \ldots, g_D] \tag{37}$$

The process of particle evolution can be expressed as follows:

$$v_{ij}(t+1) = wv_{ij}(t) + l_1 r_1(t)(p_{ij}(t) - x_{ij}(t)) + l_2 r_2(t)(p_g(t) - x_{ij}(t)) \tag{38}$$

$$x_{ij}(t+1) = x_{ij}(t) + v_{ij}(t+1) \tag{39}$$

where l_1 and l_2 are learning factors. r_1 and r_2 are uniform random numbers in the range of $[0, 1]$. p_g stands for the globally optimal particle. w is the inertial weight related to the capacity for global convergence, and weighs the capacity for global search and local search. The dynamic inertial weight is used in this paper, and is expressed as follows:

$$w = w_{\max} - \frac{(w_{\max} - w_{\min}) \cdot T}{T_{\max}} \tag{40}$$

where $w_{\max}$ and $w_{\min}$ represent the maximum and minimum inertial weights, respectively. T and $T_{\max}$ represent the current and maximum iterations, respectively.

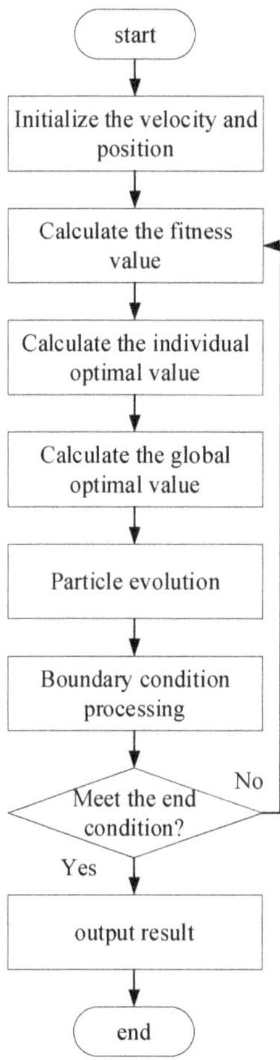

Figure 5. Particle swarm optimization flow.

The optimization problem used to obtain d_c can be expressed as follows:

$$
\begin{aligned}
&Find \quad d_l \\
&s.t. \quad |\varphi_{AIFDA}(t'_1, \theta_1, d_c, \mathbf{f}_1) - \varphi_{AIFDA}(t'_1, \theta_2, d_c, \mathbf{f}_1)| \ < b \\
&\quad\quad d_{c\_min} \leq d_c \leq d_{c\_max}
\end{aligned}
\tag{41}
$$

where d_{c_min} and d_{c_max} represent the minimum and maximum search values, respectively. b is a minimal value, which means that the phases at different angles are approximately equal.

Because d_c is a one-dimensional variable, this paper adopts the discrete PSO algorithm proposed in the literature [39] to solve it. At this time, the value of the particle state space can only be 0 or 1, and the D-dimensional binary particle corresponds to the value of d_c, one by one. The velocity update formula is still (36), and the position update formula is as follows:

$$
s(v_{ij}) = \frac{1}{1 + e^{-v_{ij}}}
\tag{42}
$$

$$
x_{ij} = \begin{cases} 1 & q < s(v_{ij}) \\ 0 & else \end{cases}
\tag{43}
$$

where q is a random number in $[0, 1]$. Finally, the expression of the fitness function $fit\_1(d_c)$

is determined as (44), according to (41):

$$fit\_1(d_c) = \left|\varphi_{AIFDA}(t'_1, \theta_1, d_c, \mathbf{f}_1) - \varphi_{AIFDA}(t'_1, \theta_2, d_c, \mathbf{f}_1)\right| \tag{44}$$

When the frequency increment $\mathbf{f}$ is determined, the phase center d_c can be obtained. Similarly, $\mathbf{f}$ can be optimized to make the direction finding of the phase interferometer point to the phase center of the virtual radiation source d_c'. The optimization problem for obtaining $\mathbf{f}$ can be expressed as follows:

$$\begin{aligned} &Find \quad \mathbf{f} \\ &s.t. \quad \left|\varphi_{AIFDA}(t'_1, \theta_1, d_c', \mathbf{f}) - \varphi_{AIFDA}(t'_1, \theta_2, d_c', \mathbf{f})\right| < b \\ &\quad f_{\min} \leq \mathbf{f}(m) \leq f_{\max}, 1 \leq m \leq M \end{aligned} \tag{45}$$

where $f_{\min}$ and $f_{\max}$ represent the minimum and maximum frequency offset search, respectively. Because $\mathbf{f}$ is a multidimensional variable, the PSO algorithm proposed in the literature [40] can be adopted. According to (45), the fitness function can be expressed as follows:

$$fit\_2(\mathbf{f}) = \left|\varphi_{AIFDA}(t'_1, \theta_1, d_c', \mathbf{f}) - \varphi_{AIFDA}(t'_1, \theta_2, d_c', \mathbf{f})\right| \tag{46}$$

Finally, when the frequency increment of the FDA is $\mathbf{f}$, it can realize angle deception and active anti-jamming.

5. Simulation Results

The numerical simulation results will be presented in this section. Unless otherwise stated, base carrier frequency $f_0 = 1$ GHz, the total number of transmitting array elements is $M = 8$, and the interval is half wavelength $d = 0.15$ m. The pulse duration $T_p = 20$ us, and the pulse repetition interval $T = 1$ ms.

5.1. PA Phase Characteristics

In order to understand the propagation law of the signal phase in space, we simulate the time–angle–phase diagram. When the signal propagates to 6 km, the propagation delay is $\tau = 20$ us. In this case, the real-time $t = 20$ us can be expressed as the propagation delay $\tau = 20$ us plus the time index within pulse $t' = 0$ us, and the real-time $t = 40$ us can be expressed as the propagation delay $\tau = 20$ us plus the time index within pulse $t' = 20$ us. When the signal propagates to 120 km, the propagation delay is $\tau = 400$ us. In this case, the real-time $t = 400$ us can be expressed as the propagation delay $\tau = 400$ us plus the time index within pulse $t' = 0$ us, and the real-time $t = 420$ us can be expressed as the propagation delay $\tau = 400$ us plus the time index within pulse $t' = 20$ us. The simulation results of the two cases are shown in Figure 6.

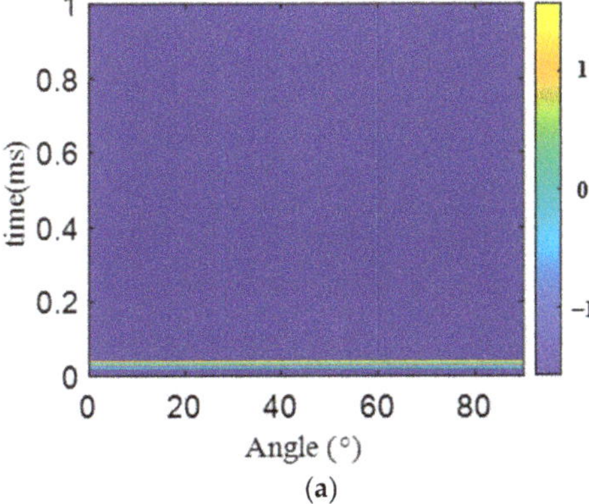

(a)

Figure 6. *Cont.*

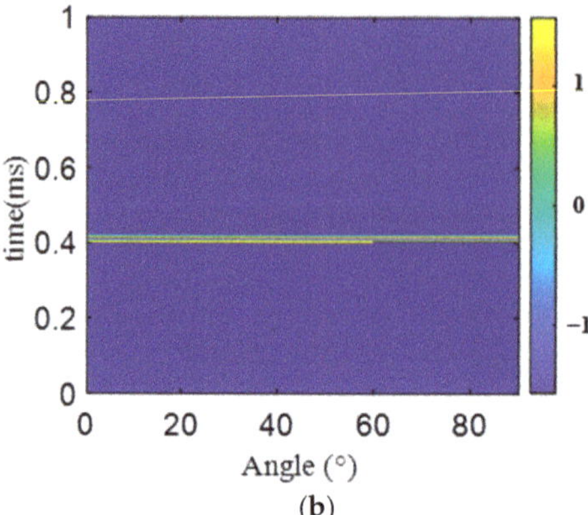

(**b**)

Figure 6. Time–angle–phase relationship of PA-transmitted signal: (**a**) $r = 6$ km; (**b**) $r = 120$ km.

As seen from Figure 6, the phase of the PA-transmitted signal propagates to the far end along with the electromagnetic wave, and each point in space traverses the phase value, independent of the propagation range. In order to understand this problem more clearly, the simulation results of the time index within pulse–angle–phase diagram are given, as shown in Figure 7.

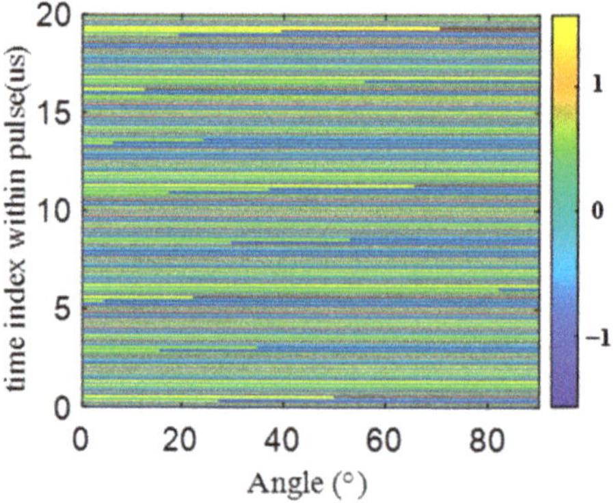

Figure 7. Time index within pulse–angle–phase relationship of the PA-transmitted signal.

It can be seen from Figure 7 that the value of the phase is related to the angle and the time index within pulse, but not to the range. The phase value of each point in space changes with the time index within pulse. The direction finding of the phase interferometer points to the phase center of the array. According to (23), the phase center of the PA-transmitted signal is $d_c = 0.525$ m. The time index within the pulse–angle–phase diagram of the PA-transmitted signal with d_c as the reference point is shown in Figure 8.

As shown in Figure 8, with the phase center as the reference point, the phase value in space has nothing to do with the angle and is only related to the time index within pulse. For the interferometer, the baseline is constant, so the time difference within the pulse of the receiving antenna is constant and the phase difference is a constant value, which ultimately enables the phase interferometer to measure the phase center of the PA signal accurately.

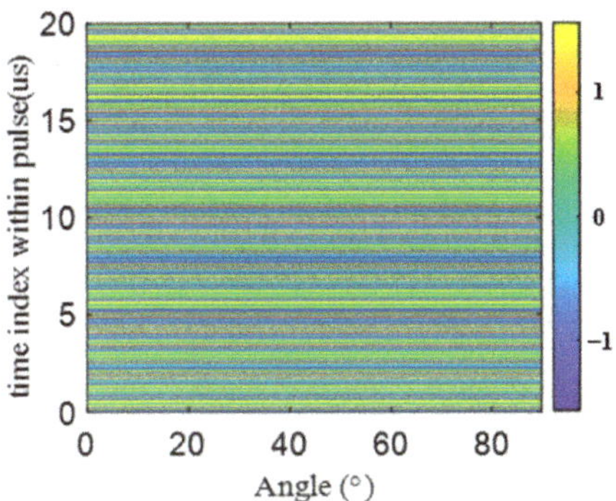

Figure 8. Time index within pulse–angle–phase relationship of the PA-transmitted signal with the phase center as the reference point.

5.2. FDA Phase Characteristics

Considering the LIFDA, its frequency increment is set as shown in (47):

$$\Delta f_m^{LI} = (m-1)\Delta f \tag{47}$$

where $\Delta f = 0.3$ MHz. After the parameters are determined, similar to the previous section, considering the signal propagation to 6 km and 120 km, the simulation results of the time–angle–phase relationship are shown in Figure 9.

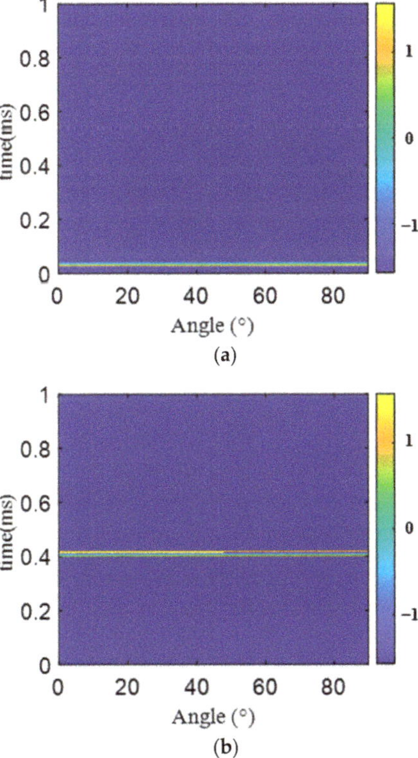

Figure 9. Time–angle–phase relationship of the LIFDA-transmitted signal: (**a**) $r = 6$ km; (**b**) $r = 120$ km.

As seen in Figure 9, the phase of the LIFDA-transmitted signal is the same as that of the PA-transmitted signal, which propagates to the far end along with the electromagnetic wave, and each point in space traverses the phase value, which is independent of the propagation range. In order to have a clearer understanding, the simulation results of the time index within the pulse–angle–phase diagram are given, as shown in Figure 10.

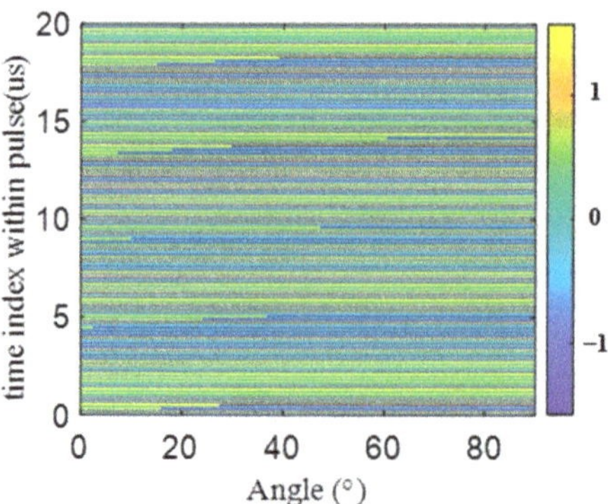

Figure 10. Time index within the pulse–angle–phase relationship of the LIFDA-transmitted signal.

It can be seen from Figure 10 that when the first element is taken as the reference point, the value of the phase is related to the angle and the time index within pulse, but not to the range, and that the phase value of each point in the space changes with the change in the time index within pulse. This paper pays attention to the location of the phase center. According to (28), the phase center of the LIFDA is $d_c = 0.5244$ m. The time index within the pulse–angle–phase diagram of the LIFDA-transmitted signal with d_c as the reference point is shown in Figure 11.

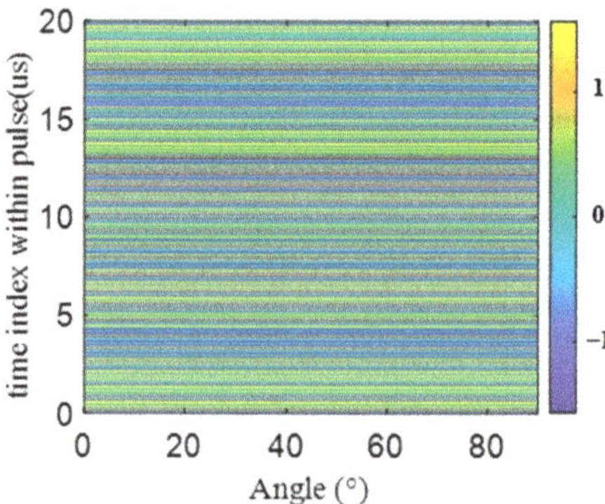

Figure 11. Time index within pulse–angle–phase relationship of the LIFDA-transmitted signal with the phase center as the reference point.

As shown in Figure 11, with the phase center as the reference point, the phase value in space is only related to the time index within the pulse, which is consistent with the theoretical analysis above. At this time, the phase interferometer direction finding is no longer pointed at the center of the array geometry, but slightly offset.

5.3. FDA Angle Deceptive

This section first calculates the phase center of the LIFDA-transmitted signal through the discrete PSO algorithm and verifies the correctness of the algorithm by comparing it with the theoretical analysis above. Then, the phase center of the virtual radiation source is set, and the PSO algorithm obtains the frequency offset of the FDA angle deception. The simulation parameters are set as follows: $N = 100$, $T_{max} = 100$, $D = 20$, $l_1 = l_2 = 1.5$, $t'_1 = 0$ us, $\theta_1 = 50°$, $\theta_2 = 20°$, $w_{max} = 0.8$, and $w_{min} = 0.4$.

The frequency increment is $\Delta f_m^{LI} = (m - 1)\Delta f$; according to (28), phase center $d_c = 0.5244$ m. Therefore, the search space is set to $d_{c\_min} = 0$ m, $d_{c\_max} = 1$ m. The fitness function is (42). The simulation results are shown in Figure 12.

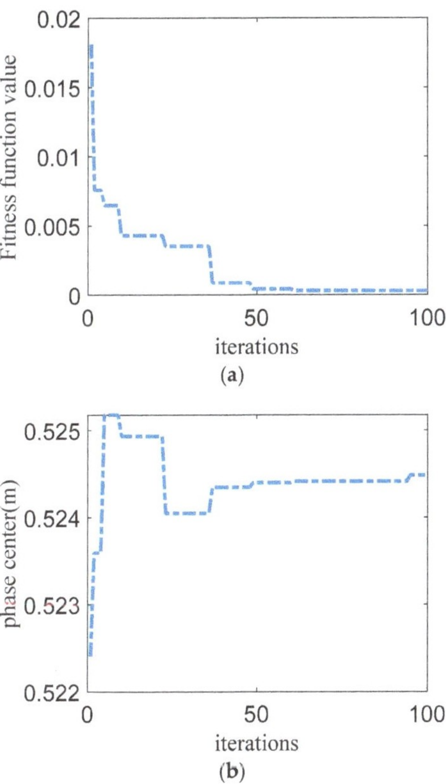

Figure 12. Relation between the LIFDA optimization results and iteration times: (**a**) fitness function value; (**b**) phase center.

It can be seen from Figure 12 that with the increase in the number of iterations, the fitness function approaches 0, indicating that the angles are different, the phases of each point with the same time index within pulse are equal, and the final phase center converges to 0.5244 m, which is consistent with the theory mentioned above.

The main lobe of the jammer is large. In order to prevent the array from being affected by the interference signals of the jammer's main lobe, the phase center of the virtual radiation source should be set as far away from the array as possible. The simulation parameters were set as follows: $d_c' = 100$ km, $f_{min} = -2$ MHz, and $f_{max} = 2$ MHz. The fitness function is (44), and the simulation results are shown in Figure 13. The frequency offsets of each element are shown in Table 1.

As can be seen from Figure 13, as the number of iterations increases, the fitness function approaches 0, and each array of the FDA adopts the frequency increment, as shown in Figure 12b, to make the phase center far away from the transmitting array, thus realizing the active angle deception of the phase interferometer.

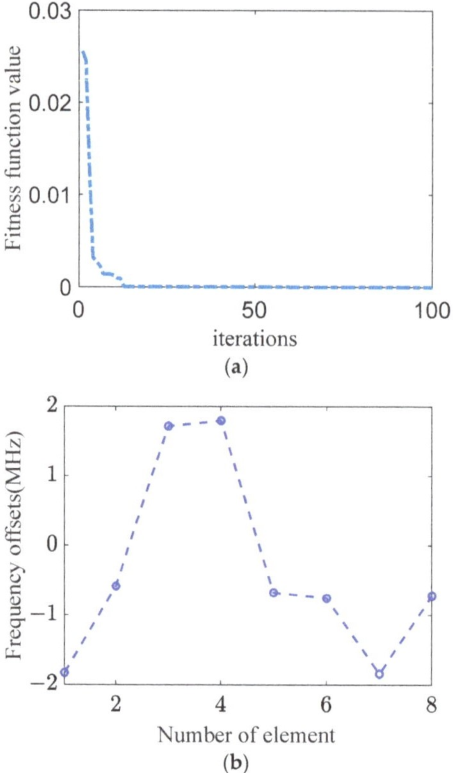

Figure 13. FDA frequency increment optimization results: (**a**) fitness function value; (**b**) frequency increment of each element.

Table 1. The frequency offsets of each element.

Number of Element	1	2	3	4	5	6	7	8
frequency offsets (MHz)	−1.83	−0.5818	1.722	1.801	−0.6737	−0.7543	−1.84	−0.7185

6. Conclusions

This paper studied the phase characteristics of FDA-transmitted signals based on the time index within pulse. In order to fully understand the phase characteristics of the array, the relationship between the time index within pulse and the phase was first analyzed, and the time index within pulse was further understood by discussing the phase characteristics and the phase center of the PA. Secondly, the phase characteristics and phase centers of the FDA-transmitted signals with linearly increasing frequency offset and arbitrary frequency offset were analyzed, and it was concluded that the phase of FDA-transmitted signals is independent of range but related to the time index within pulse, and that its phase center can deviate from the array. The potential of the FDA to cheat the phase interferometer was theoretically elaborated. Furthermore, the PSO algorithm obtained the phase center of the LIFDA, and the algorithm's effectiveness was verified. Finally, based on this algorithm, the phase center of the virtual radiation source was set, and the frequency increment of the FDA was obtained. The theoretical analysis and simulation experiments have proved that the FDA can realize angle deception and achieve active anti-interference.

Author Contributions: Conceptualization, C.Z. and C.W.; methodology, C.Z. and M.T.; software, C.Z. and M.L.; validation, C.Z., J.G. and L.B.; formal analysis, C.Z.; resources, J.G. and L.B.; writing—original draft preparation, C.Z.; writing—review and editing, C.Z., M.L., M.T. and L.B.; supervision, C.W. and J.G. All authors have read and agreed to the published version of the manuscript.

Funding: This research was funded by the Natural Science Foundation of Shanxi Province under grants 2021JM-222 and National Natural Science Funds of China under grant 62201580.

Data Availability Statement: The data and code used in this study are available upon request to the corresponding author.

Conflicts of Interest: The authors declare no conflict of interest.

References

1. Antonik, P.; Wicks, M.C.; Griffiths, H.D.; Baker, C.J. Range-dependent beamforming using element level waveform diversity. In Proceedings of the International Waveform Diversity Design Conference, Lihue, HI, USA, 22–27 January 2006; pp. 1–6.
2. Huang, J.; Tong, K.F.; Baker, C. Frequency diverse array: Simulation and design. In Proceedings of the 2009 IEEE Radar Conference, Pasadena, CA, USA, 4–8 May 2009.
3. Huang, J.; Tong, K.F.; Baker, C.J. Frequency diverse array with beam scanning feature. In Proceedings of the Antennas & Propagation Society International Symposium, San Diego, CA, USA, 5–11 July 2008.
4. Lan, L.; Marino, A.; Aubry, A.; De Maio, A.; Liao, G.; Xu, J.; Zhang, Y. GLRT-based Adaptive Target Detection in FDA-MIMO Radar. *IEEE Trans. Aerosp. Electron. Syst.* **2021**, *57*, 597–613. [CrossRef]
5. Wang, W.-Q. Overview of frequency diverse array in radar and navigation applications. *IET Radar Sonar Navig.* **2016**, *10*, 1001–1012. [CrossRef]
6. Lan, L.; Rosamilia, M.; Aubry, A.; De Maio, A.; Liao, G.; Xu, J. Adaptive Target Detection with Polarimetric FDA-MIMO Radar. *IEEE Trans. Aerosp. Electron. Syst.* **2023**, *59*, 2204–2220. [CrossRef]
7. Secmen, M.; Demir, S.; Hizal, A.; Eker, T. Frequency diverse array antenna with periodic time modulated pattern in range and angle. In Proceedings of the 2007 IEEE Radar Conference, Boston, MA, USA, 17–20 April 2007; pp. 427–430.
8. Khan, W.; Qureshi, I.M.; Basit, A.; Malik, A.N.; Umar, A. Performance analysis of MIMO-frequency diverse array radar with variable logarithmic offsets. *Prog. Electromagn. Res. C* **2016**, *62*, 23–34. [CrossRef]
9. Gao, K.; Wang, W.Q.; Cai, J.; Xiong, J. Decoupled frequency diverse array range-angle-dependent beampattern synthesis using non-linearly increasing frequency offsets. *IET Microw. Antennas Propag.* **2016**, *10*, 880–884. [CrossRef]
10. Shao, H.; Dai, J.; Xiong, J.; Chen, H.; Wang, W.-Q. Dot-shaped range-angle beampattern synthesis for frequency diverse array. *IEEE Antennas Wirel. Propag. Lett.* **2016**, *15*, 1703–1706. [CrossRef]
11. Liu, Y.; Ruan, H.; Wang, L.; Nehorai, A. The random frequency diverse array: A new antenna structure for uncoupled direction-range indication in active sensing. *IEEE J. Sel. Top. Signal Process.* **2017**, *11*, 295–308. [CrossRef]
12. Khan, W.; Qureshi, I.M. Frequency diverse array radar with time-dependent frequency offset. *IEEE Antennas Wirel. Propag. Lett.* **2014**, *13*, 758–761. [CrossRef]
13. Han, S.; Fan, C.; Huang, X. Frequency diverse array with time-dependent transmit weights. In Proceedings of the IEEE 13th International Conference on Signal Processing, Chengdu, China, 6–10 November 2016; pp. 448–451.
14. Xiong, J.; Wang, W.-Q.; Shao, H.; Chen, H. Frequency diverse array transmit beampattern optimization with genetic algorithm. *IEEE Antennas Wirel. Propag. Lett.* **2017**, *16*, 469–472. [CrossRef]
15. Chen, B.; Chen, X.; Huang, Y.; Guan, J. Transmit beampattern synthesis for FDA radar. *IEEE Antennas Wirel. Propag. Lett.* **2018**, *17*, 98–101. [CrossRef]
16. Wang, Z.; Song, Y.; Mu, T.; Luo, J.; Ahmad, Z. A short-range range-angle dependent beampattern synthesis by frequency diverse array. *IEEE Access* **2018**, *6*, 22664–22669. [CrossRef]
17. Chen, K.; Yang, S.; Chen, Y.; Qu, S.W. Accurate Models of Time-Invariant Beampatterns for Frequency Diverse Arrays. *IEEE Trans. Antennas Propag.* **2019**, *67*, 3022–3029. [CrossRef]
18. Liao, Y.; Zeng, G.; Luo, Z. Time-Variance Analysis for Frequency-Diverse Array Beampatterns. *IEEE Trans. Antennas Propag.* **2023**, *71*, 6558–6567. [CrossRef]
19. Chen, K.; Yang, S.; Chen, Y.; Qu, S.-W. Comments on "correction analysis of 'frequency diverse array radar about time'". *IEEE Trans. Antennas Propag.* **2023**, *71*, 2897–2898. [CrossRef]
20. Tan, M.; Wang, C.; Li, Z. Correction Analysis of Frequency Diverse Array Radar About Time. *IEEE Trans. Antennas Propag.* **2021**, *69*, 834–847. [CrossRef]
21. Xu, J.; Kang, J.; Liao, G.; So, H.C. Mainlobe Deceptive Jammer Suppression with FDA-MIMO Radar. In Proceedings of the 2018 IEEE 10th Sensor Array and Multichannel Signal Processing Workshop (SAM), Sheffield, UK, 8–11 July 2018; pp. 504–508.
22. Lan, L.; Liao, G.; Xu, J.; Zhang, Y.; Fioranelli, F. Suppression Approach to Main-Beam Deceptive Jamming in FDA-MIMO Radar Using Nonhomogeneous Sample Detection. *IEEE Access* **2018**, *6*, 34582–34597. [CrossRef]
23. Tan, M.; Wang, C. Reply to Comments on "Correction Analysis of 'Frequency Diverse Array Radar About Time'". *IEEE Trans. Antennas Propag.* **2023**, *71*, 2899–2902. [CrossRef]

24. Lan, L.; Xu, J.; Liao, G.; Zhang, Y.; Fioranelli, F.; So, H. Suppression of mainbeam deceptive jammer with FDA-MIMO radar. *IEEE Trans. Veh. Technol.* **2020**, *69*, 11584–11598. [CrossRef]

25. Lan, L.; Liao, G.; Xu, J.; Zhang, Y.; Liao, B. Transceive Beamforming With Accurate Nulling in FDA-MIMO Radar for Imaging. *IEEE Trans. Geosci. Remote Sens.* **2020**, *58*, 4145–4159. [CrossRef]

26. Xu, J.; Liao, G.; Zhu, S.; So, H.C. Deceptive jamming suppression with frequency diverse MIMO radar. *Signal Process.* **2015**, *113*, 9–17. [CrossRef]

27. Abdalla, A.; Wang, W.Q.; Yuan, Z.; Mohamed, S.; Bin, T. Subarray-based FDA radar to counteract deceptive ECM signals. *EURASIP J. Adv. Signal Process.* **2016**, *2016*, 104. [CrossRef]

28. Liu, W.; Liu, J.; Liu, T.; Chen, H.; Wang, Y.-L. Detector Design and Performance Analysis for Target Detection in Subspace Interference. *IEEE Signal Process. Lett.* **2023**, *30*, 618–622. [CrossRef]

29. Zhang, Z.; Wen, F.; Shi, J.; He, J.; Truong, T.K. 2D-DOA estimation for coherent signals via a polarized uniform rectangular array. *IEEE Signal Process. Lett.* **2023**, *30*, 893–897. [CrossRef]

30. Wang, X.; Guo, Y.; Wen, F.; He, J.; Truong, K.T. EMVS-MIMO radar with sparse Rx geometry: Tensor modeling and 2D direction finding. *IEEE Trans. Aerosp. Electron. Syst.* **2023**, *3297570*, 1–14. [CrossRef]

31. Wang, L.; Wang, W.-Q.; Guan, H.; Zhang, S. LPI Property of FDA Transmitted Signal. *IEEE Trans. Aerosp. Electron. Syst.* **2021**, *57*, 3905–3915. [CrossRef]

32. Hou, Y.; Wang, W.-Q. Active Frequency Diverse Array Counteracting Interferometry-Based DOA Reconnaissance. *IEEE Antennas Wirel. Propag. Lett.* **2019**, *18*, 1922–1925. [CrossRef]

33. Ge, J.; Xie, J.; Chen, C.; Wang, B. The Direction of Arrival Location Deception Model Counter Duel Baseline Phase Interferometer Based on Frequency Diverse Array. *Front. Phys.* **2021**, *9*, 598047. [CrossRef]

34. Ge, J.; Xie, J.; Wang, B. Phase characteristics of frequency diverse array radar: Phase radiation pattern, phase period, and phase centre. *IET Radar Sonar Navig.* **2022**, *16*, 759–774. [CrossRef]

35. Ge, J.; Xie, J.; Chen, C. Deceptive signal generating optimization based on frequency diverse array. *IET Radar Sonar Navig.* **2022**, *16*, 1304–1315. [CrossRef]

36. Ge, J.; Xie, J.; Wang, B.; Chen, C. The DOA location deception effect of frequency diverse array on interferometer. *IET Radar Sonar Navig.* **2021**, *15*, 294–309. [CrossRef]

37. Ge, J.; Xie, J.; Wang, B. A cognitive active anti-jamming method based on frequency diverse array radar phase center. *Digit. Signal Process.* **2021**, *109*, 102915. [CrossRef]

38. An, Z.; Zhang, Y.; Xu, L. Research on the Phase Center of Uniform Linear Array. *Adv. Mater. Res.* **2014**, *1049–1050*, 2037–2040. [CrossRef]

39. Kennedy, J.; Eberhart, R. A Discrete Binary Version of the Particle Swarm Algorithm. In Proceedings of the IEEE International Conference on Systems, Man and Cybernetics, Orlando, FL, USA, 12–15 October 1997; pp. 4104–4108.

40. Shi, Y.; Eberhart, R. A modified Particle swarm optimizer. In Proceedings of the 1998 IEEE International Conference on Evolutionary Computation, Anchorage, AK, USA, 4–9 May 1998; pp. 69–73.

MDPI

St. Alban-Anlage 66

4052 Basel

Switzerland

www.mdpi.com

Remote Sensing Editorial Office

E-mail: remotesensing@mdpi.com

www.mdpi.com/journal/remotesensing